DIGITAL ELECTRONICS
Theory, Applications, and Troubleshooting

DIGITAL ELECTRONICS
Theory, Applications, and Troubleshooting

BYRON W. PUTMAN
Computer Facilities and Communications
University of California, Berkeley

Formerly with the Heald Institute of Technology

PRENTICE-HALL, INC., Englewood Cliffs, N.J. 07632

Library of Congress Cataloging in Publication Data

Putman, Byron W., (date)
Digital electronics.

Includes index.
1. Digital electronics. I. Title.
TK7868.D5P88 1986 621.3819'58 85-9363
ISBN 0-13-212481-5

Editorial/production supervision
and interior design: *Theresa A. Soler*
Cover designer: *Lundgren Graphics, Ltd.*
Manufacturing buyer: *Gordon Osbourne*

Printed in the United States of America

10 9 8 7 6 5 4 3 2 1

ISBN 0-13-212481-5 01

Prentice-Hall International (UK) Limited, *London*
Prentice-Hall of Australia Pty. Limited, *Sydney*
Prentice-Hall Canada Inc., *Toronto*
Prentice-Hall Hispanoamericana, S.A., *Mexico*
Prentice-Hall of India Private Limited, *New Delhi*
Prentice-Hall of Japan, Inc., *Tokyo*
Prentice-Hall of Southeast Asia Pte. Ltd., *Singapore*
Editora Prentice-Hall do Brasil, Ltda., *Rio de Janeiro*
Whitehall Books Limited, *Wellington, New Zealand*

To Elaine: Who reluctantly
accepted revised priorities and unwritten
poems, but provided unlimited support
and inspiration

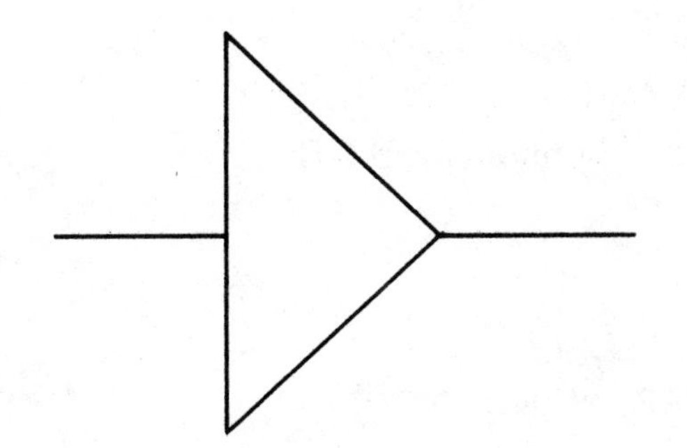

CONTENTS

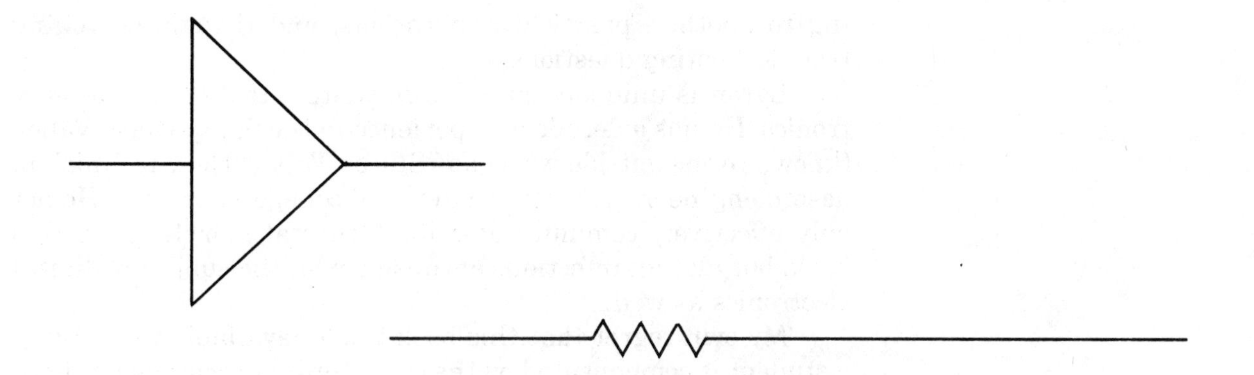

FOREWORD

During my years as a student I searched for instructors who taught in an orderly and logical fashion material that was directly applicable to my work in the electronics industry. Byron Putman was my college instructor for my digital and microprocessor courses. In his classes we learned not just what makes circuits work, but more important, what may cause them *not* to work. He taught theory that I see reflected and used everyday in the "real world." It is much easier to learn when you are taught not to memorize how something works, but to understand how it actually functions.

Understanding and good theory enable the student to "think things through." When the student runs into a new circuit or device, it is not a cause for thinking "I've never seen this before" and stopping at that point. Instead, the thought is "I've seen something similar before. If I stop and analyze this, I should be able to figure out how it works." Unknown quantities do not become brick walls impeding any progress; they are paths, new means of traveling from one point to another.

When I edited the manuscript version of this work, I was continually pleased to see exactly what I had enjoyed in Byron's lectures and lab sessions shining through in this book: a concise presentation, a logical progression, a building of ideas—one leading to another, practical applications, and thought-provoking troubleshooting questions.

Byron is uniquely qualified to write a text on digital electronics. He has a decade of experience in the Santa Clara Valley (known to the outside world as "Silicon Valley") as a technician, design engineer, technical writer, and college instructor. He not only effectively communicates his technical knowledge in this book, but also an infectious enthusiasm for the subject of digital electronics as well.

My own regret that this book was unavailable while I was a student is compensated by the opportunity to commend it here to a new generation of electronic technology students.

Ellen L. Lawson
The Thomas Engineering Company

ACKNOWLEDGMENTS

I would like to thank the following people and companies for their direct and indirect efforts to assist me in the creation of this text:

Dr. Mark Hawkins and Dr. Gene Sealbach of Foothill College, two of the very best instructors that I have had the pleasure to know. Dr. Hawkins, a model of support and enthusiasm, helped me gain the confidence required for a project of this length. Dr. Sealback introduced me to the concept of an intuitive approach to abstract subjects.

Ellen L. Lawson, who edited the manuscript and provided creative input and support.

Texas Instruments, RCA, Intel, and Data I/O for providing specifications, block diagrams, and pinouts.

Mark Diamond and Hanania Bolcover, two of my most memorable students, for providing the motivation to write this book.

Professor Roy Kratzer of Foothill College. One of the first educational professionals to fully understand the changing role of electronic technicians in the computer service industry.

Jim, Mack, Sparky, and Christopher of the CFO Engineering Lab at UC Berkeley. Simply for sharing their vast pool of technical knowledge.

Roger Lanser of Lockhead Missile and Space. For his keen technical insight and for occasionally forcing me out from in front of my computer and into Yosemite.

Peter and Linda. Peter for his technical assistance and for the photographs in this text. And Linda for her steady stream of somewhat cynical, yet very applicable criticisms.

And my cat, Hemuior, for keeping me company during those endless hours of development and execution.

A special note of thanks to the following group of reviewers for their incisive critiques and much-appreciated feedback: Robert A. Ciuffetti, GTE Sylvania Technical School; James W. Nice, ITT Technical Institute, Portland; Robert D. Carpenter, Wake Technical College; John L. Keown, Southern Technical Institute; Richard Prestopmik, Fulton-Montgomery Community College; Ralph E. Folger, Jr., Hudson Valley Community College; and Neal D. Voke, Triton College.

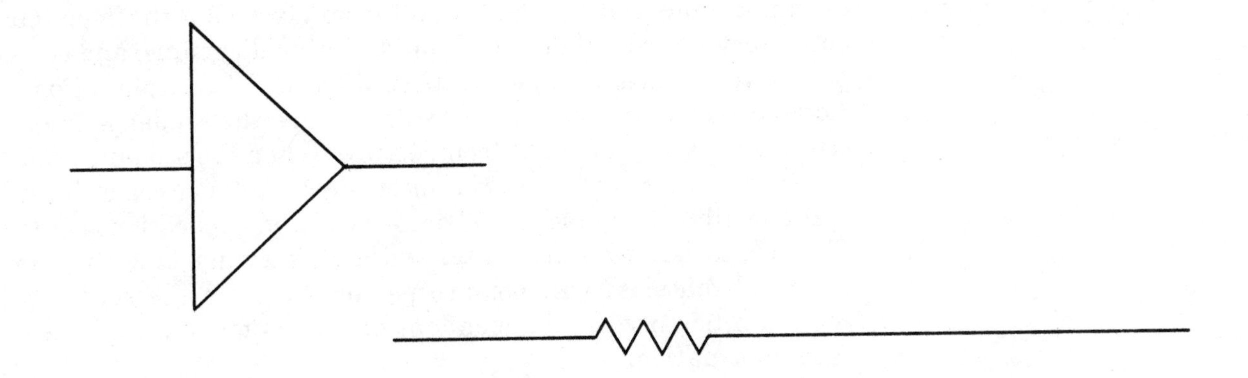

PREFACE

People seek an electronics education at technical institutes, junior colleges, and four-year colleges for a wide variety of reasons. The common motivation of these students is the desire to enter the job market with a valuable and marketable skill. This is not a book on digital circuit design; the major objective of this text is to prepare these students to enter the "real" world of modern industrial electronics with a background that will enable them to be productive electronic test technicians and troubleshooters from their first day on the job.

This text is built around the analysis, application, and troubleshooting techniques of the most widely used commercial digital integrated circuits. Flip-flop applications are constructed with MSI flip-flops, not handfuls of simple gates. Counter applications employ MSI counter ICs, not cascaded flip-flops. I have often had the frustrating experience of working with recent graduates who knew how to construct any divide-by-N counter with discrete flip-flops, but had never been exposed to a simple 74LS90 MSI counter. Full understanding and appreciation of digital devices can be accomplished only in meaningful, realistic applications and troubleshooting simulations.

This text is written in a manner that appeals to the student's intuition and previous knowledge. It was designed to be used and read, not carried to and from class as a three-pound obligatory parcel; all device pinouts, timing diagrams, and complete circuits are strongly supported by detailed explanations. Advance reading assignments will prepare the students to derive the maximum benefit from lecture. When the student graduates and enters industry, the most important sources of information will be manufacturers' data books and applications notes. The technician who graduates without knowing how to read technical material will not prosper or grow. On the contrary, that technician will be dependent on others for new and accurate information.

This text assumes that the reader has a solid background in basic electronics and transistor theory. The last chapter, which introduces data conversion techniques, also requires an elementary knowledge of op-amp theory.

Chapter 1 introduces the concepts of digital electronics and the binary number system. Number theory is held to a bare minimum. The introduction to the hexadecimal number system is deferred until Chapter 12. No concept is be introduced until it can be applied immediately to the material at hand.

Chapters 2 and 3 introduce logic gates. Most texts present this material in one chapter. They rely on truth tables and rote memorization techniques. An alternative to memorization is the concept of dynamic input and unique output levels. The logic symbol of each gate is used to describe its operation; memorization submits to understanding. Positive-logic and negative-logic symbols of each gate are equally stressed; the logic symbol selected to represent a particular gate will be that symbol which best represents the gate's function within the context of the circuit. This technique is known as "self-documenting design." It results in exceptionally readable schematics for both the engineer and the technician.

Chapter 4 illustrates applications of simple gates. Concepts of Boolean algebra and circuit-reduction techniques are not stressed. Instead, a structured intuitive approach to solving simple logic problems is developed. Several circuits are then designed, inspected, and simple troubleshooting procedures are considered. The circuits developed in Chapter 4 lay the foundation for the survey of MSI devices in Chapter 7.

Chapter 5 introduces the TTL and CMOS logic families. There is no doubt that LS TTL is still the dominate logic family in the electronics industry, but as CMOS technology continues to increase in speed, it will displace TTL in many traditionally bipolar applications. In the near future, the 74HC00 series of high-speed CMOS SSI and MSI devices and logic-gate arrays will dominate the market. This text has a balanced presentation of devices from the 74LS00 TTL series and the 4000B standard CMOS series. Most of the applications in this text use TTL devices because static sensitive CMOS ICs may not survive the rigors of the typical college electronics laboratory.

Chapter 6 is entitled "Troubleshooting Logic Levels." In an electronics production facility, well over 50% of the faults lie in printed circuit board problems or incorrectly stuffed components. This chapter closely examines the elements of the PCB and techniques of isolating shorts and finding opens. The material developed in this chapter is used as the basis of troubleshooting techniques in the remainder of the text.

Chapter 7 is a survey of the most popular MSI combinational devices. The function table and pinout of each IC presented are examined in great detail. Well-supported and highly entertaining applications are illustrated that are designed to reinforce operation of the selected devices. The chapter-end questions test the student's understanding of each circuit by considering the cause and effects of various troubleshooting problems.

Chapters 8 and 9 introduce sequential devices. The idea of feedback is employed to develop the extremely important concept of memory. Timing diagrams are the major vehicle used to illustrate the operation of sequential devices. Each event in the timing diagrams is carefully discussed. Selected applications are used to reinforce the operation of simple sequential circuits. The one-shot, an extremely important device that is often ignored in digital texts, is discussed in great detail. Once again, the chapter problems and questions revolve around troubleshooting the applications illustrated within the chapter.

Chapter 10 is entitled "MSI Counters and Shift Registers." This chapter puts it all together; simple gates, MSI combinational devices, flip-flops, and one-shots are integrated with a wide variety of counters to create moderately complex circuits. These applications are well supported with text and detailed timing diagrams. This is an extremely critical chapter. The applications in this chapter require that the students understand not only each device, but also the function that it provides within the context of the circuit system. Interesting, entertaining, and challenging applications are the keys that transform the student into a technician.

Three-state devices are used in bus systems and bus systems are controlled by microprocessors. Traditionally, the pursuit of three-bus architecture is left to a microprocessor class. This is a conceptual mistake. The stage is now set to introduce the microprocessor as the master of the bus. The microprocessor is illustrated as a device whose major function is to perform memory-read and memory-write operations. The microprocessor is introduced as a natural extension of complex MSI devices. The concept of bus systems creates the need for three-state devices. Chapter 11 establishes the three-bus architecture of microprocessor systems and surveys the major three-state devices. The hexadecimal number system is also introduced to support the memory-mapping techniques in Chapter 12.

Because Chapter 11 introduced the read and write cycles of

the microprocessor and three-state devices, complete memory systems can now be examined. Chapter 12 creates memory systems from static RAMs, dynamic RAMs, and ROMs. The techniques of memory decoding and memory mapping are studied in detail. The major specifications of RAMs are illustrated in timing diagrams. Familiar TTL components are used to create a row-and-column address multiplexing circuit for dynamic RAMs. There is absolutely no need to wait for a class on microprocessors to introduce memory systems. Students who enter a microprocessor class with a complete background in digital devices and memory systems can concentrate their efforts on understanding the software/hardware interaction of intelligent systems.

No digital text is complete without a chapter on data conversion techniques. Chapter 13 discusses the major types of DACs and A/Ds. Converters are developed using previously examined digital circuits and op amps. Reviews of the current summing op-amp and op-amp integrator are included in the chapter.

Byron W. Putman

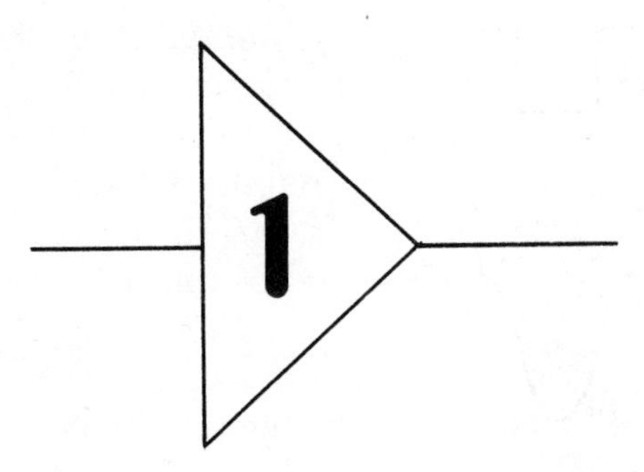

INTRODUCTION TO DIGITAL

1.1 ANALOG AND DIGITAL SIGNALS

When most people are asked to name a waveform they think of the sine wave. The sine wave is the fundamental waveform; all other waveforms are derived from various phase and frequency combinations of the sine wave. Another common waveform is the square wave. Let's take a moment to contrast these two waveforms (see Figure 1.1).

The amplitude of the sine wave is continuously changing. Following that line of thinking, we can say that the amplitude of any given sine wave has an infinite number of possible values. These values will exist between the positive and negative peaks. The sine wave is an *analog* signal. The term "analog" refers to a waveform whose amplitude changes in a continuous fashion.

The amplitude of a square wave consists of only two values, a high value and a low value. These amplitude changes occur discretely, that is, in steps. The square wave is called a *digital* waveform. This book is devoted to the study of electronic circuits whose outputs have only two possible values. The use of the term "digital" indicates that a numerical value can be associated with each level of the waveform. Our decimal number system is based on the fact that human beings have 10 fingers. In essence, a square wave has only two "fingers." It should seem logical that digital electronics supports a base 2 number system. The digits

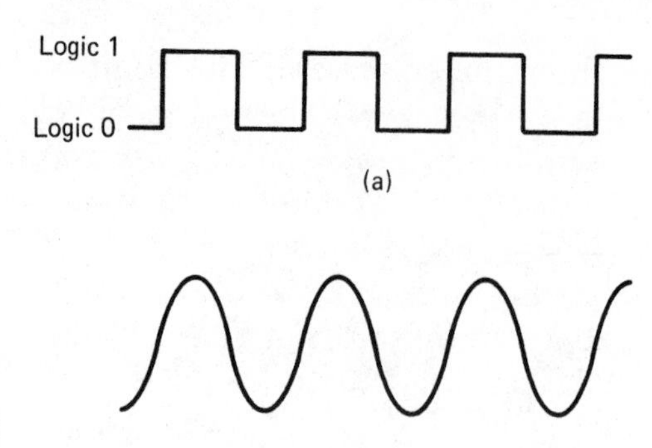

Figure 1.1 (a) Square wave; (b) sine wave.

in this system will be 0 and 1. The digit 0 will be assigned to the low part of the square wave and the digit 1 to the high part of the square wave.

The term "digital" is commonly used to describe a device that utilizes a numerical display. Watches, voltmeters, and almost any other device that one can imagine can be found with digital displays. Devices that use digital displays always contain digital electronic circuitry.

1.2 THE BINARY NUMBER SYSTEM

1.2.1 Number Bases and Counting

Because a square wave has two levels, digital electronics can be described by a base 2 number system. The base 2 system is called the *binary number system.* The base of a number system describes the number of unique digits in that system. In our decimal number system there are 10 unique digits, 0 through 9. In the binary number system there are two unique digits: 0 and 1.

All number systems, regardless of the base, function identically. We should take a moment and review the decimal number system. Most number systems operate with a *weighted code.* Weighted codes use place values. Refer to the following example.

What does the number 537 really represent?

10^2	10^1	10^0
5	3	7

$$\begin{aligned} 10^0 \times 7 &= 7 \\ 10^1 \times 3 &= 30 \\ 10^2 \times 5 &= \underline{500} \\ & \quad 537 \end{aligned}$$

Notice that the place value of the rightmost column is equal to the base raised to the 0 power. The next column's place value is equal to the base raised to the 1st power; the value of the last column is the base raised to the 2nd power. This process could continue indefinitely to represent a number of any size.

Consider the act of counting. We start with the first digit

in the number system. This digit is always the number 0. Each time we increment the count, we will use the next digit in the system to represent the new, updated count. What happens when we run out of unique digits? When we use up all the digits in our number system, we must restart the counting procedure with the number zero. To assure that the objects we have already counted are not lost or forgotten, a *carry* will be generated. This carry will be added to the value in the next column.

Counting Table

	Comments:				
	2^3	2^2	2^1	2^0	*Decimal equivalent*
0000					0
+ 1					
0001				0 + 1 = 1	1
+ 1					
0010			0 + 1 = 1	1 + 1 = 0	2
+ 1					
0011				0 + 1 = 1	3
+ 1					
0100		0 + 1 = 1	1 + 1 = 0	1 + 1 = 0	4
+ 1					
0101				0 + 1 = 1	5
+ 1					
0110			0 + 1 = 1	1 + 1 = 0	6
+ 1					
0111				0 + 1 = 1	7
+ 1					
1000	0 + 1 = 1	1 + 1 = 0	1 + 1 = 0	1 + 1 = 0	8
+ 1					
1001				0 + 1 = 1	9
+ 1					
1010			0 + 1 = 1	1 + 1 = 0	10
+ 1					
1011				0 + 1 = 1	11
+ 1					
1100		0 + 1 = 1	1 + 1 = 0	1 + 1 = 0	12
+ 1					
1101				0 + 1 = 1	13
+ 1					
1110			0 + 1 = 1	1 + 1 = 0	14
+ 1					
1111				0 + 1 = 1	15

A decimal number is referred to as a *digit.* The term digit indicates one of the 10 fingers. A binary number is called a *bit.* The word "bit" is derived from the phase "*b*inary dig*it.*" Notice the counting pattern. On every count the least significant bit toggles, 0 to 1 or 1 to 0. The next column toggles on every other count; the column after that, on every four counts. Finally, the most significant column toggles every eight counts. The frequency at which the column toggles indicates its place value. You should memorize the binary bit patterns for the quantities 0 through 15.

1.2.2 Binary Addition

From time to time you will need to add or subtract two binary numbers. Addition is as easy as counting. All you must remember is these four rules:

$$0 + 0 = 0$$

$$1 + 0 = 1$$

$$1 + 1 = 0 \text{ and a carry of } 1$$

$$1 + 1 + 1 = 1 \text{ and a carry of } 1$$

Consider these addition examples:

(a)	(b)	(c)	(d)
1001	1110	1010	0011
+ 0011	+ 1000	+ 0111	+ 0111
1100	1 0110	1 0001	1010

Examine addition example (d). The least significant column creates a carry. This carry is added to the next column, where two 1's are already to be summed. This is a situation where the last addition rule is applied:

$$1 + 1 + 1 = 1 \text{ and a carry of } 1$$

1.2.3 Binary Subtraction

Before we attempt binary subtraction, let's review the methods that we employ in decimal subtraction. The act of subtraction is closely tied to the concept of a *borrow.* The borrow in subtraction is analogous to the carry in addition. The major conceptional problem that people experience with a borrow is; How much are we really borrowing? Refer to the following subtraction problem:

$$\begin{array}{r} 54 \\ -\ \underline{8} \\ 46 \end{array}$$

When we attempt the subtraction it is obvious that we must borrow from the 10's column. The value that we borrow is therefore equal to 10. Adding our borrow to the value in the 1's column yields a value of 14. Eight subtracted from 14 gives us six. This process may seem painfully obvious, but it is important that you consciously understand the details of a borrow operation. Consider the following subtraction problem:

$$\begin{array}{r} 504 \\ -\ \ 8 \\ \hline 496 \end{array}$$

This time we cannot borrow from the adjacent digit. We must first borrow from the 100's place to the 10's place, and then borrow from the 10's place to the 1's place.

We have seen that one of the rules of binary addition is 1 + 1 = 0 plus a carry. The operation of borrowing is the inverse operation of a carry. The following subtraction problem illustrates the manner in which you should approach a binary borrow.

$$\begin{array}{r} 10 \\ -\ \ 1 \\ \hline 01 \end{array} = \begin{array}{r} \not{1}0^{1+1} \\ -\ \ 1 \\ \hline 01 \end{array}$$

The first binary subtraction problem does not require a borrow:

$$\begin{array}{r} 1101 \\ -\ \ 100 \\ \hline 1001 \end{array}$$

The following binary subtraction problems require only elementary borrows.

$$\begin{array}{r} 1010 \\ -\ \ \ 1 \\ \hline 1001 \end{array} \qquad \begin{array}{r} 1101 \\ -\ \ 11 \\ \hline 1010 \end{array} \qquad \begin{array}{r} 1000 \\ -\ 100 \\ \hline 0100 \end{array} \qquad \begin{array}{r} 1110 \\ -\ 1101 \\ \hline 0001 \end{array}$$

This subtraction problem illustrates a complex binary borrow:

$$\begin{array}{r} 1100 \\ -\ \ \ 1 \\ \hline 1011 \end{array}$$

1.3 DIGITAL CODES

All numbers exist in some form of code. In most number systems, the place value of any given column is equal to the base raised to the number of the column minus 1. Once again, let's examine the decimal number system to help illustrate this concept. The rightmost column of any base 10 number has the place value of 1. Ten raised to the 0 power is equal to 1. The next column has the place value of 10, which is 10 raised to the 1st power; the next column 100, which is 10 raised to the 2nd power.

1.3.1 The 8421 Code

Binary numbers can also be expressed in a form that is analogous to the decimal system. This binary code is called the *8421 code.* You should assume that all binary numbers are in 8421 code unless you are directed otherwise. The base of the binary number system is 2. Two raised to the 0 power is equal to 1. The place value of the rightmost column in the binary number system is equal to 1. The place value of the next column is equal to 2 raised to the 1st power, or 2. The next column is 2 raised to the 2nd power, or 4. The next, 2 raised to the 3rd power, or 8. As in the decimal number system, this method of assigning place values to columns can continue indefinitely. The following formula will produce the place value of any given column:

$$\text{place value} = 2^{(\text{column number} - 1)}$$

1.3.2 The BCD Code

The circuitry in calculators, digital watches, or voltmeters is digital. These devices "think" in binary. Human beings, on the other hand, think in decimal. A digital watch with a binary output would be difficult for us to read. We must have a method of representing decimal numbers in a binary form. That is the function of the *binary-coded decimal* (BCD) *code.* These digital devices will manipulate decimal information that is in an encoded binary form. Before this information is output to the display, it will first be routed through a decoder. It will be the job of this decoder to transform the BCD number back into its decimal equivalent. Following is a complete list of all the numbers in the BCD code.

BCD number	*Decimal equivalent*	*BCD number*	*Decimal equivalent*
0000	0	1000	8
0001	1	1001	9
0010	2	1010	Undefined
0011	3	1011	Undefined
0100	4	1100	Undefined
0101	5	1101	Undefined
0110	6	1110	Undefined
0111	7	1111	Undefined

You should notice that the BCD code is nothing more than the binary equivalent of the decimal digit that it represents. Because there are only 10 decimal digits, the last six combinations in BCD are not used. These combinations are considered to be undefined. Four bits are required for each decimal digit that we wish to represent.

Decimal number	*BCD representation*
835	8 3 5
	1000 0011 0101
29	2 9
	0010 1001
337	3 3 7
	0011 0011 0111

1.3.3 Other Codes

There are many other binary codes. If you own a microcomputer, you have surely heard of the ASCII code. This code will be introduced in a later chapter. Of the remaining codes, few are in common use. The only other code that we will examine is the *Gray code.*

The Gray code is an *unweighted code.* This means that the columns of a binary number encoded in Gray code do not have a place value. Each number in the Gray code differs from the preceding number by only 1 bit. The Gray code is used in situations where the position of a mechanical shaft will be interpreted into a binary form. Another application of the Gray code is in Karnaugh mapping. In Chapter 4 we will use Karnaugh maps to analyze digital circuits.

Decimal	*Gray code*	*Decimal*	*Gray code*
0	0000	8	1100
1	0001	9	1101
2	0011	10	1111
3	0010	11	1110
4	0110	12	1010
5	0111	13	1011
6	0101	14	1001
7	0100	15	1000

1.4 CONVERTING BETWEEN DECIMAL AND BINARY

Occasionally when working with digital circuits, you will be required to convert back and forth between decimal and binary. Many digital devices manipulate information in groups of four bits. Because of this fact, the binary representations of the decimal numbers 0 through 15 should be memorized. Infrequently, you will be required to work with numbers outside this range. When this situation occurs, you will need an algorithm to accomplish the conversion. An algorithm is nothing more than a procedure, or recipe, for solving a mathematical problem.

1.4.1 Binary-to-Decimal Conversion

We have stated that the place value of any column is equal to

$$\text{place value} = \text{number's base}^{(\text{column number} - 1)}$$

The place values of the first eight columns of a binary number are expressed in the following table.

Column number	*Formula*	*Place value*
1	2^0	1
2	2^1	2
3	2^2	4
4	2^3	8
5	2^4	16
6	2^5	32
7	2^6	64
8	2^7	128

Take a moment to consider this table. An important skill that a digital technician should possess is the ability to recognize patterns. You may have noticed that the place value doubled for each successive column. In the decimal system the place value increases by a factor of 10 for each successive column. This is a key concept in understanding number systems. How do we apply this information to the problem of binary-to-decimal number conversion? We merely sum all the place values for those columns in the binary number that contains a 1. For example, convert the binary number 1001 1101 into decimal. The 8-bit binary number was broken into two groups of 4 bits to improve readability.

Column values	128	64	32	16	8	4	2	1
Binary number to convert	1	0	0	1	1	1	0	1

$$\begin{array}{r} 128 \\ +\ 16 \\ +\ \ 8 \\ +\ \ 4 \\ +\ \ \underline{1} \\ 157 \end{array}$$

Convert 0011 0110 into decimal.

$$\begin{array}{r} 32 \\ +16 \\ +\ 4 \\ +\ \underline{2} \\ 54 \end{array}$$

1.4.2 Decimal-to-Binary Conversion

It is important that you be able to perform decimal-to-binary conversion. Many employers include number-conversion questions on their entrance exams. The problem is that technicians perform these conversions so infrequently that they often forget the algorithm. There are many fast and efficient methods for accomplishing decimal-to-binary conversion. These methods usually contain one or more "tricks' that are quickly forgotten. The conversion method that will be illustrated is perhaps the slowest and most inefficient, but it is also the most intuitive. Once learned, it is easily remembered.

Following is a word description of our *algorithm:* The first step we will perform is to find the largest place value that is less than or equal to the number that we wish to convert. We will assign a 1 to that bit position and subtract the place value from the number. We will repeat this process until the number we wish to convert is reduced to zero.

As an example, let us convert the number 89 into binary. First, we list the binary column values for reference:

Place Values

128 64 32 16 8 4 2 1

Step 1 64 is the largest column value that is less than or equal to 89. We will place a 1 in that bit position and subtract 64 from 89.

The intermediate binary result is 0100 0000.
The decimal value left to convert is 89 − 64 = 25.

Step 2 16 is the next column value that is less than or equal to the value to convert. We place a 1 in that bit position and subtract 16 from 25.

The intermediate binary result is now 0101 0000.
The decimal value left to convert is 25 − 16 = 9.

Step 3 The next place value that is less than or equal to the value to convert is 8.

The intermediate binary result is now 0101 1000.
The decimal value left to convert is 9 − 8 = 1.

Step 4 The decimal value of 1, obviously, will fit into the first column of the binary number.

The final binary result is 0101 1001.

As a second example, convert the value 167 into binary. This example will be worked through with a minimum of details.

Step number	*Intermediate binary result*
1. 167 − 128 = 39	1000 0000
2. 39 − 32 = 7	1010 0000
3. 7 − 4 = 3	1010 0100
4. 3 − 2 = 1	1010 0110
5. 1 − 1 = 0	1010 0111

The final binary result is 1010 0111.

The other important number system used in digital electronics is called *hexadecimal.* The hexadecimal system employs a base of 16 and is used as a shorthand method of representing long binary numbers. The hexadecimal number system will be introduced in Chapter 13.

SUMMARY

The output of a digital circuit can take on only one of two possible voltage levels. The higher voltage level is assigned the numerical value of 1, and the lower voltage level is assigned the numerical value of 0. A square wave is an example of a digital signal. Digital electronics is described by the binary number system.

Most students find their first class in digital electronics to be an extremely enjoyable experience. Digital circuits employ few discrete devices. A typical digital printed circuit board may contain 50 or more integrated circuits without a single discrete transistor. Because digital electronics use so many complex integrated circuits, sophisticated systems are easily designed and constructed. Take care not to underestimate the level of difficulty involved in your first exposure to digital. The material may seem easy at first, but it quickly accelerates in difficulty. The field of electronics is constantly evolving. Any technician who desires to prosper must keep current with the present state of the technology. A characteristic that most successful technicians share is that they enjoy electronics. Do they enjoy electronics because they are good at it, or are they good at electronics because they enjoy it? In most cases the latter is true.

QUESTIONS AND PROBLEMS

1.1. Perform the following binary additions.

(a) 1010 0111	(b) 0101 0101	(c) 1111 0001
+ 0011 0011	+ 0011 1100	+ 0100 1111

1.2. Perform binary subtractions on the three parts of problem 1.1.

1.3. Convert the following binary numbers into decimals.

(a) 0011 1110	(b) 1111 0001	(c) 0101 1010
(d) 0001 1110	(e) 0011 0011	(d) 1111 1111

1.4. Convert the following decimal numbers into binary.

(a) 127	(b) 96	(c) 231
(d) 37	(e) 51	(f) 255

1.5. Indicate the BCD representation for the following decimal numbers.

(a) 13 (b) 693 (c) 39

1.6. What decimal numbers do the following BCD numbers represent?

(a) 1001 0010 (b) 0001 0101 0111 (c) 1100 0100

1.7. How many unique binary numbers can be created with:

(a) 3 bits?	(b) 4 bits?
(c) 8 bits?	(d) 16 bits?

1.8. Name three everyday events that are digital in nature.

1.9. Name three everyday events that are analog in nature.

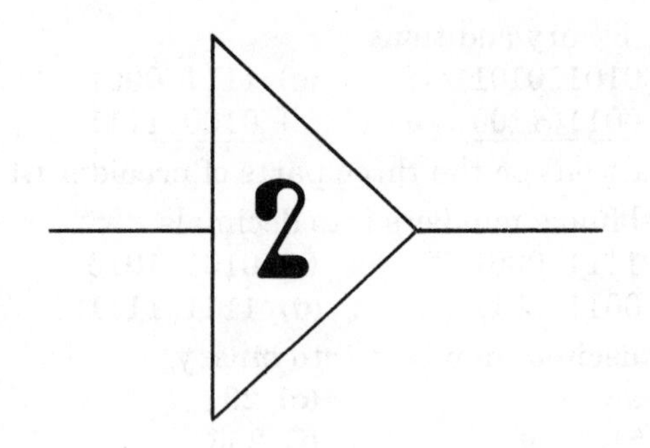

INTRODUCTION TO LOGIC GATES

2.1 REVIEW OF DIGITAL LEVELS

A digital signal will be either a logic 1 or a logic 0. In most systems a *logic 1* will refer to the higher voltage, whereas a *logic 0* will refer to the lower voltage. A square wave is a perfect example of a digital signal; at any instant in time, it is either high or low. In this chapter we will refer to a logic 1 as +5 V and a logic 0 as 0 V. We will also associate a logic 0 with a false condition and a logic 1 with a true condition.

Logic 0	*Logic 1*
0 V False	5 V True

2.2 THE LOGIC GATE

The term "logic" is a common word in conversational English. Have you ever considered the formal definition of "logic"? It is derived from the Greek root "logs" and the suffix "ic." "Logs" refers to words or, in this case, reason. The suffix "ic" means "relating to" or "the study of." Therefore, logic means the study of reason.

"logs"—words, reason
"ic"—relating to, the study of
logic—"the study of reason"

The Greek philosophers developed many simple laws of logic that were applied to the study of natural phenomena. In the nineteenth century, George Boole developed a branch of mathematics especially designed to handle these logical arguments. This math, called *Boolean algebra*, is an important tool used in the development of digital circuits.

"Gate" is another word in common use. In your study of electronics, you have surely noticed that in technical English, common words are often taken and given very specific, rigorous definitions. In common usage, "gate" means

an opening in a wall or fence, a means of entrance or exit

The technical definition of "gate" is a natural extension of its common form:

a device that outputs a signal when specified input conditions are met

We now know that a *logic gate* is

a digital device that performs a predetermined process

2.3 THE PHYSICAL REALIZATION OF LOGIC GATES

The logic gates that we will study in this book are contained in *integrated circuits* (ICs). In a later chapter we will examine the internal workings of ICs. Until then, we will view logic gates as functional blocks. Do not concern yourself with the details of how these circuits are realized; simply concentrate on understanding the basic logic functions.

2.4 A STRUCTURED INTRODUCTION TO LOGIC GATES

Each type of gate will be introduced by noting the following characteristics:

1. A conversational example of how we use this logic function in everyday conversation. This English version will define both the positive and negative approaches to each logic function.
2. A simplified circuit schematic using diodes, resistors, and transistors to illustrate a first approximation of how the circuit could be constructed using common electronic components. In these circuits all the components will be considered "ideal." That is, a forward-biased diode will look like a closed switch and have no forward voltage drop accross it; a reverse-biased diode will look like an open switch (see Figure 2.1). All transistors will be in either cutoff or saturation. Transistors will not be used as linear amplifiers but as electronic switches (Figure 2.2). Remember that when a transistor is cutoff, no collector current flows and the collec-

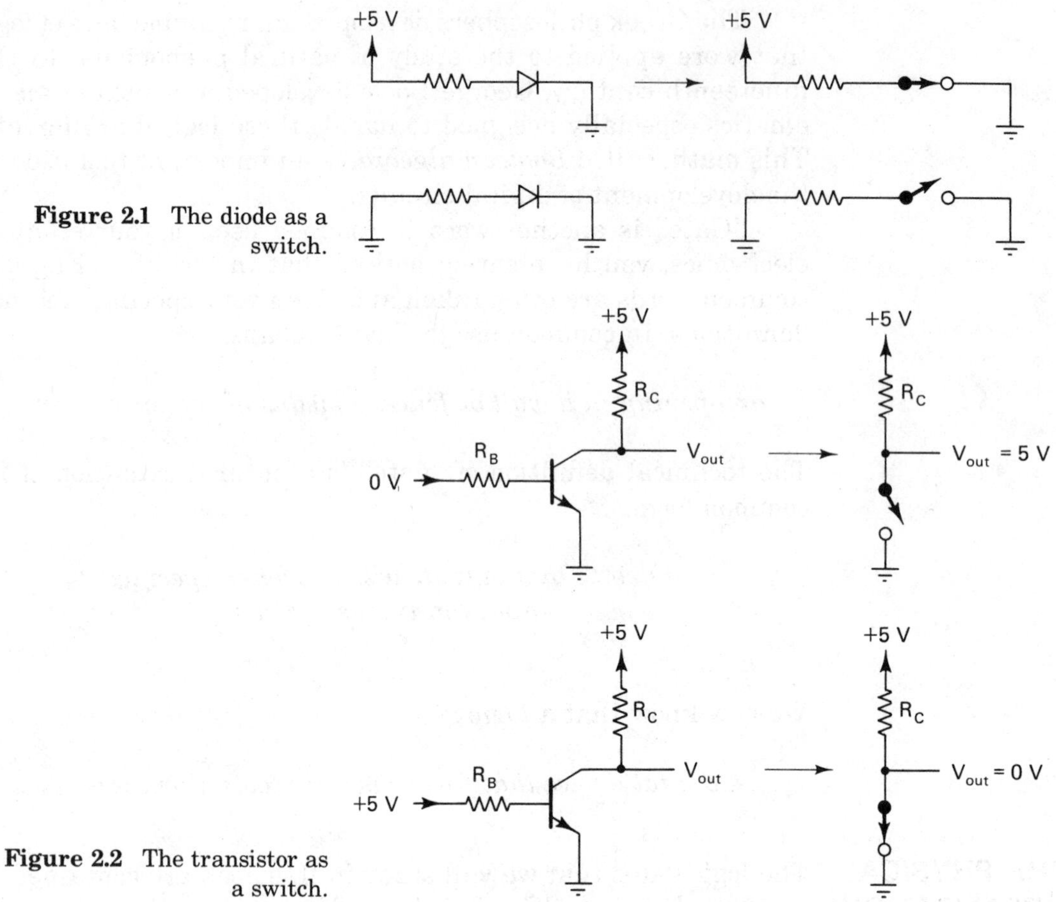

Figure 2.1 The diode as a switch.

Figure 2.2 The transistor as a switch.

tor–emitter junction appears to be open. This means that during cutoff the transistor looks like an open switch. On the other hand, when the transistor is turned on with a positive voltage on its base, it will go into saturation. A transistor is in *saturation* when

the collector current no longer follows the base current by a factor of beta

When a transistor is in saturation, the collector–emitter junction offers a minimum impedance and looks like a closed switch.

3. A truth table. A *truth table* contains all the possible combinations of input values and the output value that corresponds to each input combination. Because digital signals are either low or high, the total number of input combinations possible for any gate will be equal to

$$\text{number of unique input combinations} = 2^n$$

where n is the number of inputs. For example:

A two-input gate will have $2^2 = 4$ *unique input combinations.*

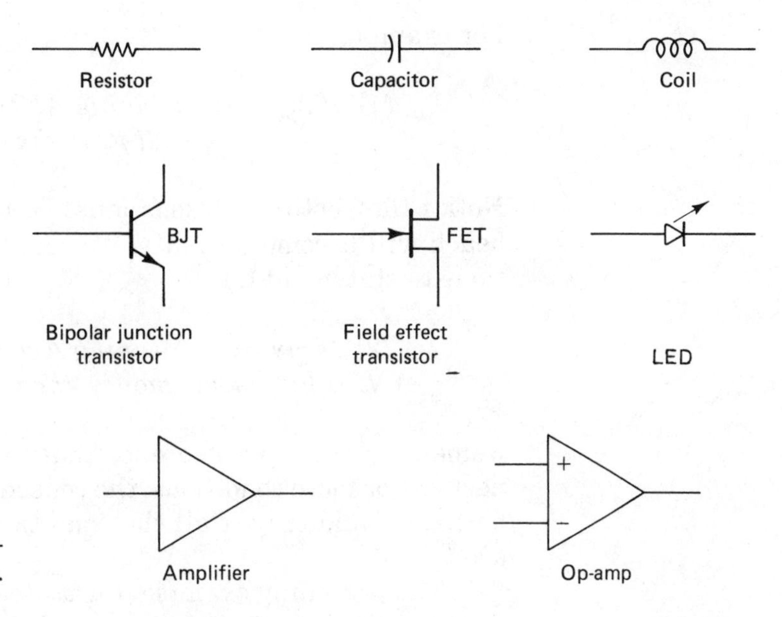

Figure 2.3 Common schematic symbols.

A three-input gate will have 2^3 = 8 unique input combinations.

4. Every gate has a dynamic input level and a unique output level. A *dynamic input level* is

   an input level that forces the output of a gate to a particular value, regardless of the other input values applied to the gate

   A *unique output level* is

   that output level of a gate that occurs exclusively for only one combination of inputs

Figure 2.3 illustrates some common schematic symbols that you should already know. Each logic gate will have two symbols that can be used to represent it in a schematic diagram. The symbol that is chosen will be the one that best describes the function of the gate in the context in which it is applied. Schematics should be more than just "road maps" of components; they should help the technician to understand how the circuit functions. This is why great care should be taken in choosing the proper logic symbol to represent a gate.

2.5 THE AND GATE

2.5.1. The AND Function

We often use sentences that fit into the following form:

If ⟨first condition is true⟩ AND ⟨second condition is true⟩, then ⟨consequence is true⟩.

For example:

(1) If my car is running AND I get the day off, then I'll go to the beach.

Notice that both conditions must be true before the day at the beach will become a reality. We could easily add further conditions to statement (1)

(2) If my car is running AND I get the day off AND I have some money, then I'll go to the beach.

Remember that in a sentence that uses the word AND to connect two or more conditions, the consequence of the sentence will be true if and only if all the conditions stated in the sentence are true.

The preceding example illustrated a positive-logic approach to the AND function. A true condition AND a true condition yielded a true consequence. The next example will illustrate a negative-logic version of the AND function.

If ⟨first condition is false⟩ OR ⟨second condition is false⟩, then ⟨consequence will be false⟩.

All the terms in the preceding statement were negative terms. This is just another way of expressing the original AND statement.

(3) If my car isn't running OR I can't get the day off, then I won't go to the beach.

Statements (1) and (3) say exactly the same thing. There are two different methods of expressing an AND statement. It all depends on one's point of view. Statement (1) has positive overtones; all its elements refer to true states. Statement (3) has negative overtones; every part of statement (3) is negative. Often, the form of AND that we choose to use depends on whether or not we believe that an event is likely to occur.

2.5.2 A Diode Model of an AND Gate

We can build a crude form of AND gate using two diodes and a resistor (Figure 2.4). The truth table in Figure 2.4 depicts all the possible input combinations of V_{in1} and V_{in2}. Let's follow each line of the truth table step by step:

Line 1: V_{in1} = 0 V, V_{in2} = 0 V Applying 0 V to each input essentially places a ground on the cathode of D1 and D2 (Figure 2.5). We can see that D1 and D2 are both forward biased. Therefore, +5 V (remember that we are considering the diodes as ideal

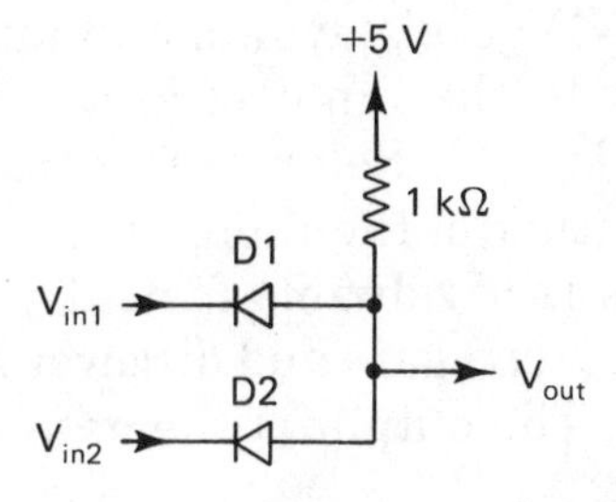

V_{in}		V_{out}
2	1	
0 V	0 V	0 V
0 V	5 V	0 V
5 V	0 V	0 V
5 V	5 V	5 V

Figure 2.4 Diode AND gate and truth table.

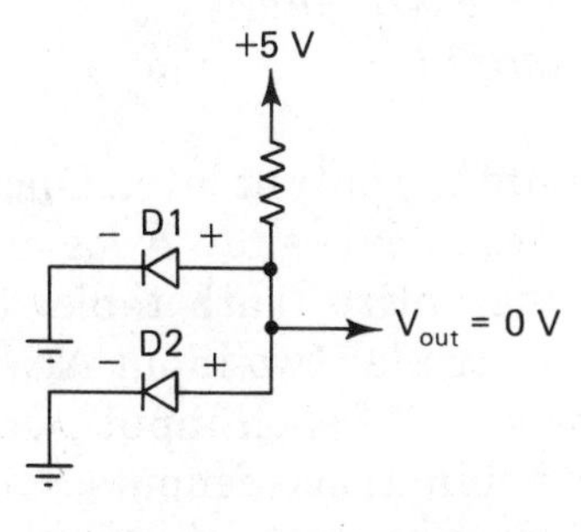

Figure 2.5 Both inputs low.

devices) is dropped across the 1-kΩ resistor and the output is 0 V.

Line 2: V_{in1} = 5 V, V_{in2} = 0 V Here D1 is not forward biased because it has +5 V applied to its cathode. But D2 is still forward biased and conducting, with 0 V on its cathode. Current finds a path through D_2 to ground and the output voltage is once again equal to 0 V.

Line 3: V_{in1} = 0 V, V_{in2} = 5 V Diodes D1 and D2 have switched the roles they played in line 2 of the truth table. With a path to ground through D1, current flows through the resistor. The +5 V is dropped across the resistor and the output is still equal to 0 V.

Line 4: V_{in1} = 5 V, V_{in2} = 5 V This is the case that we have been waiting for: both inputs are true (Figure 2.6). Both D1 AND D2 are reverse biased. The current has no place to flow. We know that *Ohm's law* states:

The voltage drop across a resistor is equal to the value of the resistor times the current flowing through it.

Because there is no current flowing through the resistor, there is no voltage drop across it. That is, the voltages at both ends of the resistor are the same. Therefore, the output is equal to +5 V: a logic 1. If either input is a logic 0, current will flow through the resistor and the output voltage will be pulled down to 0 V.

2.5.3 Dynamic Input and Unique Output Levels

Each of the four basic gates has a dynamic input level. No matter how many inputs a gate may have, if its dynamic input level is applied to the gate, the output will be forced to a particular level, independent of the other inputs.

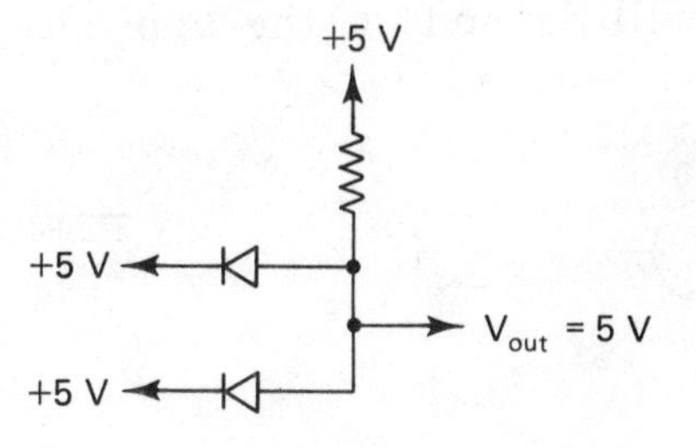

Figure 2.6 Both inputs high.

The AND gate's dynamic input level is a logic 0. If a logic 0 is applied to the input of an AND gate, the output will be forced to a logic 0. Refer to the truth table of the AND gate. Whenever either of the inputs is at logic 0, the output is also at logic 0. This is a valuable fact. If you are checking the inputs of a 10-input AND gate and discover a logic 0, you need not check any further. The output of the gate will be a logic 0.

A logic 0 on any input of an AND gate will force a logic 0 output.

Each of the basic gates also has a unique output level. Once again, let's examine the truth table of the AND gate. A logic 1 on the output appears only once in the entire truth table. If someone told you that the output of a particular two-input AND gate was logic 0, could you predict the state of each input? Obviously not. There are three different combinations of inputs that can result in a logic 0 on the output. Now, compare this to the situation where you were informed that the output of the AND gate was equal to logic 1. Could you now accurately predict the inputs? Of course. All the inputs must be logic 1. Therefore, a logic 1 on the output of an AND gate is a unique event that can result from only one combination of inputs.

A logic 0 on the output of an AND gate is a unique event.

The dynamic input level and the unique output level of a gate are useful characteristics to know. You will see later that these two parameters actually define each of the four basic gates.

2.5.4 The Schematic Symbols of the AND Gate

We have seen that there are two different possible approaches to the concept of an AND gate.

1. The positive approach: If all the inputs are true, the output will be true.

2. The negative approach: If any of the inputs are false, the output will be false.

Each of the different views of an AND gate has its own circuit symbol. Which symbol is chosen for any particular situation will depend on the application of the gate within the

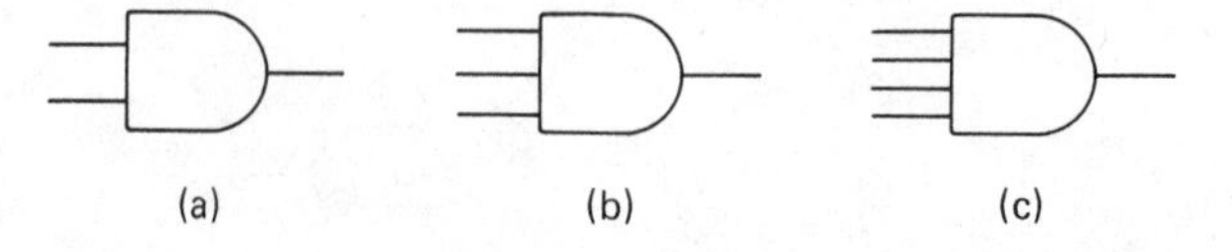

Figure 2.7 Positive-logic symbols for the AND gate.

context of the other circuitry. A full explanation of this idea will be presented in Chapter 3. The symbol that looks like half of an ellipse represents the AND function (Figure 2.7). By adding more inputs we can represent AND gates of any size. The proper way to interpret this symbol is (Figure 2.8):

If input (a) is high AND input (b) is high, then the output will be high; otherwise, the output will be low.

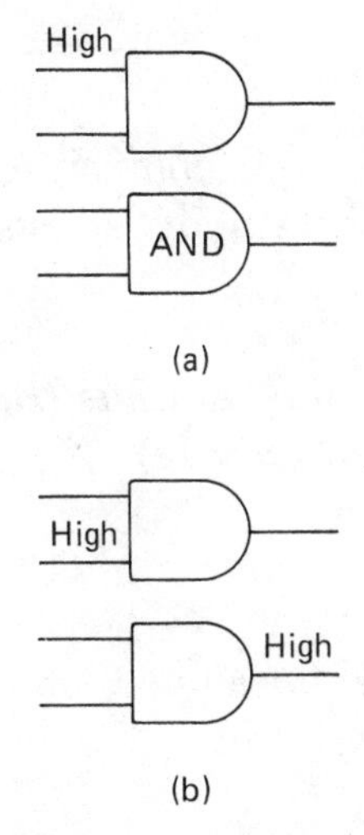

Figure 2.8 Interpretation of the AND symbol.

An alternative symbol for the AND gate is shown in Figure 2.9. This symbol represents the negative-logic approach to an AND function. The symbol that consists of the three curved lines represents the OR function. The small circles on the inputs and output are called *bubbles*. A bubble is really a small zero. A bubble on the input of a gate means to expect a logic 0 on this input, whereas a bubble on the output of a gate means to expect a logic 0 output. This may sound a bit confusing, but an illustration of how to interpret the symbol should clarify this situation (Figure 2.10):

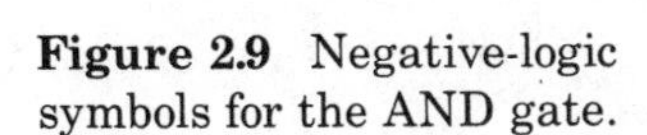
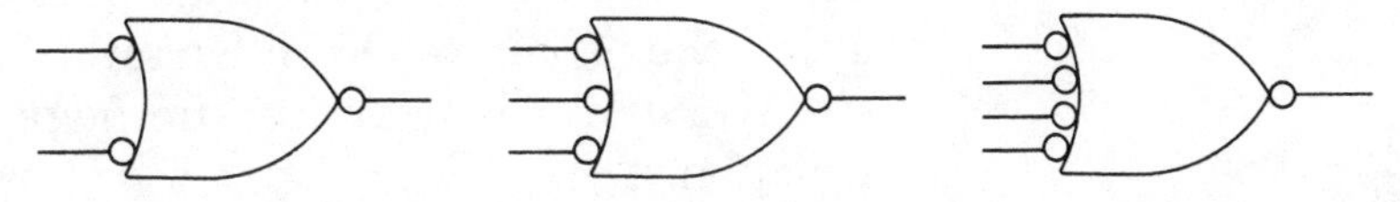

Figure 2.9 Negative-logic symbols for the AND gate.

If input (a) is low OR input (b) is low, then the output will be low; otherwise, the output will be high.

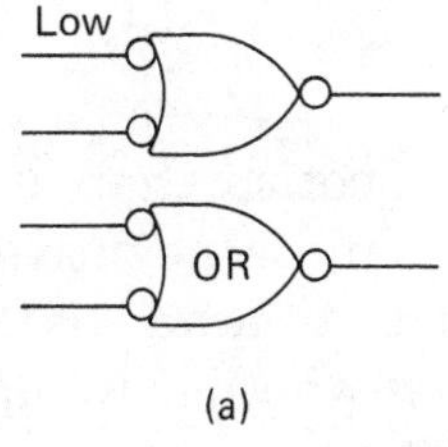

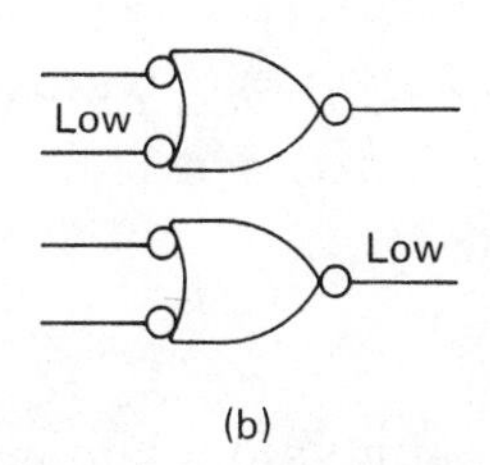

Figure 2.10 Interpretation of negative-logic symbol of an AND gate.

Notice that wherever we had a bubble we referred to that pin in its logic 0 state.

2.5.5 Summary of the AND Function

Review the summary of the AND gate given in Figure 2.11 before proceeding to the next section. Do not try to memorize all these facts. You are already familiar with the AND function. All you have to do is understand the AND function in its context of digital electronics.

B	A	Out
0	0	0
0	1	0
1	0	0
1	1	1

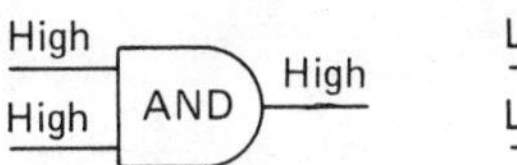

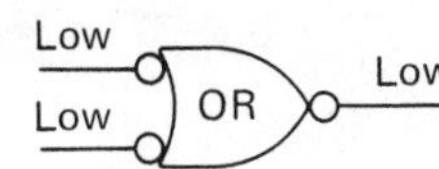

Figure 2.11 Summary of the AND function.

2.6 THE OR GATE

2.6.1 The OR Function

The other common connector that we use in compound sentences is the term OR. The following statement shows the typical structure of an OR sentence:

If ⟨condition 1 is true⟩ OR ⟨condition 2 is true⟩, then ⟨consequence will be true⟩.

For example:

If I get a raise OR the bank gives me a loan, then I can buy a new car.

If either or both conditions are true, the consequence will be true. This type of OR is called an *inclusive Or,* because it includes the case when both statements are true. An exclusive OR does not include the case when both conditions are true. We will study the exclusive OR in Chapter 3.

We can now examine the negative-logic version of the OR function.

If ⟨condition 1 is false⟩ AND ⟨condition 2 is false⟩, then ⟨consequence will be false⟩.

Important Observation. Did you notice that a negative-logic AND uses an OR as a connector and a negative-logic OR uses an AND as a connector? This proves that an AND is an OR and an OR is an AND. It all depends on whether we approach a certain situation from a positive- or a negative-logic point of view.

If I don't get a pay raise AND the bank won't give me a loan, then I can't buy a new car.

2.6.2 Diode Model of an OR Gate

The circuit shown in Figure 2.12 is a simple implementation of a two-input OR gate using two diodes and a resistor. Did you notice that this OR gate uses the same components as the diode AND gate? The diodes are turned around so that their anodes are connected to the input signals. Also, the resistor is con-

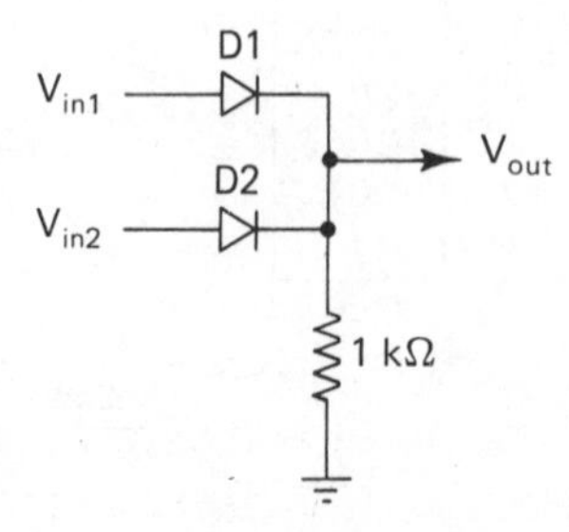

V_{in} 2	V_{in} 1	V_{out}
0 V	0 V	0 V
0 V	5 V	5 V
5 V	0 V	5 V
5 V	5 V	5 V

Figure 2.12 Diode OR gate and truth table.

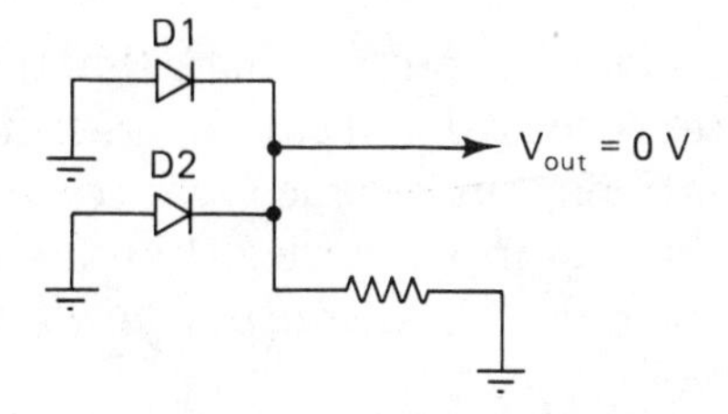

Figure 2.13 Both inputs low.

nected to ground instead of to +5 V. It seems that the diode OR gate is the diode AND gate flipped upside down and inside out. This should give you some insight into the relationship between the AND and OR functions.

The truth table in Figure 2.12 lists all the possible input combinations, together with their respective output levels. Let's follow through each line in the truth table as it relates to the schematic.

Line 1: V_{in1} = 0 V, V_{in2} = 0 V With both inputs at logic 0, we can create the equivalent circuit shown in Figure 2.13. Because neither diode is conducting, the output will be pulled down to 0 V via the resistor.

Line 2: V_{in1} = 5 V, V_{in2} = 0 V Although D2 is still not conducting, D1 is forward biased and 5 V will appear at the output.

Line 3: V_{in1} = 0 V, V_{in2} = 5 V This is just the opposite situation of line 2. This time, D2 is conducting and the output will also be 5 V.

Line 4: V_{in1} = 5 V, V_{in2} = 5 V Both D1 and D2 are conducting. The output is, once again, equal to 5 V.

In summary, if either D1 OR D2 OR both D1 and D2 are conducting, the output will be 5 V; otherwise, the output will be equal to 0 V.

2.6.3 Dynamic Input and Unique Output Levels

Just like the AND gate, the OR gate also has a dynamic input level. By examining the truth table for the OR gate, it is obvious that the dynamic input level of the OR gate must be a logic 1. By applying a logic 1 to any input of an OR gate, the output will be forced to logic 1, independent of the other inputs. Again we see that the AND and OR functions have opposite characteristics. As you will remember, the dynamic input level of an AND gate is a logic 0, which forces a logic 0 output. The OR gate has a dynamic logic 1 input, which forces a logic 1 output.

A logic 1 on the input of an OR gate
forces the output to logic 1.

The truth table also reveals the unique output level of the OR gate. We see that a logic 0 appears on the output for only one

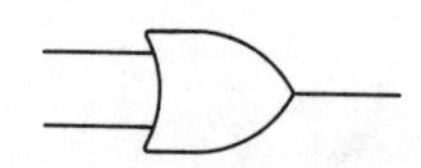

Figure 2.14 Positive-logic OR-gate symbol.

combination of inputs. A logic 0 is the unique output level of the OR gate. A logic 1 output can result from three separate input combinations, but a logic 0 can result from only one unique set of inputs. Again we notice that the AND and OR gates are opposite in their unique output levels.

A logic 0 on the output of an OR gate is a unique event.

2.6.4 The Schematic Symbols of the OR Gate

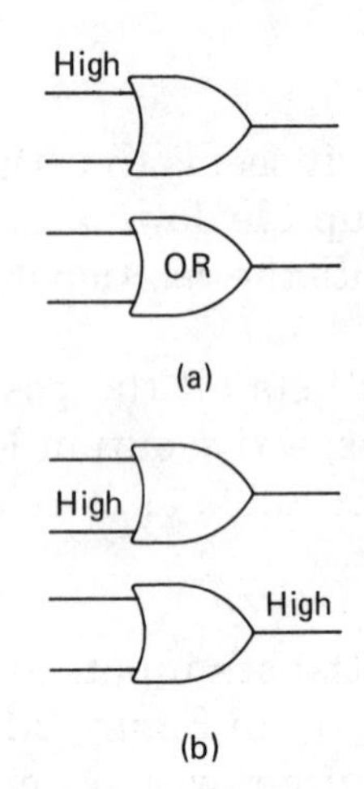

Figure 2.15 Interpretation of a positive logic OR gate symbol.

The OR gate will also have two different symbols. Just like the AND gate, the OR-gate symbols will be either positive- or negative-logic representations of the OR function. The positive-logic symbol is shown in Figure 2.14. You already know the meaning of the symbol composed of the three curved lines. This is an OR symbol, representing the OR function. We have seen that the negative-logic representation of the AND function uses an OR symbol with bubbled inputs and outputs. This OR symbol has no bubbles. That means that the inputs and output are expected to be high levels for the gate to be active. You should interpret this OR symbol as follows (Figure 2.15):

If input (a) is high OR input (b) is high, the output will be high; otherwise, the output will be low.

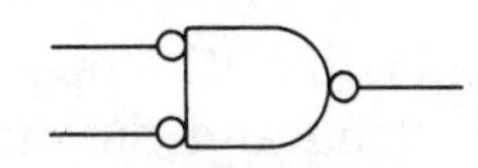

Figure 2.16 Negative-logic symbol of an OR gate.

The negative-logic symbol (Figure 2.16) should be interpreted as (Figure 2.17);

If input (a) is low AND input (b) is low, the output will be low; otherwise, the output will be high.

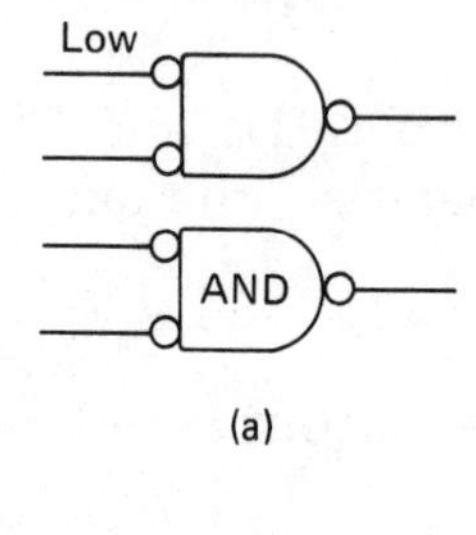

This negative-logic OR symbol is used when we want to output a logic 0 in response to all logic 0's on the inputs. In the next section we will learn some simple applications that will give a practical meaning to each of the four logic symbols that we have encountered so far.

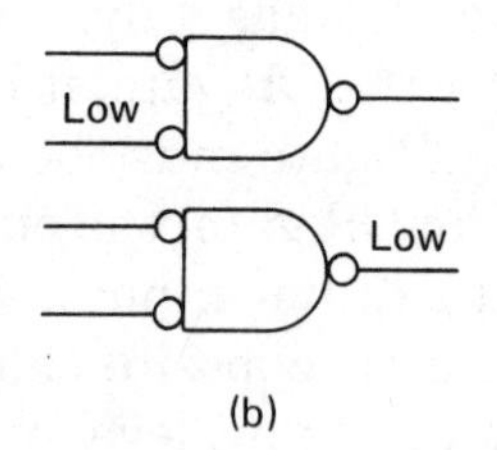

Figure 2.17 Interpretation of a negative-logic OR gate symbol.

2.6.5 Summary of the OR function

Review the summary in Figure 2.18 to test your understanding of the OR function.

2.7 REVIEW OF THE LOGIC SYMBOLS FOR AND AND OR

Figure 2.19 illustrates each logic symbol that we have covered. The output of each gate is driving an LED. It is standard practice to place a current-limiting resistor in series with each LED. Without this resistor, the LED would quickly be destroyed by

B	A	Out
0	0	0
0	1	1
1	0	1
1	1	1

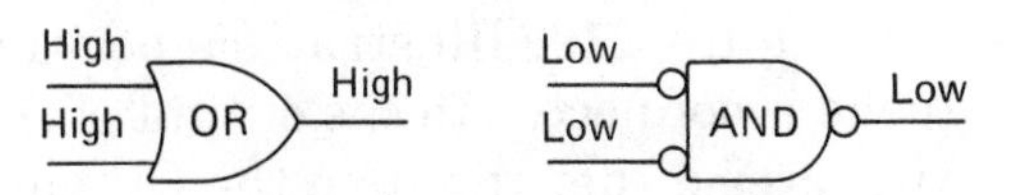

Figure 2.18 Summary of the OR gate: dynamic input level is a logic 1; unique output level is a logic 0.

excessive current flow. Let's examine each circuit to consider how it functions and why a particular symbol was used.

Figure 2.19a shows the positive-logic symbol for an AND function. If both input 1 and input 2 are high, the output will go high. This will place +5 V on the anode of the LED and the LED will illuminate. This circuit can be best summarized as being

an active-high-input, active-high-output AND gate

By the term *active* we are referring to

the level that causes a desired circuit action to occur

In this case, the desired circuit action is to illuminate the light-emitting diode (LED). This is accomplished by applying logic 1's to each input of Figure 2.19a.

Figure 2.19b uses the negative-logic symbol for an AND gate. We know that the unique output level of an AND gate is logic 1. Here there are three different input combinations that will illuminate the LED. A logic 0 on the output of this AND gate will effectively place a ground on the cathode of the LED. This will enable it to illuminate. The circuit can be summarized as

an active-low-input, active-low-output OR gate

Positive-logic active-high output:

Negative-logic active-low output:

+5 V

220 Ω

AND

220 Ω

(a)

(b)

+5 V

220 Ω

OR

220 Ω

(c)

(d)

Figure 2.19 Active levels in the AND and OR gates.

Even though this circuit is an AND gate, in this application it is being used as an OR gate. We are saying that the job of this gate is to light the LED whenever either input is low. The concepts of AND and OR are not rigid concepts but rather, flexible tools used to process digital signals.

Figure 2.19c illustrates a positive-logic OR gate. The LED's anode is connected to the output of the OR gate. This means that whenever either input on the OR gate is at logic 1, the LED will light. This can be summarized as

an active-high-input, active-high-output OR gate

Figure 2.19d represents the negative-logic symbol of an OR function. The cathode of the LED is pointed toward the output of the gate. This tells us whenever the output of this gate goes low, the LED will light. This can be summarized as

an active-low-input, active-low-output AND gate

As in Figure 2.19b, this AND function appears to be used out of place. In fact, we are using an OR gate to provide an AND function. We want the LED to light when both inputs are low. This gate provides that function.

Compare Figures 2.19a and c. Both these gates provide an active-high output response. The LED at the output of each of these gates will light with a logic 1 output of each gate. Contrast this idea with Figures 2.19b and d. These gates provide an active-low output level. A low on the output of these two gates will illuminate the LEDs.

2.8 THE INVERTER

2.8.1 The Not Function

The NOT logic function is simple. It inputs a logic level and outputs the logic level that the input is not. If the operand for the NOT function was logic 0, the output would be a logic 1—the logic value that the input is not. On the other hand, if the operand for a NOT function was a logic 1, the output would be a logic 0.

2.8.2 A Transistor Inverter

Up to this point, each logic function that we have studied has had two or more operands. A one-input AND gate or a one-input OR gate makes absolutely no sense at all. But the NOT logic function operates on only one operand. The gate that provides this function is called an *inverter* (Figure 2.20). This transistor configuration should remind you of a common-emitter amplifier. We can see that the base is driven with V_{in} and the output is equal to the voltage drop across the collector–emitter junction of the transistor. Another thing we should consider is

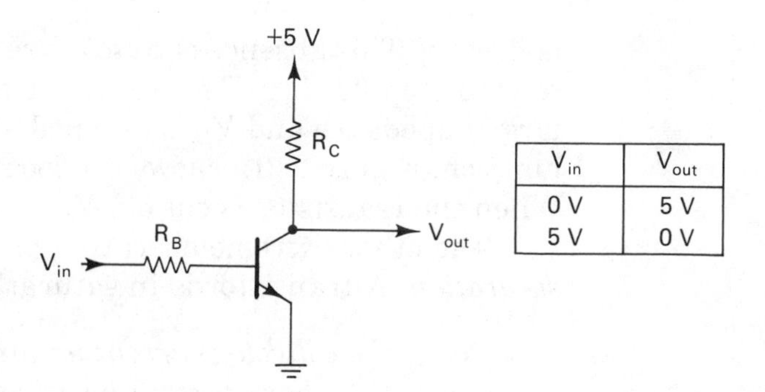

V_{in}	V_{out}
0 V	5 V
5 V	0 V

Figure 2.20 Transistor inverter and truth table.

that the output voltage in a common–emitter amplifier is 180 degrees out of phase with the input voltage; that is, the output is phase inverted. We can use this characteristic to create a logic gate that will invert a logic signal. The truth table for the inverter illustrates the simplicity of this logic function. A logic 0 input will output a logic 1. On the other hand, a logic 1 input will output a logic 0. The input logic level will be inverted—flipped upside down and inside out.

Up to this point in your study of electronics, you have used transistors as linear devices. A transistor amplifier is a good example of a linear device. The output of an amplifier should be an exact copy of the input signal, the only difference being that the amplitude of the output voltage will be equal to the input voltage times the voltage gain of the amplifier. In digital electronics, transistors are not used as linear devices, that is, as amplifiers, but as electronic switches.

Transistor fundamentals are reviewed in Figure 2.21. Figure 2.21b shows a typical way to bias a transistor. Resistors R_{B1} and R_{B2} act as a voltage divider. This causes a calculated value of base current to flow. This base current is multiplied by a factor of beta to set the amount of collector current. The collector current, in turn, sets the quiescent operating point in the amplifier. This resting point of the amplifier is called the *Q point.* Figure 2.21a shows that the *Q* point is typically set halfway between cutoff and saturation.

A transistor is in a *cutoff* condition when no base current

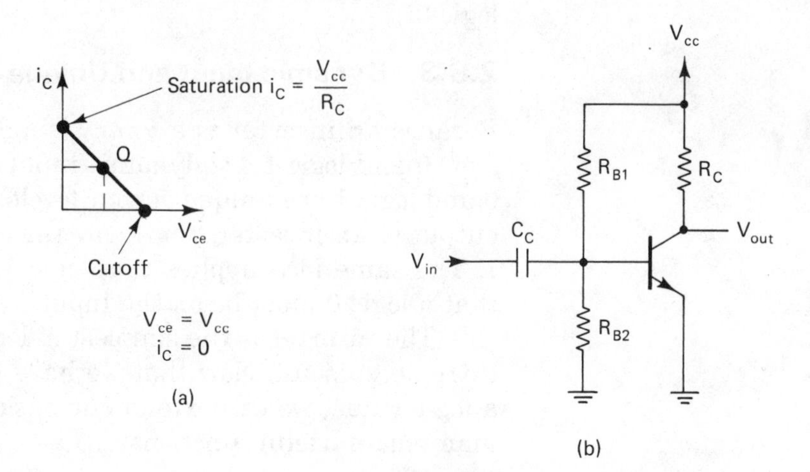

Figure 2.21 Review of transistor fundamentals.

is flowing. This absence of base current causes the collector current to stop flowing. The transistor appears to be an infinitely large impedance and V_{cc} is dropped across the collector–emitter junction. Figure 2.21a shows the location of cutoff on the graph. When the transistor is cut off, V_{out} will equal V_{cc}.

The other extreme operating point of a transistor is called *saturation.* A transistor is in saturation when

the collector current no longer follows the base current by a factor of beta

At saturation the transistor's collector–emitter junction will offer minimum impedance. In fact, the transistor is virtually a short circuit. Therefore, collector current will be maximum and equal V_{cc}/R_C, but the output voltage will be equal to the voltage from collector to emitter, which is zero volts.

It appears that if we wanted to use a transistor as an electronic switch it would operate at cutoff when it represented an open switch and at saturation when it represented a closed switch. When the transistor is cut off, the output voltage is "pulled up" to V_{cc} via the collector resistor. When the transistor is in saturation, the output voltage is "pulled down" via the transistor to ground.

Refer to the schematic of the transistor inverter circuit, Figure 2.20.

Case 1: V_{in} = 0 V The transistor is receiving no base drive current. The transistor will be in cutoff. V_{out} will be pulled up to +5 V via the collector resistor.

Case 2: V_{in} = 5 V The value of the base resistor was chosen to saturate the transistor with a logic 1 input. The collector current will be maximum and V_{out} will be pulled down via the transistor to 0 V.

The circuit acts as a digital inverter. It is said to *complement* the input logic level. To complement a digital level means to replace it with its opposite digital level. That is, complementing a logic 0 yields a logic 1, and complementing a logic 1 yields a logic 0.

2.8.3 Dynamic Input and Unique Output Levels

Because an inverter is a *unary* (single operand) function, both logic 0 and logic 1 are dynamic input levels. Further, both logic 0 and logic 1 are unique output levels. If you see a logic 0 on the output of an inverter, you know that the input must be a logic 1. The same idea applies to a logic 1 on the output: you know that a logic 0 must be on the input.

The inverter is the simplest of logic functions, but it is also extremely useful. Now that we have the ability to complement a logic value, we can extend our basic gates to provide us with many more useful functions.

2.8.4 Logic Symbols of the Inverter

We have seen that a bubble on the input or output of a logic gate implies a low level. We can use a bubble and the universal symbol of an amplifier to create a logic symbol for the inverter. Figure 2.22a represents the positive-logic symbol for the inverter. The triangle represents an amplifier. We know that the inverter is a digital device, and one may wonder why the symbol for an amplifier was chosen to represent it. In your study of digital electronics, you will discover that an inverter is often used to *buffer* an output from another logic gate. A buffer provides current gain. For this reason an amplifier symbol was chosen for an inverter: it symbolizes current amplification. For now, just accept the symbol at face value. A logic 1 will be input and the bubble symbolizes the inversion to a logic 0. A logic 0 can be input and the bubble will invert it into a logic 1. This is a positive-logic symbol because it "expects" to input a logic 1 and output a logic 0.

The negative-logic symbol (Figure 2.22b) simply moves the bubble to the input of the inverter. This symbol "expects" to input a logic 0 and output a logic 1. The symbol that is chosen for any particular application will depend on whether an active-high or active-low output is expected from the inverter. Figure 2.23 illustrates active-high and active-low outputs of the inverter and the proper symbol for each.

2.8.5 Summary of the Inverter

The inverter is summarized in Figure 2.24.

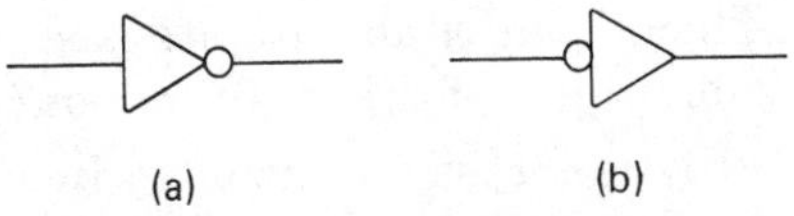

(a) (b)

Figure 2.22 Logic symbols of the inverter: (a) positive logic; (b) negative logic.

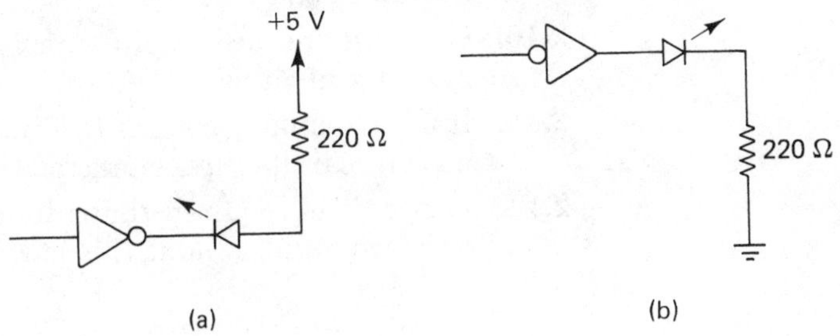

+5 V
220 Ω
(a)
220 Ω
(b)

Figure 2.23 Active low (a) and high (b) inverter circuits.

Input	Output
0	1
1	0

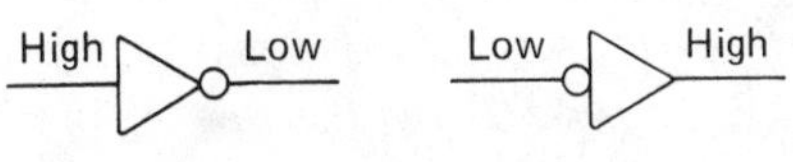

High Low Low High

Figure 2.24 Summary of the inverter: because the inverter function is unary, both logic 0 and logic 1 are dynamic input levels; each output is a unique output.

QUESTIONS AND PROBLEMS

2.1. Define the following terms.
(a) Dynamic input level
(b) Unique output level
(c) Active level of a digital input or output

2.2. What is the dynamic input level of:
(a) An AND gate?
(b) An OR gate?
(c) An inverter?

2.3. What is the unique output level of:
(a) An AND gate?
(b) An OR gate?
(c) An inverter?

2.4. Match the logic symbol with the proper word description.

(a) 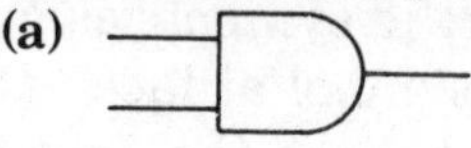 (1) Active-high-input, active-high-output OR gate

(b) 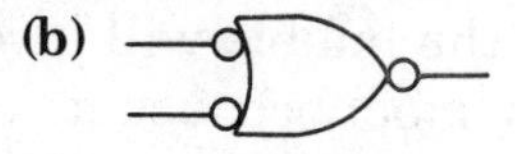 (2) Active-low-input, active-low-output OR gate

(c) 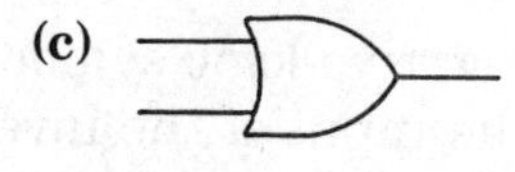 (3) Active-low-input, active-low-output AND gate.

(d) 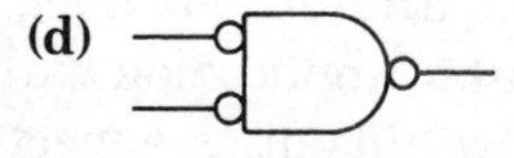 (4) Active-high-input, active-high-output AND gate

2.5. What logic body symbol is used to describe a gate in its:
(a) Dynamic input state?
(b) Unique output state?

2.6. Draw a circuit that illuminates an LED whenever all three inputs are low. Take care to use the logic symbol that best illustrates the function of this circuit.

2.7. The output of an AND gate is stuck low. What could be the possible cause of this malfunction?

2.8. The output of an OR gate is stuck high. What could be the possible cause of this malfunction?

2.9. Why is a resistor included in series with every LED? If a particular LED requires 12 mA of forward current, what value should this resistor be?

2.10. Contrast the fashion in which transistors are used in analog and digital circuits.

2.11. If a particular gate has five inputs, how many possible combinations can its inputs take on?

2.12. A circuit is required that illuminates an LED when inputs A or B are high or input C is low. Illustrate this circuit.

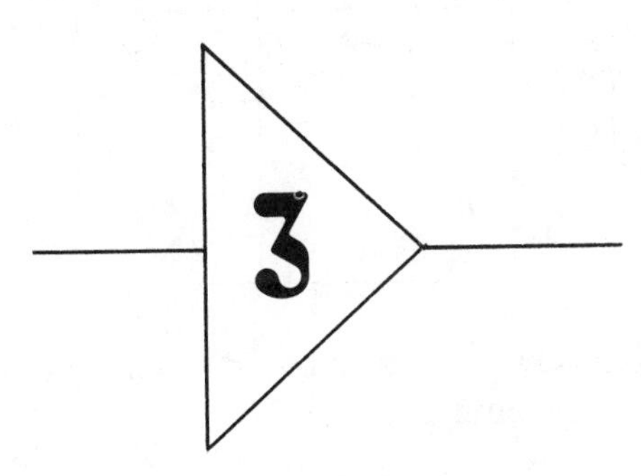

ADVANCED LOGIC GATES: NAND, NOR, AND EXCLUSIVE OR

3.1 THE NAND GATE

3.1.1 The NAND Function

The NAND gate is really a combination of the NOT and AND functions. The word "NAND" stands for NOT-AND. A NAND gate can be modeled as an AND with an inverter on its output. The truth table in Figure 3.1 compares the AND and NAND functions. This truth table follows directly from the model. You may wonder why the NAND function is explicitly defined. Why not just use an AND gate and an inverter whenever we want the function of NOT-AND? There are two answers to this question.

1. In the integrated-circuit technology called TTL (transistor-to-transistor logic), the natural logic function is NAND. In TTL all gates are really constructed using NAND gates as the basic building block. In the next chapter we will study the TTL logic family in depth.
2. The NAND function is called a *universal building block.* This means that all the logic functions that we have introduced—AND, OR, and NOT—can be implemented using the NAND gate.

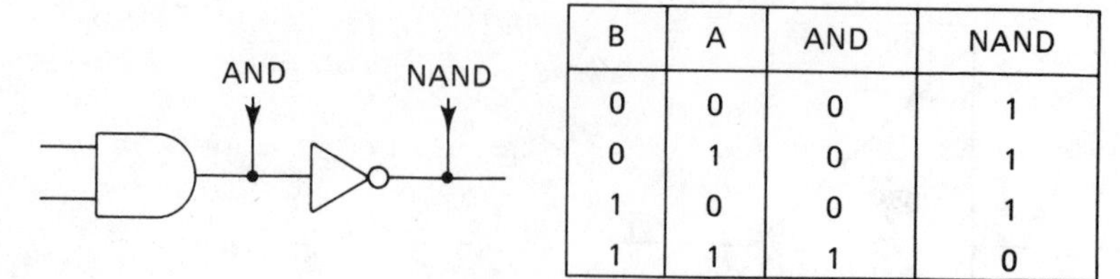

B	A	AND	NAND
0	0	0	1
0	1	0	1
1	0	0	1
1	1	1	0

Figure 3.1 NAND model and truth table.

For these two reasons, the NAND gate may be the most widely used logic function that you will learn.

3.1.2 Dynamic Input and Unique Output Levels

You now know that the NAND gate is an embodiment of the AND and NOT functions, but you may wonder how this affects the dynamic input and unique output levels. Just like the AND function, the NAND function has a dynamic input level of logic 0. A logic 0 on any of the inputs will force the output to a particular level, regardless of the other inputs. But unlike the AND, where a logic 0 in forces a logic 0 out, the NAND reacts with a logic 1 output to a logic 0 input.

A logic 0 on the input of a NAND gate forces a logic 1 output.

This should make perfect sense when you consider that an inverter would take a logic 0 on the output of the AND and turn it into a logic 1. If you are probing the inputs of a NAND gate and you come across a logic 0, you need not look any further. The output of the gate should be a logic 1.

By referring to the truth table in Figure 3.1, we can find the unique output level of a NAND gate. This unique output level is logic 0. For only one combination of inputs (all logic 1's) will a NAND gate output a logic 0. This is the opposite unique output level of an AND gate. Picture an AND with all 1's on its inputs, outputting a unique logic 1. Invert this logic 1 and you will have a logic 0: the unique output level of a NAND gate for the same combination of inputs.

The unique output level of a NAND gate is a logic 0.

The unique output level of an OR gate is also a logic 0. But a logic 0 occurs on the output of the OR gate when its inputs are all at logic 0. We can see that the NAND gate shares characteristics of both the AND gate and the OR gate.

3.1.3 The Logic Symbols of the NAND Gate

Like the previous three logic gates that we have encountered, the NAND gate has two different logic symbols: one symbol for its positive-logic applications, and another symbol for negative-logic applications.

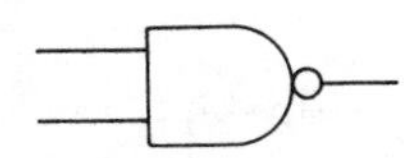

Figure 3.2 Positive-logic NAND symbol.

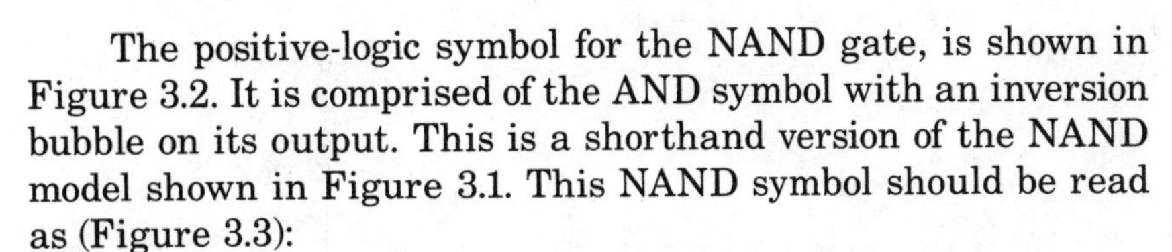

The positive-logic symbol for the NAND gate, is shown in Figure 3.2. It is comprised of the AND symbol with an inversion bubble on its output. This is a shorthand version of the NAND model shown in Figure 3.1. This NAND symbol should be read as (Figure 3.3):

If input (a) is high AND input (b) is high, then the output will be low; otherwise, the output will be high.

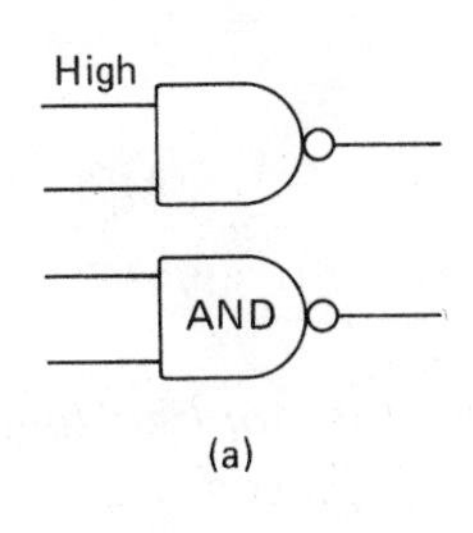

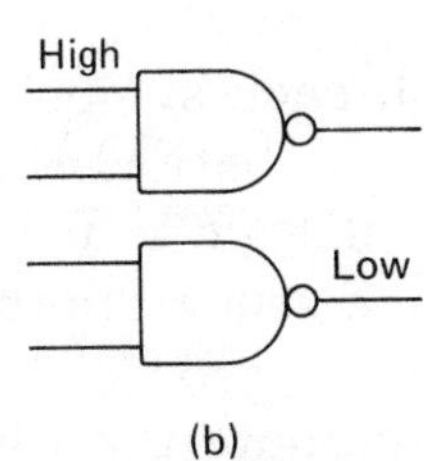

Figure 3.3 Interpretation of the NAND symbol.

The inputs are not bubbled, so they can be thought of as being active high. The main body of the symbol is an AND gate, and the output is bubbled. Therefore, this positive-logic symbol of the NAND gate can be thought of as

An active-high-input, active-low-output AND gate

The second NAND symbol, representing its negtive-logic application, is illustrated in Figure 3.4. You can picture this symbol as having an inverter in series with each input of an OR gate. This symbol should be interpreted as saying (Figure 3.5):

If input (a) is low OR input (b) is low, then the output will be high; otherwise, the output will be low.

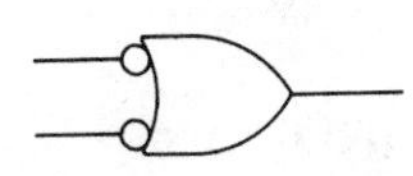

Figure 3.4 Negative-logic NAND symbol.

We see that the inputs are bubbled, indicating an active-low-input level. The main body of the symbol is an OR gate and the output is not bubbled, indicating an active-high-output level. With these thoughts in mind, we can think of this negative-logic symbol of the NAND gate as representing

An active-low-input, active-high-output OR gate

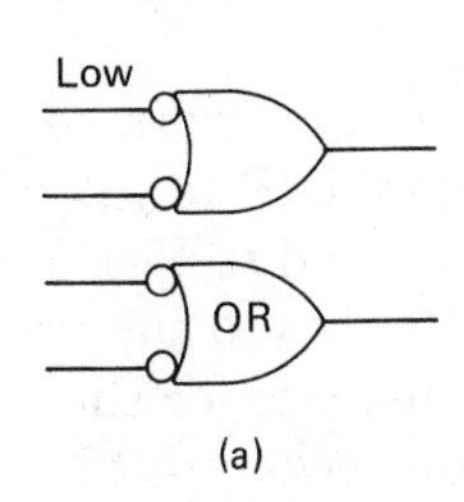

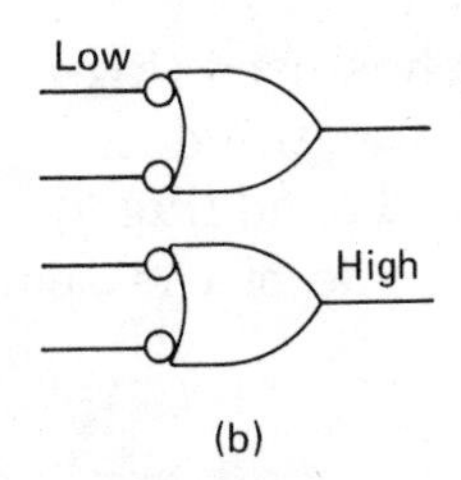

Figure 3.5 Interpretation of negative-logic NAND symbol.

Another way to think of these two symbols is in the context of unique output and dynamic input levels. Figure 3.6a shows that this representation of the NAND gate refers to its unique output level. This symbol says:

To get a unique output of logic 0, both inputs must be at logic 1.

On the other hand, Figure 3.6b represents the NAND gate from

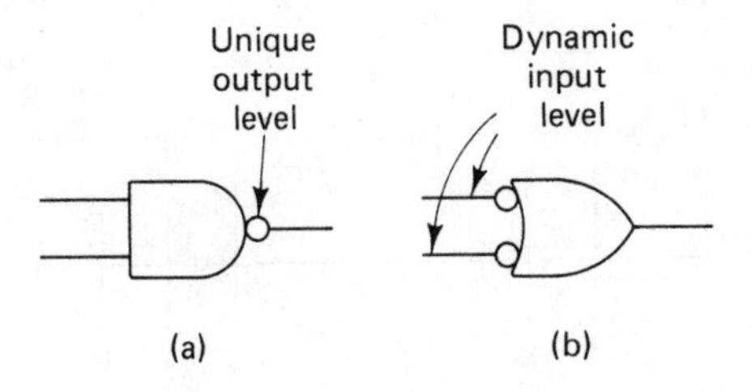

Figure 3.6 Unique output and dynamic input levels of the NAND gate.

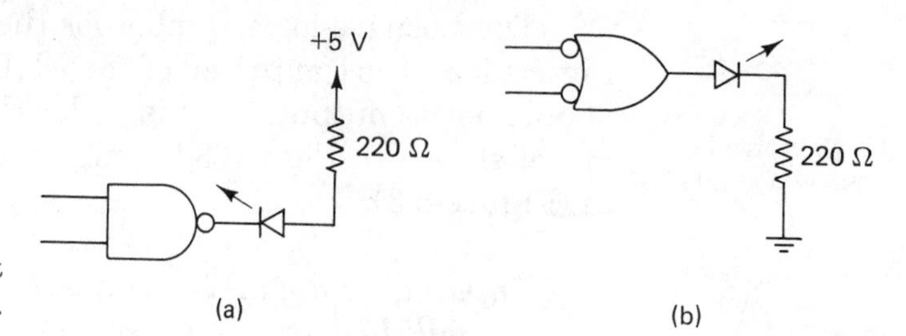

Figure 3.7 Active output levels of the NAND gate.

its dynamic input point of view. Figure 3.6b says:

If the dynamic input level of logic 0 is applied to either input, then the output will be forced high.

Figure 3.7 shows the context in which each symbol should be used. Figure 3.7a indicates that to illuminate the LED, both the first AND the second inputs must be at logic 1. This, in turn, causes a logic 0 output, which supplies a ground to the cathode of the LED.

Figure 3.7b indicates that to illuminate the LED, either input 1 OR input 2 must be at logic 0; this will force the output to logic 1. This logic 1 will supply current to the anode of the LED.

3.1.4. Creating Inverters from NAND gates

We have stated that the NAND gate is a universal building block. By using only NAND gates, we can perform any logic function: AND, OR, and invert.

It is common practice for IC manufacturers to place four two-input NAND gates into one 14-pin DIP (dual in-line pin) package (Figure 3.8). With one pin dedicated to *Vcc*, and another pin to ground, a 14-pin dip will have 12 pins to support the logic gates. Each gate will require three pins, two for the inputs and one for the output. That means that four two-input NAND gates can be encapsulated in one 14-pin DIP.

Let's say that a circuit designer has used three of the four NAND gates in this DIP package. Later the design may need the addition of a single inverter. Instead of adding another IC to the circuit, it would be convenient if one of the spare NAND

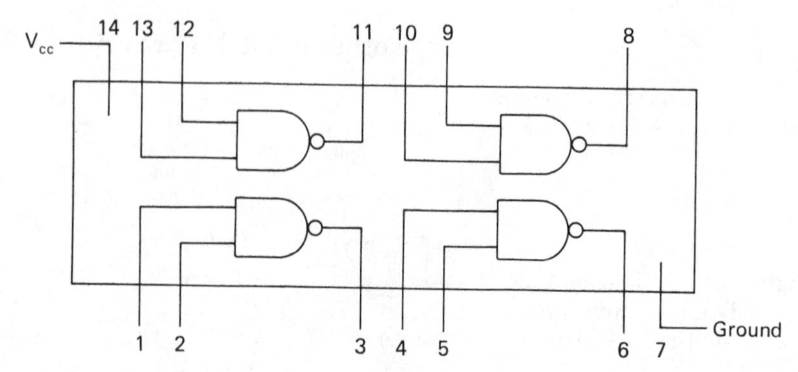

Figure 3.8 NAND gates within an integrated circuit.

TABLE 3.1 Deriving an Inverter from a NAND Gate

B	A	NAND
0	0	1
0	1	1
1	0	1
1	1	0

A = B in lines 1 and 4

gates could be used as an inverter. How would we go about making this two-input NAND gate behave as if it were an inverter?

Let's review the truth table (Table 3.1) of the NAND gate and see if it will provide us with some insight into this problem. Given the constraint that an inverter has only one input, we see a good possibility in the truth table. In lines 1 and 4, both input values are the same. Furthermore, when both inputs are equal to 0, the output will be high; when both inputs are equal to logic 1, the output will be low.

If we tie all the inputs of a NAND gate together, it will act as an inverter.

Figure 3.9 shows both the positive- and negative-logic methods of using a two-input NAND gate as an inverter.

Figure 3.10 illustrates common alternative symbols to Figure 3.9. A designer may want to stress that the NAND gate is functioning as an inverter, but the designer also wants to provide the information that the inverter will (physically) be found in a NAND-gate package. The solution is to show a multiple-input inverter. We know that a multiple-input inverter makes absolutely no sense. Therefore, this is a perfect way of showing an inverter that is really constructed from a NAND gate.

There is also one other way of forcing a NAND gate to function as an inverter. Think of the NAND gate in its negative-logic terms: as an active-low-input, active-high-output OR gate (Figure 3.11). We know that the dynamic input level of a NAND gate is a logic 0. That is, a logic 0 on an input will force a logic 1 on the output, regardless of the other input levels. Consider tying one input of this NAND gate to its nondynamic input level of

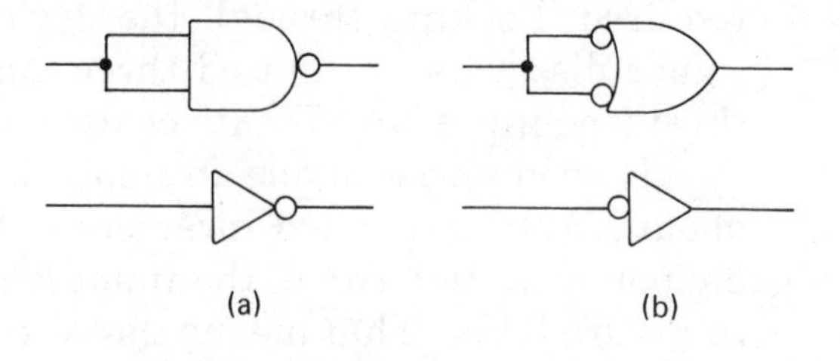

Figure 3.9 NAND gates employed as inverters.

Figure 3.10 Alternative symbols for NAND gates used as inverters.

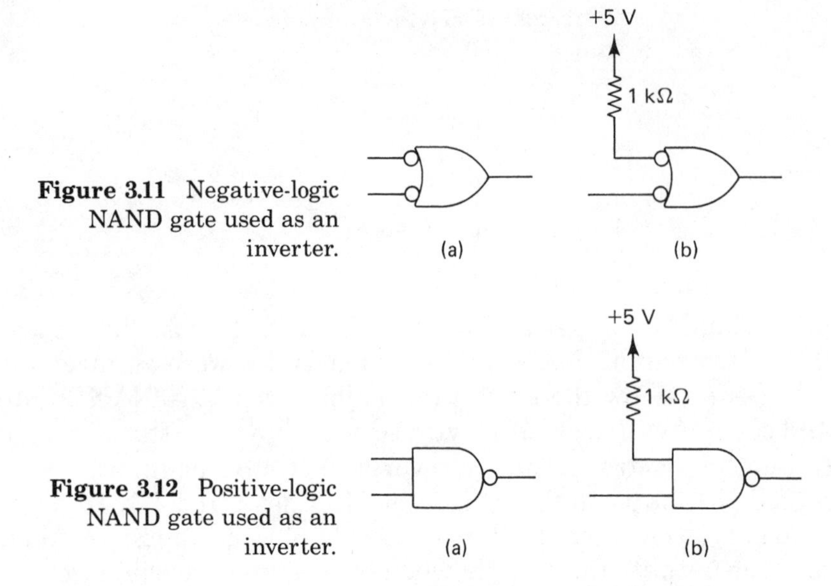

Figure 3.11 Negative-logic NAND gate used as an inverter.

Figure 3.12 Positive-logic NAND gate used as an inverter.

logic 1. We know that a logic 1 on the input of a NAND gate will not force the output to any particular level. That means that the output level of this gate will depend solely on its other input. If the input of Figure 3.11 goes low, the output will go high. That is what this symbol tells us. On the other hand, if the input goes high, neither input will have a logic 0, therefore, the output will go low. The NAND gate now acts as an inverter.

To use a NAND gate as an inverter, tie all but one input to the nondynamic input level of logic 1 (Figure 3.12). This method is not preferred to the first method of tying all the inputs together, because it requires the addition of an extra resistor. You may wonder why the input cannot be directly tied up to +5 V. From your study of power supplies, you should know that the instant a power supply is turned on, a voltage spike may occur that can damage semiconductor devices. This resistor is usually between 1 and 10 kΩ. Its function is to protect the internal circuitry of the gate.

3.1.5 Reducing the Number of Inputs on a NAND Gate

This brings up another problem that we should consider. A circuit designer discovers that another two-input NAND gate is required. Looking through the devices in the circuit, the designer discovers an unused three-input NAND gate. How can this three-input NAND gate be used as a two-input NAND gate?

It seems reasonable to apply what we have just learned about NAND gates and inverters to this problem. What would happen if we tied two of the input leads together, as illustrated in Figure 3.13a? That means that inputs A and B will always be of the same value. Inspect the truth table of the three-input NAND gate in Figure 3.13. Each row in which inputs A and B are equal has been circled. We can make a new truth table to

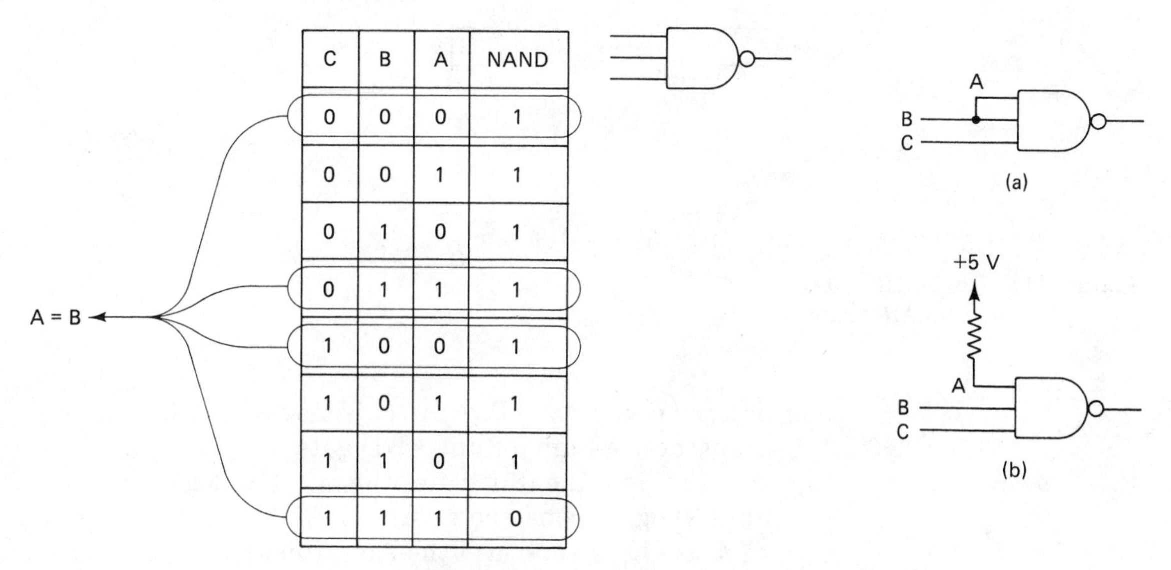

C
B
A
NAND
0 0 0 1
0 0 1 1
0 1 0 1
0 1 1 1
1 0 0 1
1 0 1 1
1 1 0 1
1 1 1 0
A = B
A
B
C
(a)
+5 V
A
B
C
(b)

Figure 3.13 Using a three-input NAND gate as a two-input NAND gate.

TABLE 3.2 NAND Truth Table Where Input A Is Equal to Input B

C	B/A	NAND
0	0	1
0	1	1
1	0	1
1	1	0

reflect only those parts of the first truth table which are applicable to Figure 3.13a (Table 3.2). The new truth table should be familiar. It is the truth table for a two-input NAND gate. We see that this approach works perfectly well.

To reduce the number of inputs of a NAND gate, tie any extra inputs to inputs that are being used.

The alternative method of tying all unused inputs to logic 1, the nondynamic input level, also works. But an extra resistor must be added to protect the gate.

3.1.6 Using the NAND Gate as an AND Gate

If you had only two-input NAND gates and you needed a two-input AND gate, how could you go about making NAND gates appear to function as AND gates?

When the NAND gate was introduced, it was modeled as an AND gate followed by an inverter. With this approach in mind, we can use two two-input NAND gates to function as a two-input AND gate. Figure 3.14a illustrates a positive-logic AND gate constructed from NAND gates. Notice that the output bubble on the first NAND gate is canceled by the input bubbles

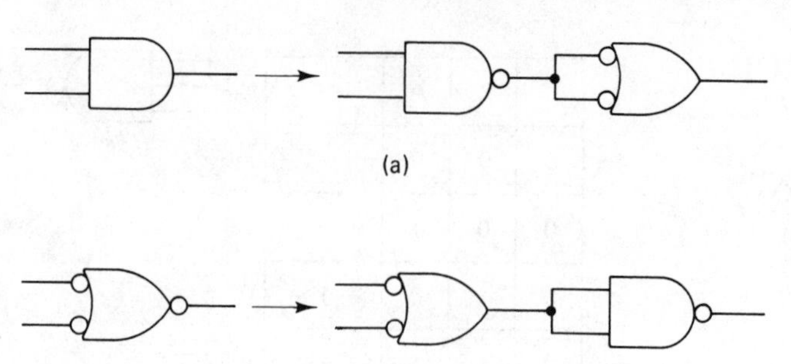

Figure 3.14 Two NAND gates used as an AND gate.

on the second NAND gate. This leaves us with an active-high-input, active-high-output AND gate.

Figure 3.14b illustrates the negative-logic representation of an AND gate constructed from NAND gates. As you know, the negative-logic version of the AND function appears as an active-low-input, active-low-output OR gate. This is precisely what Figure 3.14b represents.

3.1.7 Summary of the NAND Gate

Figure 3.15 summarizes the NAND gate.

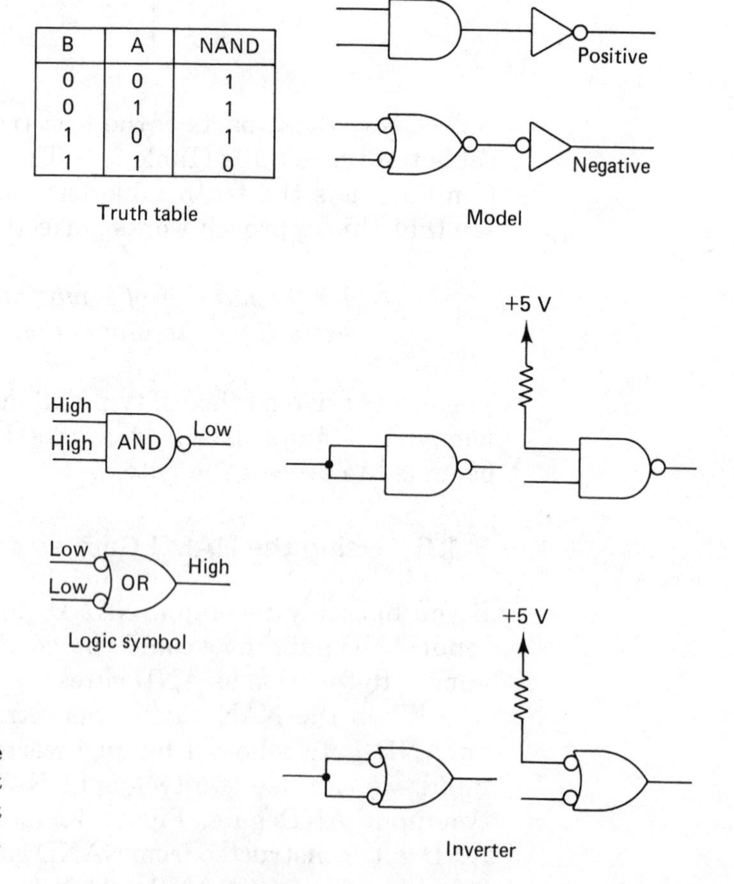

B	A	NAND
0	0	1
0	1	1
1	0	1
1	1	0

Truth table

Figure 3.15 Summary of the NAND gate: dynamic input level is logic 0; unique output level is logic 0.

3.2 THE NOR GATE

3.2.1 The NOR Function

The NOR gate is a combination of the OR and NOT functions: NOT OR (Figure 3.16). The NOR gate shares the same relationship to the OR gate that the NAND gate shares with the AND gate. The NOR gate is also a universal building block. We will learn to use the NOR gate as an inverter and an OR gate. We can model the NOR gate as an OR gate followed by an inverter. Therefore, the output values of the truth tables of the OR and NOR gates will be complements of each other.

3.2.2 Dynamic Input and Unique Output Levels

In review, the dynamic input level of both the AND and NAND functions is logic 0. A logic 0 on the input of an AND gate forces a logic 0 on the output, whereas a logic 0 on the input of a NAND gate forces a logic 1 on the output. We noticed this to be an interesting relationship between the AND and NAND gates. The OR and NOR gates share a similar relationship.

The dynamic input level of both the OR and NOR is a logic 1. We can see from the truth table in Figure 3.16 that a logic 1 on the input of the OR forces a logic 1 on the output, but a logic 1 on the input of the NOR forces a logic 0 on the output. Considering the model of the NOR gate, this should not be a surprising fact.

A logic 1 on the input of a NOR gate forces a logic 0 on the output.

Referring once again to the truth table in Figure 3.16, we can establish the unique output level of the NOR gate. We see that the unique output level of the NOR gate is a logic 1. A logic 1 output occurs for only one combination of inputs. This unique output level is the same for both the AND gate and the NOR gate. We have observed that the NAND gate and the OR gate both had unique output levels of logic 0, so it seems that the NOR gate is also a combination of the AND and OR functions.

A logic 1 on the output of a NOR gate is a unique event.

Table 3.3 should give you a great deal of insight into the four logic functions that we have examined. It is important to notice that each of the four logic functions can be identified by

B	A	OR	NOR
0	0	0	1
0	1	1	0
1	0	1	0
1	1	1	0

OR NOR

Figure 3.16 Model and truth table of the NOR gate.

TABLE 3.3 Summary of Dynamic Input and Unique Output Levels

Dynamic input	Unique output	Function
0	0	NAND
0	1	AND
1	0	OR
1	1	NOR

its dynamic input and unique output levels. Take a moment before proceeding and consider the interwoven relationships among these functions.

3.2.3 The Logic Symbols of the NOR Gate

Figure 3.16 leads us to believe that the symbols shown in Figure 3.17 should describe the positive- and negative-logic NOR function. Figure 3.17a can be taken to represent an active-high-input, active-low-output OR gate. Figure 3.17b can be interpreted as an active-low-input, active-high-output AND gate.

Figure 3.18a should be read as:

If input (1) is high OR input (2) is high, then the output will be low; otherwise, the output will be high.

Figure 3.18b should be read as:

If input (1) is low AND input (2) is low, then the output will be high; otherwise, the output will be low.

The OR-body logic symbol can be thought of as representing the active input levels, and the AND-body logic symbol as representing the unique output level (Figure 3.19).

The logic symbol chosen to represent the gate will usually

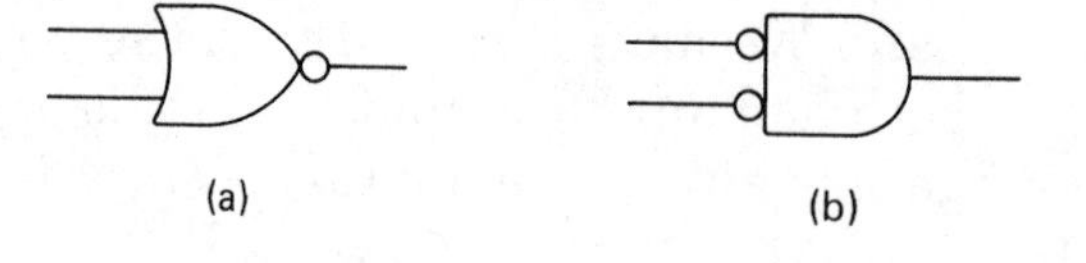

Figure 3.17 Logic symbols of the NOR gate: (a) positive; (b) negative.

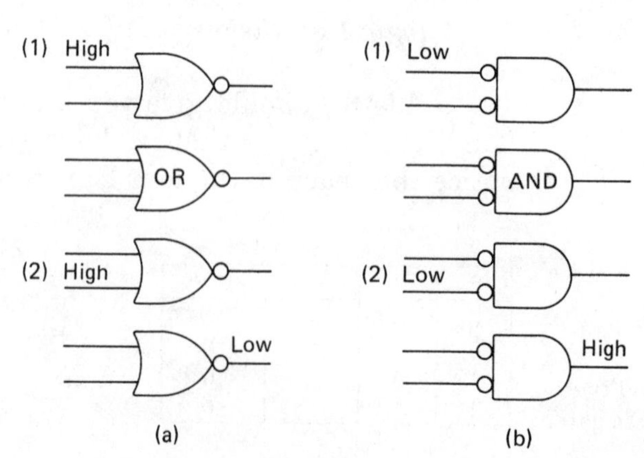

Figure 3.18 Interpretation of NOR-gate symbols.

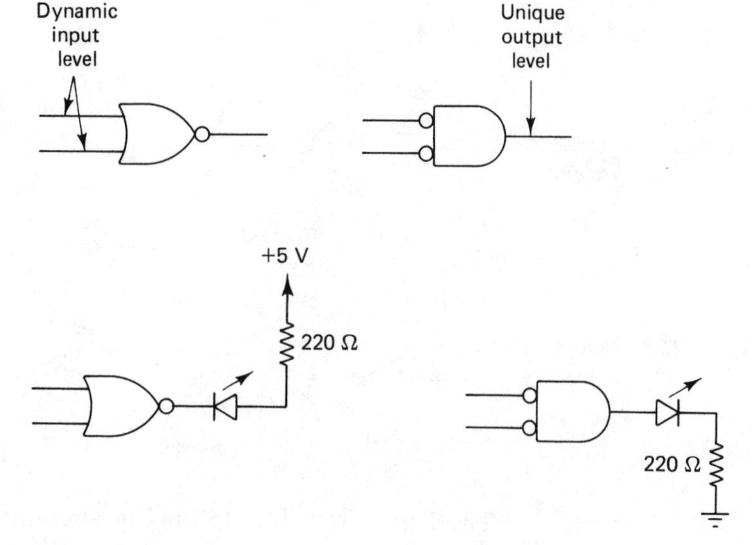

Figure 3.19 Dynamic and unique output levels of the NOR gate.

Figure 3.20 Active low (a) and high (b) levels of the NOR gate.

indicate the active level of the output. By *active* is meant:

> *the output level which will be an indication that predetermined conditions have been met*

In Figure 3.20, an illuminated LED will indicate an active output.

3.2.4 Using the NOR Gate as an Inverter

We discovered that one way to use a NAND gate as an inverter is to tie all the inputs together. Let's refer to the truth table (Table 3.3) to see if this approach will also work for the NOR gate (see Figure 3.21). In the first and last lines of the truth table, the inputs of A and B are equal. This would symbolize the inputs as being tied together. With logic 0's on both inputs, the output will be a logic 1; with logic 1's on both inputs, the output will be a logic 0. This appears to satisfy our requirements for an inverter.

We also use a NAND gate as an inverter by tying one input to a logic 1 level. Would this approach also work for a NOR-gate inverter? (See Figure 3.22). Obviously not. A logic 1 is the dynamic input level of a NOR gate. That means if we tied one input to logic 1, the output would be stuck at a logic 0 level. Let's take the NOR's nondynamic input level of logic 0 and tie the extra input to that level (Figure 3.23). This configuration appears to function correctly. Figure 3.23a shows the positive-logic NOR inverter, and Figure 3.23b illustrates the negative-logic

Figure 3.21 Using the NOR gate as an inverter.

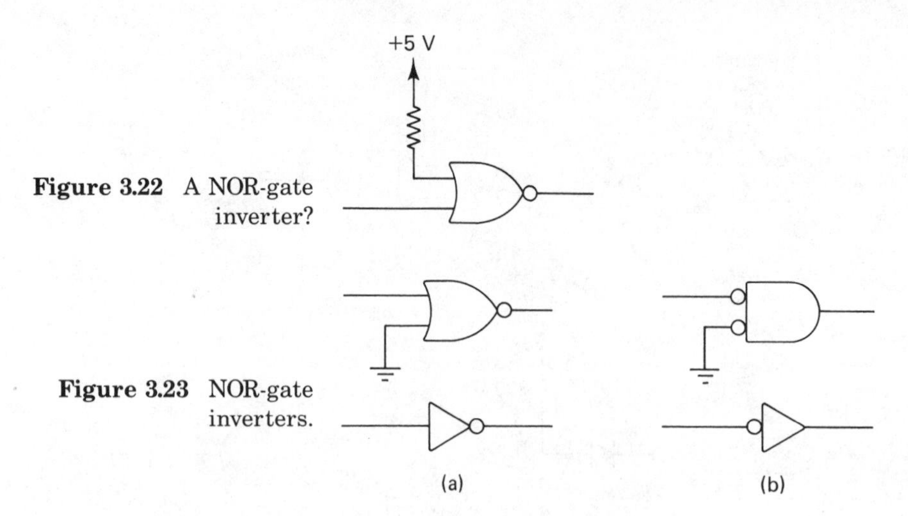

Figure 3.22 A NOR-gate inverter?

Figure 3.23 NOR-gate inverters.

NOR inverter. The following statement sums up the technique to use when implementing NANDs and NORs as inverters:

To use a NAND or NOR gate as an inverter, tie all but one input to the nondynamic input level.

3.2.5 Reducing the Number of Inputs on a NOR Gate

The preceding statement concerning the use of NOR gates as inverters can be slightly modified to include the concept of reducing the number of inputs to any gate: AND , OR, NAND, and NOR. As we have noted previously, sometimes a situation occurs where a designer needs an extra gate. Instead of adding another IC package, the designer may choose to modify a surplus gate from another IC. It appears to be a simple matter to use a NOR or NAND as an inverter. What if a two-input NOR gate is needed and the only surplus gate is a four-input NOR gate? The following statement applies to all gates that need to have their number of inputs reduced.

To reduce the number of inputs on any gate, tie the unused inputs
1. to other inputs that will be used,
or,
2. to the nondynamic input level of the particular gate.

Figure 3.24 shows examples of reduced input gates.

3.2.6 Using the NOR gate as an OR Gate

Just as two NAND gates can be used to simulate an AND gate, two NOR gates can be used to simulate an OR gate. The idea is still the same: NOR and OR truth tables are complements of each other. We can invert the output of a NOR gate to get a

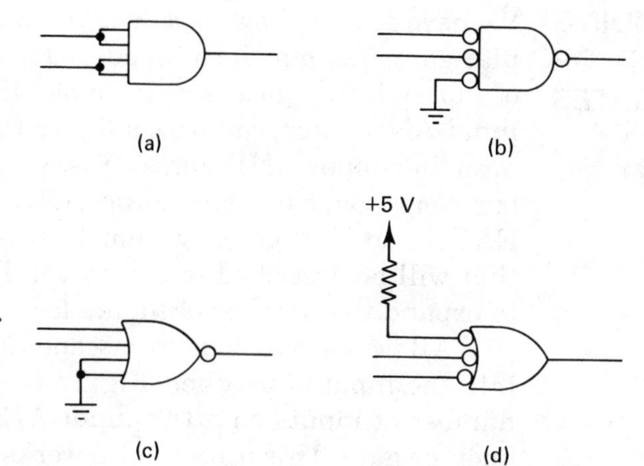

Figure 3.24 Examples of reduced input gates: (a) two-input AND; (b) two-input OR; (c) two-input NOR; (d) two-input NAND.

circuit that behaves like an OR gate (Figure 3.25). Figure 3.25a illustrates the combination of NOR gates that create a positive-logic OR gate. Notice how the output bubble from the first NOR gate effectively gets canceled by the input bubbles of the second NOR gate. We are left with a positive-logic OR gate. On the other hand, Figure 3.25b pictures a negative-logic OR gate constructed from two NOR gates. Here the input bubbles and output bubble remain, because there is nothing to cancel them.

3.2.7 Summary of the NOR Function

Figure 3.26 summarizes the NOR function.

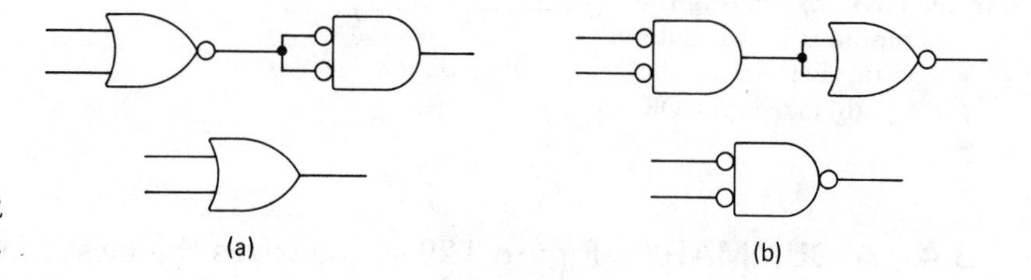

Figure 3.25 Using two NOR gates as an OR gate.

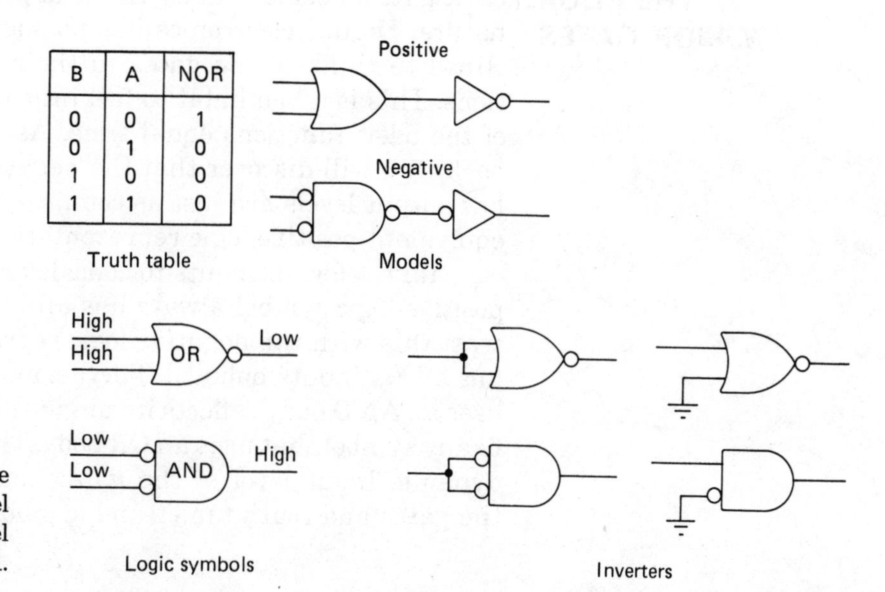

B	A	NOR
0	0	1
0	1	0
1	0	0
1	1	0

Figure 3.26 Summary of the NOR gate: dynamic input level is a logic 1; unique output level is a logic 1.

3.3 EXPANDING AND AND OR GATES

We have studied how to reduce the number of inputs of a particular gate. You may have wondered if we can expand the number of inputs into a gate. For example, if we have a few extra two-input AND gates, can we configure them to operate as three- or even four-input AND gates? Yes, we can. The process is simple but works only for expanding AND and OR gates. To expand NAND and NOR gates, we must use a more complicated process that will be described in Chapter 4. For now, we will learn how to expand the number of inputs for AND and OR gates.

All we have to do is to cascade the output of one AND gate into the input of another (Figure 3.27). This expands by 1 the number of inputs on a two-input AND gate each time we add another gate. This process also works perfectly well for expanding the inputs of OR gates (Figure 3.28).

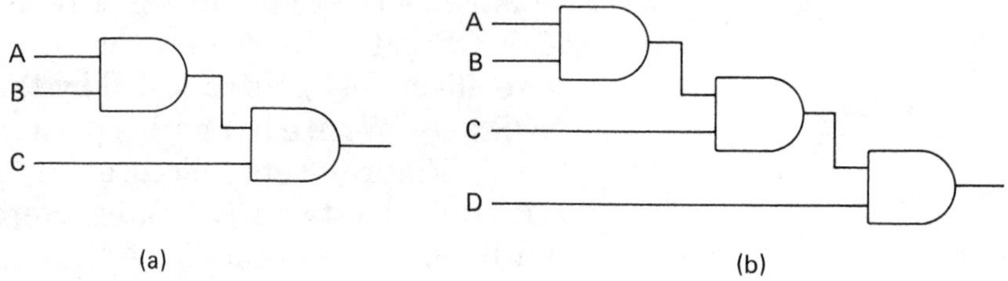

A
B
C
(a)
A
B
C
D
(b)

Figure 3.27 Expanding the number of inputs of an AND gate: (a) three-input AND; (b) four-input AND.

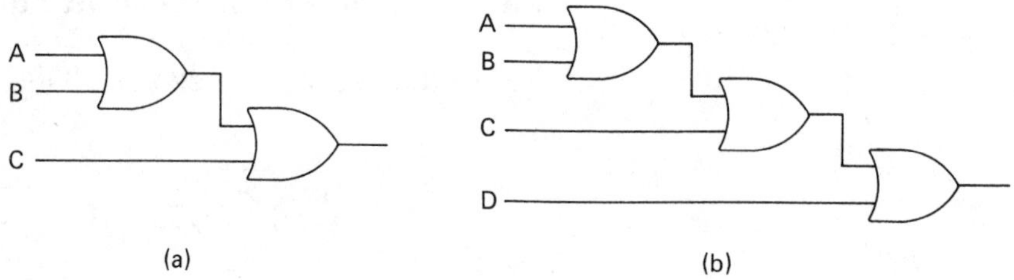

A
B
C
(a)
A
B
C
D
(b)

Figure 3.28 Expanding the inputs of an OR gate: (a) three-input OR; (b) four-input OR.

3.4 A SUMMARY OF THE FOUR MAJOR GATES

Figure 3.29 summarizes the most important aspects of the four basic gates. People usually think in positive logic; it is our basic nature. Digital electronics has no such nature. You will be inclined to think of the gates in their positive-logic representations. This is a bad habit to fall into. Give both representations of the basic functions equal time. As you proceed through this book, you will discover that the negative-logic gates and active-low output levels are just as common, if not more so, than their equivalent positive-logic representations.

Take a few moments to consider Figure 3.29. Notice that a positive-logic symbol always has all its inputs unbubbled. Contrast this with the negative-logic representation, which always has all its inputs bubbled. Furthermore, any gate symbol that uses an AND body reflects its unique output level. Compare this to any symbol that uses an OR body. This symbol will reflect the dynamic input level of the gate. You need not memorize all the gates and truth tables and symbols. If you understand the

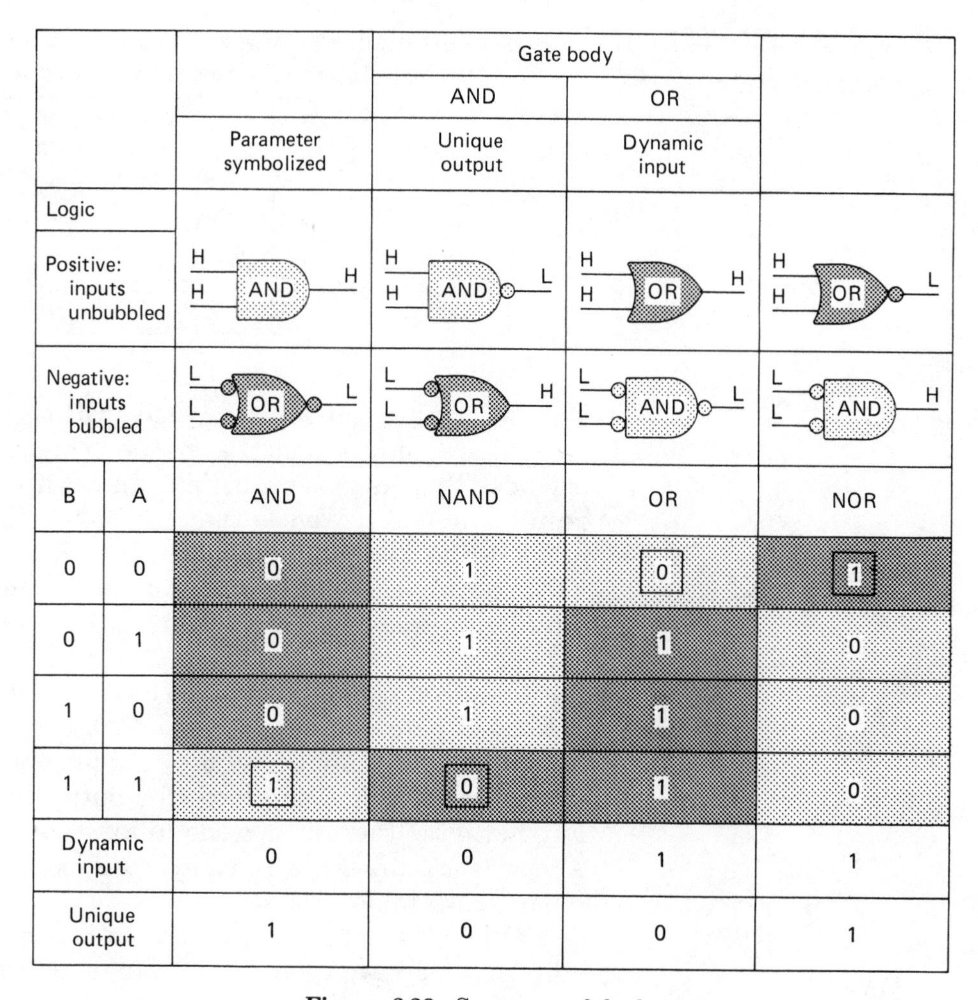

			Gate body		
			AND	OR	
		Parameter symbolized	Unique output	Dynamic input	
Logic					
Positive: inputs unbubbled		H, H → AND → H	H, H → AND → L	H, H → OR → H	H, H → OR → L
Negative: inputs bubbled		L, L → OR → L	L, L → OR → H	L, L → AND → L	L, L → AND → H
B	A	AND	NAND	OR	NOR
0	0	0	1	0	1
0	1	0	1	1	0
1	0	0	1	1	0
1	1	1	0	1	0
Dynamic input		0	0	1	1
Unique output		1	0	0	1

Figure 3.29 Summary of the basic gates.

meaning of the AND and OR symbols, you should be able to derive the truth table of any gate:

1. Place a high on an unbubbled input or a low on a bubbled input.
2. Connect the input values with the name of the gate.
3. If the output is unbubbled when the conditions in steps 1 and 2 are met, the output will go high; otherwise, the output will be low.
4. If the output is bubbled when the conditions in steps 1 and 2 are met, the input will go low; otherwise, the output will be high.

3.5 THE EXCLUSIVE-OR GATE

3.5.1 The Exclusive-OR Function

The OR function that we have studied is true if

A is true OR B is true OR both A and B are true.

TABLE 3.4 Truth Table of the Exclusive-OR Function

B	A	OR	XOR
0	0	0	0
0	1	1	1
1	0	1	1
1	1	1	0

This is formally called an inclusive OR because it includes the case where both A and B are true. The exclusive-OR function excludes the case where both A and B are true. Compare the results Table 3.4. We see that the only difference in the truth table is the last line. An exclusive-OR function, abbreviated XOR, outputs a logic 0 whenever both inputs have the same value. It outputs a logic 1 whenever both inputs have different values.

The XOR function seems to be different from the other four basic logic gates that we have studied. For one thing, it does not have a dynamic input level or a unique output level. Neither a logic 1 nor a logic 0 can force the output to a particular level. Furthermore, there is not one unique output level that occurs. A logic 0 appears twice and a logic 1 appears twice on the output in the truth table.

Other than the inverter, which is a unary function, every gate that we have seen can be expanded to any number of inputs. The XOR gate comes in only one form, and that has two inputs.

The XOR is a special gate, used only for special applications. It is widely employed in digital ICs that perform math functions. In Chapter 4 you will learn that the XOR gate is really composed of two inverters, two AND gates, and an OR gate.

3.5.2 The Logic Symbol of the Exclusive-OR Gate

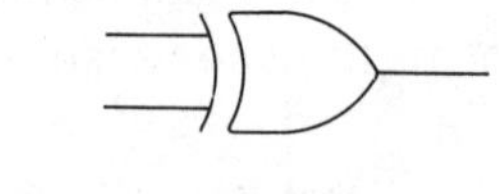

Figure 3.30 XOR gate symbol.

Another different characteristic of the XOR gate is that it has only one logic symbol (Figure 3.30). Every other gate that we have studied has both a negative- and a positive-logic symbol. This looks like the positive-logic symbol of an OR gate with a curved line across the inputs.

3.5.3 An Application of the XOR Gate

One circuit that we can construct with the XOR gate is illustrated in Figure 3.31. If the control line is at a logic 0, the value on the other input of the gate passes through unchanged. If the control line is held at logic 1, the value on the other input of the XOR gate is complemented. That is why this circuit is referred to as a *controlled complementer.*

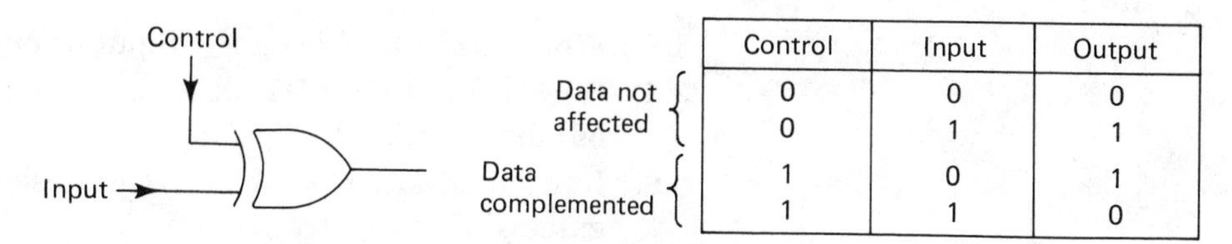

Control	Input	Output
0	0	0
0	1	1
1	0	1
1	1	0

Figure 3.31 The XOR gate as a controlled complementer.

SUMMARY At this point in your study of digital electronics, you have learned the concepts of digital signals and how they apply to logic gates. You should feel comfortable with the logic functions of invert, AND, OR, NAND, and NOR in both their positive- and negative-logic interpretations. Do not try to memorize all the gate symbols. Instead, you should learn to understand their meanings. The next chapter deals with a special branch of mathematics called Boolean algebra. This math describes logic circuits. You will learn to design simple logic circuits to solve practical problems. Without a good understanding of the four basic logic functions, you will have a difficult time grasping new material.

QUESTIONS AND PROBLEMS

3.1. Create a word description for each of the following logic symbols.

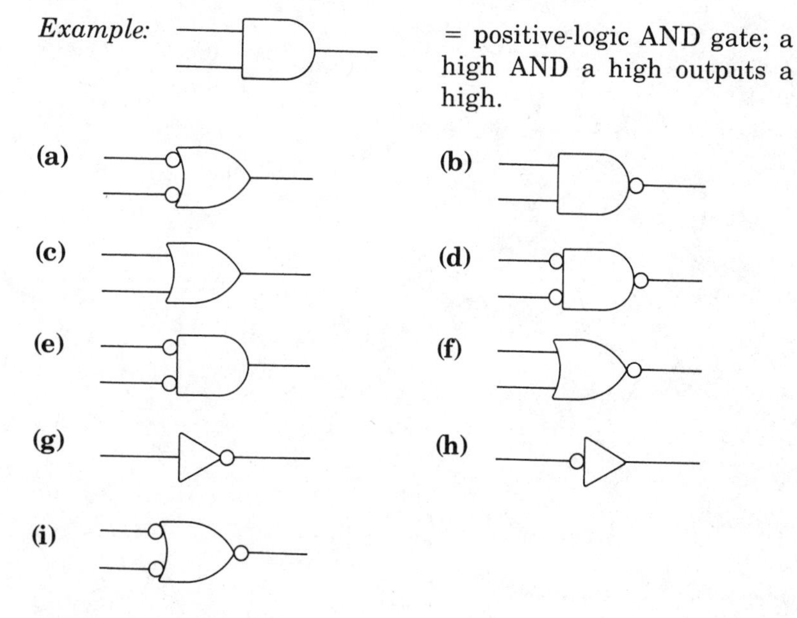

3.2. What is the dynamic input level of:
(a) A NAND gate?
(b) A NOR gate?
(c) An XOR gate?

3.3. What is the unique output level of:
(a) A NAND gate?
(b) A NOR gate?
(c) An XOR gate?

3.4. What does a bubble on the input or output of a gate indicate?

3.5. Why are the NAND and NOR gates referred to as "universal building blocks"?

3.6. In what situation would a circuit designer use a NAND or NOR gate as an inverter?

3.7. If the output of a NAND gate is stuck high, what could be the possible problem?

3.8. If the output of a NOR gate is stuck low, what could be the possible problem?

3.9. How does the XOR function differ from all the other logic functions?

3.10. If the output of an XOR gate is always equal to the A input, what can you conclude about the B input?

3.11. A circuit is required that illuminates an LED whenever all three of its inputs are at logic 0. Illustrate the two circuits that could accomplish this task. Remember that an LED can be illuminated on either a logic 1 or a logic 0, depending on the LED's orientation.

3.12. A circuit is required that illuminates an LED whenever any of its three inputs are at a logic 0. Illustrate the two circuits that accomplish this function.

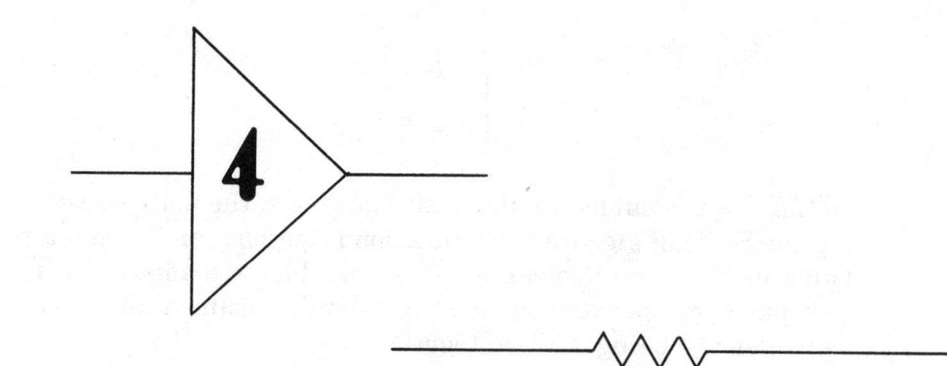

APPLICATIONS OF GATES

4.1 FUNDAMENTALS OF BOOLEAN ALGEBRA

In the mid-nineteenth century, George Boole wrote an essay entitled "An Investigation of the Laws of Thought." As a mathematician, Boole was concerned with statements that were digital in nature: either true or false. The symbolic notation that is used to describe digital circuitry is called *Boolean algebra.* Circuit designers translate their ideas into Boolean equations and then transform these equations into logic circuits. Technicians, on the other hand, most often perform the reverse process: they start with schematics of the logic circuits and translate them back into Boolean equations. These equations can then be used as tools for troubleshooting or to analyze circuit action.

4.1.1 The Boolean Operators

Take a moment to draw the truth tables for the AND and OR functions on a piece of scratch paper. Examine the AND truth table closely. Is there a common arithmetic function that describes this truth table? Yes, there is: simple multiplication. For that reason the logic AND function is represented by the multiplication symbol—a dot (·).

$$0 \cdot 0 = 0$$
$$0 \cdot 1 = 0$$
$$1 \cdot 0 = 0$$
$$1 \cdot 1 = 1$$

NOTE: As is common in algebraic equations, the dot representing the Boolean algebra AND function is optional and most often is not used. If two Boolean algebra variables are adjacent with no separating operation symbol, you should assume that the two variables are being ANDed together.

Now examine the OR truth table. The simple math function that best describes logical OR is addition. The last line in the truth table, where both inputs are equal to logic 1, is the only case that differs from arithmetic addition. The plus sign (+) is used to symbolize the logical OR function.

$$0 + 0 = 0$$
$$0 + 1 = 1$$
$$1 + 0 = 1$$
$$1 + 1 = 1$$

The NOT function is indicated with a bar ($\overline{\ }$). This bar symbolizes inversion. It can extend over a single operand or a group of operands.

$$\overline{0} = 1$$
$$\overline{1} = 0$$

Figure 4.1 illustrates the relationships between the truth table, logic symbol, and Boolean equation of each function. Notice that the Boolean equations that describe the active-low output representation of the logic gates all have a long inversion bar over them. This inversion bar is analogous to the bubble on the output of the negative-logic symbols, and can be thought of as implying that the active output value of the equation is a logic 0.

Refer to Figure 4.1. Let's see how the Boolean equation for the negative-logic AND function works. To test this equation we will substitute each line of the truth table into the input variables.

$A = 0, B = 0$:

1. $\overline{A}$ is equal to 1 and $\overline{B}$ is equal to 1.
2. $1 + 1 = 1$.
3. $\overline{1} = 0$.

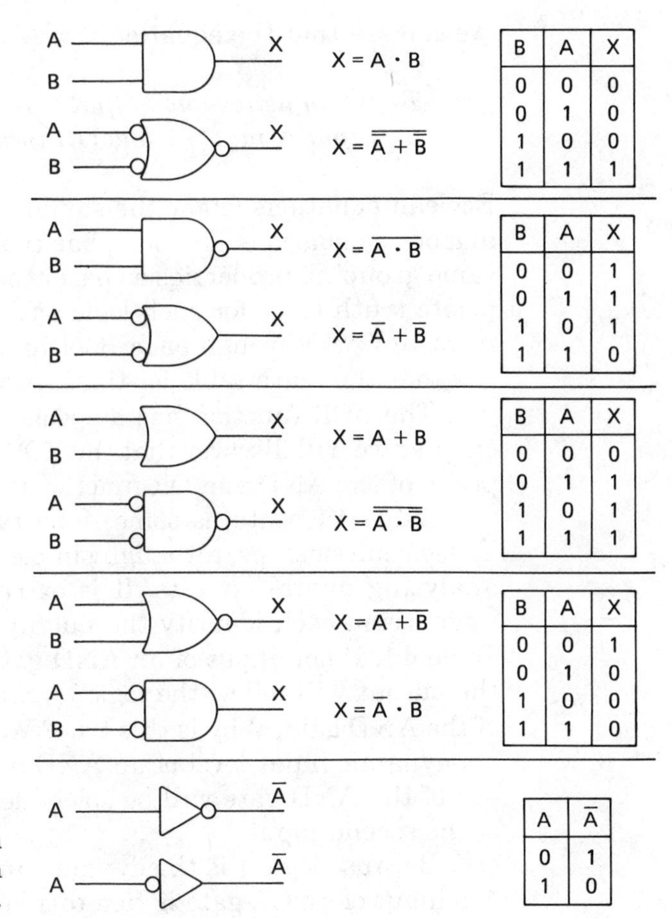

B	A	X
0	0	0
0	1	0
1	0	0
1	1	1

B	A	X
0	0	1
0	1	1
1	0	1
1	1	0

B	A	X
0	0	0
0	1	1
1	0	1
1	1	1

B	A	X
0	0	1
0	1	0
1	0	0
1	1	0

A	$\overline{A}$
0	1
1	0

Figure 4.1 Logic gates, truth tables, and their Boolean equations.

A = 1, B = 0:

1. $\overline{A}$ is equal to 0 and $\overline{B}$ is equal to 1.
2. 0 + 1 = 1.
3. $\overline{1} = 0$.

A = 0, B = 1:

1. $\overline{A}$ is equal to 1 and $\overline{B}$ is equal to 0.
2. 1 + 0 = 1.
3. $\overline{1} = 0$.

A = 1, B = 1:

1. $\overline{A}$ is equal to 0 and $\overline{B}$ is equal to zero.
2. 0 + 0 = 0.
3. $\overline{0} = 1$.

We can see that the equation x = A + B actually says:

To get an active-low output from an AND gate, either input A must be low OR input B must be low.

Boolean equations follow the same rules as those for any other algebraic equations. The long bar that covered A and B has the same grouping properties as parentheses. Substitute the appropriate truth table for each logic gate in Figure 4.1. Take a moment to work through each Boolean equation and assure yourself that the truth table for that function is satisfied.

The XOR function has a special symbol (⊕). Later in this chapter we will discover that the XOR function is actually a composite of the AND and OR functions.

Table 4.1 contains some important Boolean relationships. These relationships, although simple, play an important role in analyzing digital circuits. It is extremely important that you understand each identity thoroughly. Consider the first line of Table 4.1. If one input of an AND gate is tied to a logic 1 level, the output will follow the logic level applied to the other input of the AND gate. Why is this true? We know that a logic 1 is the nondynamic input level of an AND gate; consequently, the output of the AND gate will be solely dependent on the logic level of the second input.

Because logic 1 is the dynamic input level of an OR gate, if one input of an OR gate is tied to a logic 1 level, the output will be stuck at logic 1, regardless of the value on the other input.

The second line in Table 4.1 shows the AND gate with a constant dynamic input level and the OR gate with a constant nondynamic input level. If one input of an AND gate is tied to a logic 0 level, the output will be logic 0, regardless of the other inputs. If one input of an OR gate is tied to a nondynamic input level of logic 0, the output will be solely dependent on the logic level of the other input.

In the latter part of this chapter we will use the Boolean identities illustrated in the first two lines of Table 4.1 to implement simple AND and OR gates as electronic switches.

The third line of Table 4.1 illustrates the process that we used in Chapter 3 to reduce the number of inputs on AND and

Table 4.1 Boolean Identities

AND	NOT	OR
$X \cdot 1 = X$		$X + 1 = 1$
$X \cdot 0 = 0$		$X + 0 = X$
$X \cdot X = X$		$X + X = X$
$X \cdot \overline{X} = 0$		$X + \overline{X} = 1$
	$\overline{\overline{X}} = X$	

OR gates. If the two inputs of an AND gate or OR gate are tied together, the output of the gate will be equal to the logic level on the single input. In this manner AND and OR gates can be used as noninverting buffers.

The fourth line in Table 4.1 illustrates a formal statement concerning the concept of dynamic input levels. If a logic level and its complement (as if a line driving one input of a gate were routed through an inverter and then applied to the other input) are both applied to the inputs of a gate, then logically one of the inputs must be a dynamic input level. If the dynamic input level of logic 0 is applied to an AND gate, the output will be forced to logic 0. If the dynamic input level of logic 1 is applied to an OR gate, the output will be forced to a logic 1 level.

The last line in Table 4.1 illustrates the concept of double inversion. If a signal is passed through two inverters, it will be the original logic level input into the first inverter. Two inverters can be used as a noninverting buffer. We will discover in Chapter 5 that a signal can be delayed a specified length of time by running it through two inverters.

Table 4.1 should not contain any surprises; it only formalizes the properties of logic gates that we have already examined in depth in the first three chapters.

4.1.2 De Morgan's Law

De Morgan's law is a simple algorithm that concerns the relationship between positive and negative logic (Figure 4.2). As was true with Table 4.1, De Morgan's law should not tell you anything new. We have already discovered both the positive and negative equivalents of the basic logic functions by simple inspection.

De Morgan's law implies that if you desire to change a positive-logic symbol into a negative-logic symbol, or vice versa, you should execute these steps:

1. Interchange the basic logic symbol. An AND becomes an OR; an OR becomes an AND.
2. Add a bubble to each input and output of the new symbol.
3. If any input or output has two bubbles, they cancel each other and should be omitted.

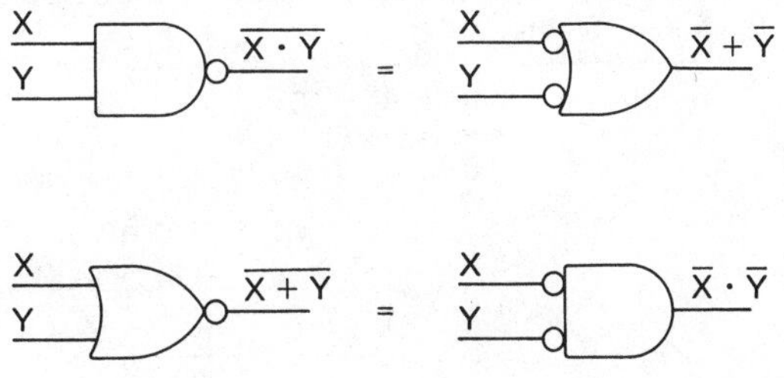

X
Y
$\overline{X \cdot Y}$
=
X
Y
$\overline{X} + \overline{Y}$

X
Y
$\overline{X + Y}$
=
X
Y
$\overline{X} \cdot \overline{Y}$

Figure 4.2 Positive- and negative-logic duals.

4.1.3 Creating an Equation and Truth Table from a Logic Diagram

When you are troubleshooting or analyzing digital circuits it is often helpful to create a Boolean equation describing the circuit. You can then substitute values for the inputs and find out what the output and all intermediate logic levels should be. Consider Figure 4.3, which illustrates a simple circuit composed of five gates. Because this circuit has two inputs, there are four possible input combinations. To establish what output logic level should exist for each of these four input combinations, we must follow these steps:

1. Construct a Boolean equation from the logic diagram.
2. Create a truth table and substitute the input values of each row into the equation.

Step 1 Construct the Boolean equation. Figure 4.1 indicates the procedure for establishing the equation. Start at the inputs on the left side of the schematic. Follow each input or group of inputs as it is applied to the various gates. At the output of each gate write the Boolean equation that describes the action of the gate. This output equation will now be used as an input to the next gate. This process will continue until you reach the final output.

Step 2 Construct a truth table. This truth table should contain the input variables, each intermediate equation, and the final equation for the output of the circuit. Fill in the truth table from left to right. Each intermediate answer will become the input value for the next equation. When the truth table is complete

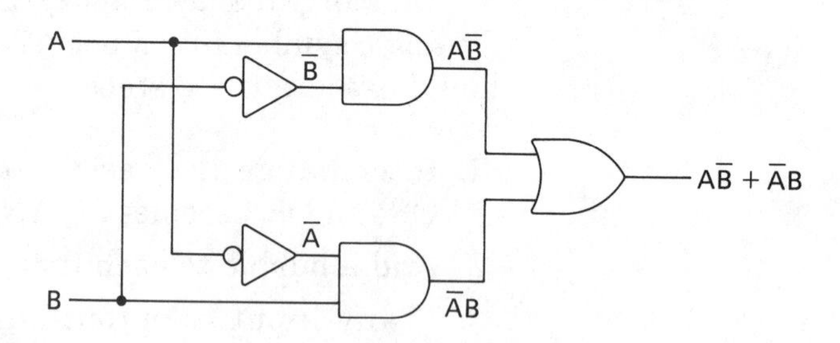

Inputs		Intermediate values				Output
B	A	$\bar{B}$	$\bar{A}$	$A\bar{B}$	$\bar{A}B$	$A\bar{B} + \bar{A}B$
0	0	1	1	0	0	0
0	1	1	0	1	0	1
1	0	0	1	0	1	1
1	1	0	0	0	0	0

Figure 4.3 Logic diagram and truth table.

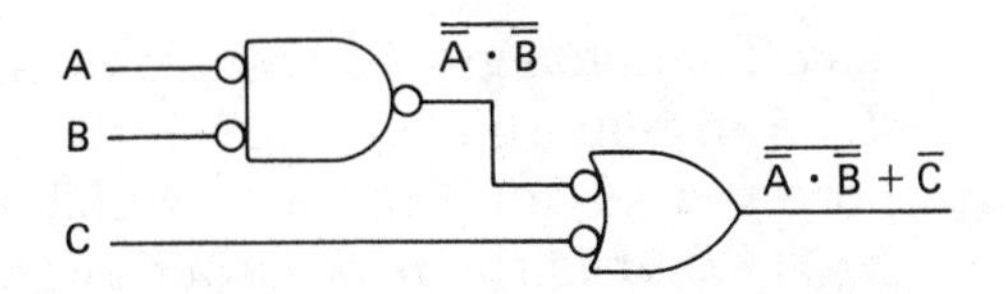

C	B	A	$\overline{A} \cdot \overline{B}$	$\overline{A} \cdot \overline{B} + \overline{C}$
0	0	0	0	1
0	0	1	1	1
0	1	0	1	1
0	1	1	1	1
1	0	0	0	1
1	0	1	1	0
1	1	0	1	0
1	1	1	1	0

Figure 4.4 Logic diagram, truth table, and Boolean equation.

you will know what each logic level in the circuit should be for any set of input values.

Examine Figure 4.3 closely. On a scratch sheet of paper insert each pair of input values into the equation and check to be sure that you come up with the same results. On a passing note, do you recognize this truth table? You should. It is the truth table for XOR. Although a designer will never use an XOR gate constructed like Figure 4.3, this circuit should help you get a good feeling for how the XOR function operates.

Figure 4.4 is another example of constructing an equation and a truth table from a logic diagram. The output goes high whenever:

1. A AND B are both low

   OR

2. C is low.

Compare the derived Boolean equation with the logic-symbol description of the circuit. The logic symbols say that there are two ways to make the output of the negative-logic NAND gate go high: both inputs A AND B must be low OR input C must be low.

4.2 PRACTICAL APPLICATIONS OF LOGIC GATES

Digital technology has progressed at an almost exponential rate. New designs employ complex ICs that contain whole subsystems and systems. The majority of applications use simple gates to interconnect these complex ICs. Nonetheless, there are still times when simple gates are the best solution for a particular application. Learning digital electronics is a progressive process. This section will illustrate a few applications of simple gates. It is urgent that you understand these simple applications before you advance to more complex devices and circuitry.

What is the process a designer engages in whenever a new circuit is required? You must:

1. Write down a concise statement of the problem, defining all inputs and outputs.
2. Fill in a truth table with the proper output values for each combination of inputs.
3. Create a Boolean equation by using the *sum-of-products method.* The sum-of-products method of creating Boolean equations is extremely simple and intuitive. We stated in the beginning of this chapter that the Boolean AND function is analogous to multiplication and the Boolean OR function is analogous to addition. The term "sum of products" refers to ORing together (summing) all the lines in a truth table that contain a logic 1 output level (products). A product will be created by ANDing together the variables in any row whose output is a logic 1. After you have all the products, they will be ORed together to create the final equation: thus the name "sum of products."
4. The final step will be the translation of the Boolean equation into the logic diagram.

4.2.1 The Three Judges: A Majority Voting Circuit

Step 1 State the problem. This circuit (Figure 4.5) will analyze the votes of three judges and indicate whether a majority of the judges has voted in favor for the motion in question. A majority will occur whenever two or more judges vote yes on an issue.

Inputs: three—one from each judge. The input device will be a single-throw, single-pole (SPST) switch.

*Outputs: two—*one LED to indicate a majority vote, another to indicate a nonmajority vote.

Step 2 Complete the truth table. Each judge will constitute one input into our circuit. With three inputs there will be eight possible voting combinations.

C	*B*	*A*	*Vote*
0	0	0	0
0	0	1	0
0	1	0	0
0	1	1	1
1	0	0	0
1	0	1	1
1	1	0	1
1	1	1	1

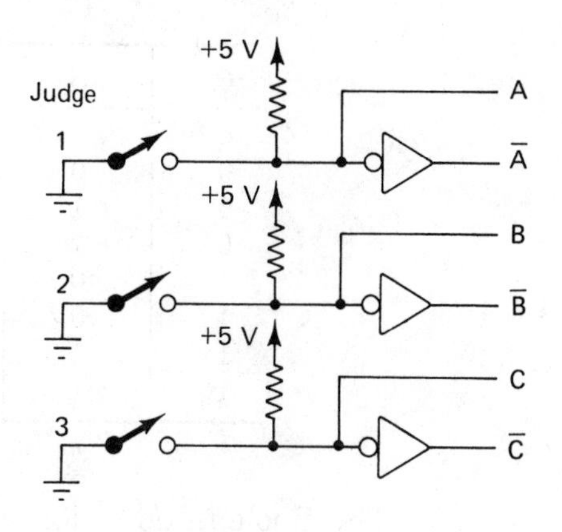

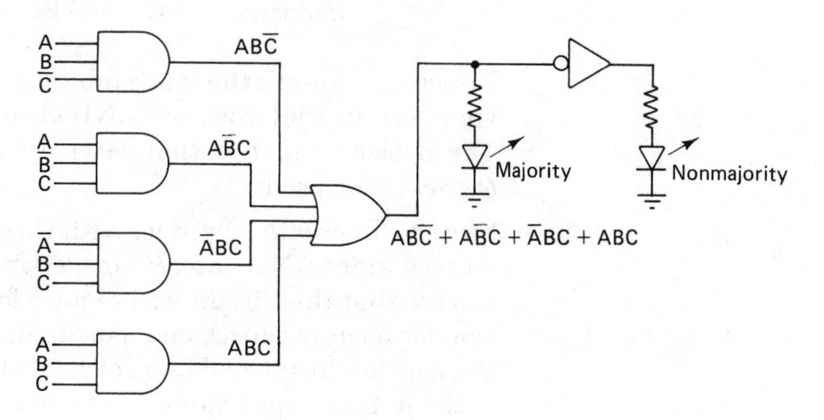

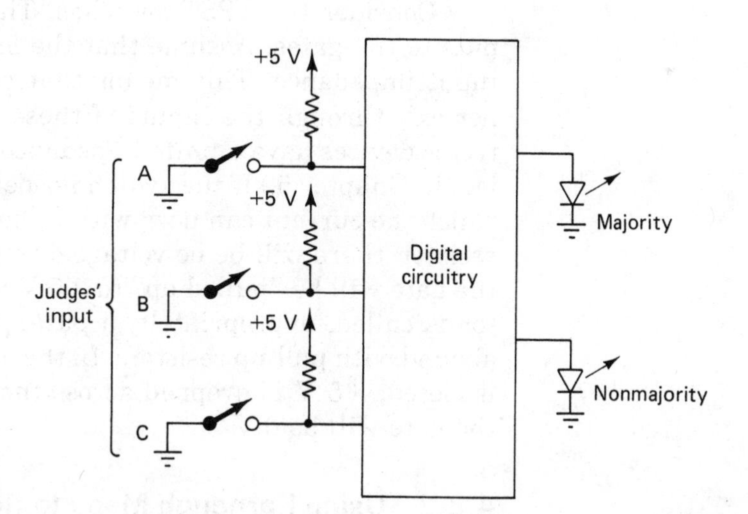

Figure 4.5 Block diagram and logic diagram for the three-judges problem.

Inspect each row of the truth table. A logic 0 indicates a no vote and a logic 1 indicates a yes vote. Whenever two or more judges vote yes, the output will go high to indicate a majority vote of yes.

Step 3 Create a Boolean equation by the sum-of-products method.

C	B	A	Vote	Product
0	0	0	0	
0	0	1	0	
0	1	0	0	
0	1	1	1	$AB\overline{C}$
1	0	0	0	
1	0	1	1	$A\overline{B}C$
1	1	0	1	$\overline{A}BC$
1	1	1	1	ABC

The Boolean equation is

$$\text{majority} = AB\overline{C} + A\overline{B}C + \overline{A}BC + ABC$$

Notice each row in the truth table that yields a logic 1. The three variables in that row are ANDed together to form a product. The Boolean equation that describes the truth table is the sum of these four products.

Step 4 To create a product with three input variables requires a three-input AND gate. By inspecting the Boolean equation we can see that this circuit will require four three-input AND gates (one for each product), one four-input OR gate (to sum together the four products), and four inverters (to create the complements of the judges' input values and drive the nonmajority LED).

Consider the SPST switches. They provide the digital inputs to the gates. Assume that the logic gates have an infinite input impedance. This means that current will neither enter nor exit through the inputs of these gates. (In reality all electronic devices have a finite impedance. We will discuss that subject in Chapter 5.) If the switch is open, there will be no path in which the current can flow; with no current flowing through the resistor, there will be no voltage dropped across it. The input to the gate will be "pulled up" to +5 V via the resistor. This resistor is called, appropriately, a *pull-up resistor*. Digital circuits abound with pull-up resistors. In the other case, when the switch is closed, +5 V is dropped across the resistor and the input to the gate will be 0 V.

4.2.2 Using Karnaugh Maps to Reduce the Number of Gates

The circuit in Figure 4.5 accomplishes its function, but one may wonder if it can be redesigned using fewer logic gates. In the early days the technology of digital integrated-circuit manufacturing was at a crude level. Complex circuits had to be constructed from discrete gates and gates were expensive. Because of this, it was important that a circuit be optimally designed using the minimum number of gates. As integrated-circuit manufacturing technology improved, the price of gates fell dramatically and the complexity of integrated circuits increased tremendously.

Circuit-reduction techniques are no longer a major concern of designers. In this section we will cover some general gate reduction concepts. If you need in-depth knowledge of circuit reduction, there are many classic texts available on the subject.

Consider the simple Boolean equation

$$y = ABC + AB\overline{C}$$

Does the logic value of variable C really have any affect on the value of the function? If both A and B are logic 1's, the value of C does not matter; if C is high, the first product is true; if C is low, the second product is true. Therefore, we can state that

$$ABC + AB\overline{C} = AB$$

We will use repeated applications of this idea to reduce the number of gates and the number of inputs required for each gate in the three-judges circuit. Our first step will be to rewrite the Boolean equation describing the three-judges circuit:

$$AB\overline{C} + A\overline{B}C + \overline{A}BC + ABC = AB\overline{C} + ABC + A\overline{B}C + ABC + \overline{A}BC + ABC$$

Notice that the term "ABC" is repeated three times in our new version of the three-judges Boolean equation. This is a perfectly acceptable action because of the Boolean identity "$A + A = A$." We have not really changed the meaning of the equation, only modified it to make the reduction much easier to see.

Now all we have to do is apply the concept that $ABC + AB\overline{C} = AB$ to the modified three-judges equation.

$$AB\overline{C} + ABC + A\overline{B}C + ABC + \overline{A}BC + ABC$$

$$AB \quad + \quad AC \quad + \quad BC$$

$$\text{vote} = AB + AC + BC$$

Closely examine the reduced three-judges equation. It says:

Whenever two judges vote yes, the vote of the third judge does not matter because a majority has already been reached.

All sum-of-products equations can be examined for reduction potential by applying this simple method. However, in a sum-of-products equation that has more than just a few products it can be difficult to pick out the reducible products. We need a visual aid to assist us in finding these products. That is the function of a Karnaugh map.

A *Karnaugh map* is a modified representation of a conventional truth table. It was invented by Maurice Karnaugh of Bell Labs. Karnaugh maps are designed so that along any row or column only the state of one variable changes with each cell. (This counting scheme uses the Gray code, which we examined in Chapter 1.) Adjacent cells arranged in such a manner will contain reducible products. This eases the effort required to do circuit reductions.

Figure 4.6 illustrates the basic form for two-, three-, and four-variable Karnaugh maps. (Karnaugh maps for equations of five or more variables are extremely ungainly and complex and are not examined in this text.) It is important to remember that these maps are just a visual aid for finding reducible products. Refer to Figure 4.6. Because the cells of the Karnaugh map are arranged to display reducible products, they do not have a sequential translation to the rows of their corresponding truth table. The cells on the left column of the Karnaugh map are considered to be adjacent to the cells in the right column, and the cells in the top row are considered to be adjacent to the cells in the bottom row. Although the Karnaugh map is illustrated as a two-dimensional object, it is much easier to think of as a sphere with the sides and tops rolled back to touch each other. If you ever question whether two cells are supposed to be adjacent, just examine them and determine if they are indeed reducible products (i.e., do they differ by only one variable?). Remember that the Karnaugh map does not work by magic; it is just a visual aid.

Figure 4.7 shows the truth table-to-Karnaugh map translation for the problem of the three judges. The output logic levels from each row in the truth table have been transferred to their corresponding cells in the Karnaugh map. Notice that we can find three pairs of adjacent cells that contain logic 1 levels. The one product that is common to all three pairs is ABC. That is the product we used to accomplish the Boolean algebra reduction of this circuit (Figure 4.8).

The following list indicates each pair of reducible products and the resultant product.

Reducible product	Reduced term
$AB\overline{C} + ABC$	AB
$\overline{A}BC + ABC$	BC
$A\overline{B}C + ABC$	AC

Notice that a product (in this case ABC) can be used any number of times. Now we sum these reduced products together to arrive at our new equation:

$$\text{vote} = AB + BC + AC$$

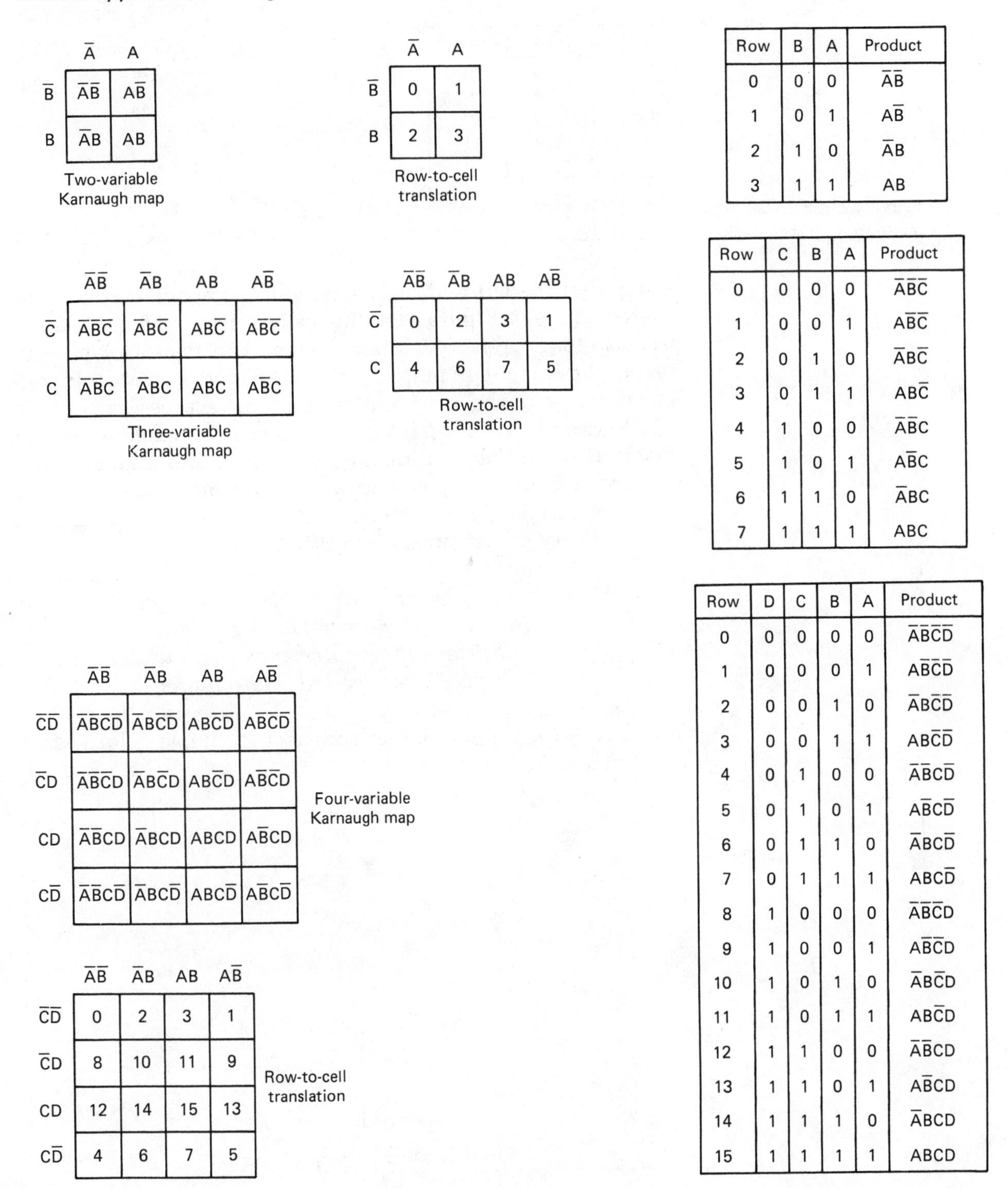

Two-variable Karnaugh map

	$\bar{A}$	A
$\bar{B}$	$\bar{A}\bar{B}$	$A\bar{B}$
B	$\bar{A}B$	AB

Row-to-cell translation

	$\bar{A}$	A
$\bar{B}$	0	1
B	2	3

Row	B	A	Product
0	0	0	$\bar{A}\bar{B}$
1	0	1	$A\bar{B}$
2	1	0	$\bar{A}B$
3	1	1	AB

Three-variable Karnaugh map

	$\bar{A}\bar{B}$	$\bar{A}B$	AB	$A\bar{B}$
$\bar{C}$	$\bar{A}\bar{B}\bar{C}$	$\bar{A}B\bar{C}$	$AB\bar{C}$	$A\bar{B}\bar{C}$
C	$\bar{A}\bar{B}C$	$\bar{A}BC$	ABC	$A\bar{B}C$

Row-to-cell translation

	$\bar{A}\bar{B}$	$\bar{A}B$	AB	$A\bar{B}$
$\bar{C}$	0	2	3	1
C	4	6	7	5

Row	C	B	A	Product
0	0	0	0	$\bar{A}\bar{B}\bar{C}$
1	0	0	1	$A\bar{B}\bar{C}$
2	0	1	0	$\bar{A}B\bar{C}$
3	0	1	1	$AB\bar{C}$
4	1	0	0	$\bar{A}\bar{B}C$
5	1	0	1	$A\bar{B}C$
6	1	1	0	$\bar{A}BC$
7	1	1	1	ABC

Four-variable Karnaugh map

	$\bar{A}\bar{B}$	$\bar{A}B$	AB	$A\bar{B}$
$\bar{C}\bar{D}$	$\bar{A}\bar{B}\bar{C}\bar{D}$	$\bar{A}B\bar{C}\bar{D}$	$AB\bar{C}\bar{D}$	$A\bar{B}\bar{C}\bar{D}$
$\bar{C}D$	$\bar{A}\bar{B}\bar{C}D$	$\bar{A}B\bar{C}D$	$AB\bar{C}D$	$A\bar{B}\bar{C}D$
CD	$\bar{A}\bar{B}CD$	$\bar{A}BCD$	$ABCD$	$A\bar{B}CD$
$C\bar{D}$	$\bar{A}\bar{B}C\bar{D}$	$\bar{A}BC\bar{D}$	$ABC\bar{D}$	$A\bar{B}C\bar{D}$

Row-to-cell translation

	$\bar{A}\bar{B}$	$\bar{A}B$	AB	$A\bar{B}$
$\bar{C}\bar{D}$	0	2	3	1
$\bar{C}D$	8	10	11	9
CD	12	14	15	13
$C\bar{D}$	4	6	7	5

Row	D	C	B	A	Product
0	0	0	0	0	$\bar{A}\bar{B}\bar{C}\bar{D}$
1	0	0	0	1	$A\bar{B}\bar{C}\bar{D}$
2	0	0	1	0	$\bar{A}B\bar{C}\bar{D}$
3	0	0	1	1	$AB\bar{C}\bar{D}$
4	0	1	0	0	$\bar{A}\bar{B}C\bar{D}$
5	0	1	0	1	$A\bar{B}C\bar{D}$
6	0	1	1	0	$\bar{A}BC\bar{D}$
7	0	1	1	1	$ABC\bar{D}$
8	1	0	0	0	$\bar{A}\bar{B}\bar{C}D$
9	1	0	0	1	$A\bar{B}\bar{C}D$
10	1	0	1	0	$\bar{A}B\bar{C}D$
11	1	0	1	1	$AB\bar{C}D$
12	1	1	0	0	$\bar{A}\bar{B}CD$
13	1	1	0	1	$A\bar{B}CD$
14	1	1	1	0	$\bar{A}BCD$
15	1	1	1	1	$ABCD$

Figure 4.6 Two-, three-, and four-variable Karnaugh maps.

This reduced equation agrees with the equation that we derived using Boolean algebra reduction techniques. Many people find that the graphic approach of Karnaugh maps leads to much easier discovery of all possible groups of reducible products.

We now examine three examples of finding groups of reducible products within the Karnaugh map (Figure 4.9). Remem-

C	B	A	Vote
0	0	0	0
0	0	1	0
0	1	0	0
0	1	1	1
1	0	0	0
1	0	1	1
1	1	0	1
1	1	1	1

	$\overline{A}\overline{B}$	$\overline{A}B$	AB	$A\overline{B}$
$\overline{C}$	0	0	1	0
C	0	1	1	1

Figure 4.7 Truth table to Karnaugh map translation.

bering that the left and right sides of a Karnaugh map are considered adjacent, Figure 4.9a illustrates a group of four reducible products, often referred to as a *quad*. The Boolean equations next to the Karnaugh map illustrate the process of reducing the four products in the quad to a two-variable expression.

Notice the two variables that were dropped, A and D. Within the quad both A and D occurred in both true and complemented form, whereas within the same quad, both B and C were expressed only in complemented form. From this example we can derive an extremely important rule:

Within any group of reducible products in a Karnaugh map, those products that appear in both true and complemented form will be dropped from the reduced equation.

This rule is a direct consequence of the simple relationship: $AB + A\overline{B} = A$.

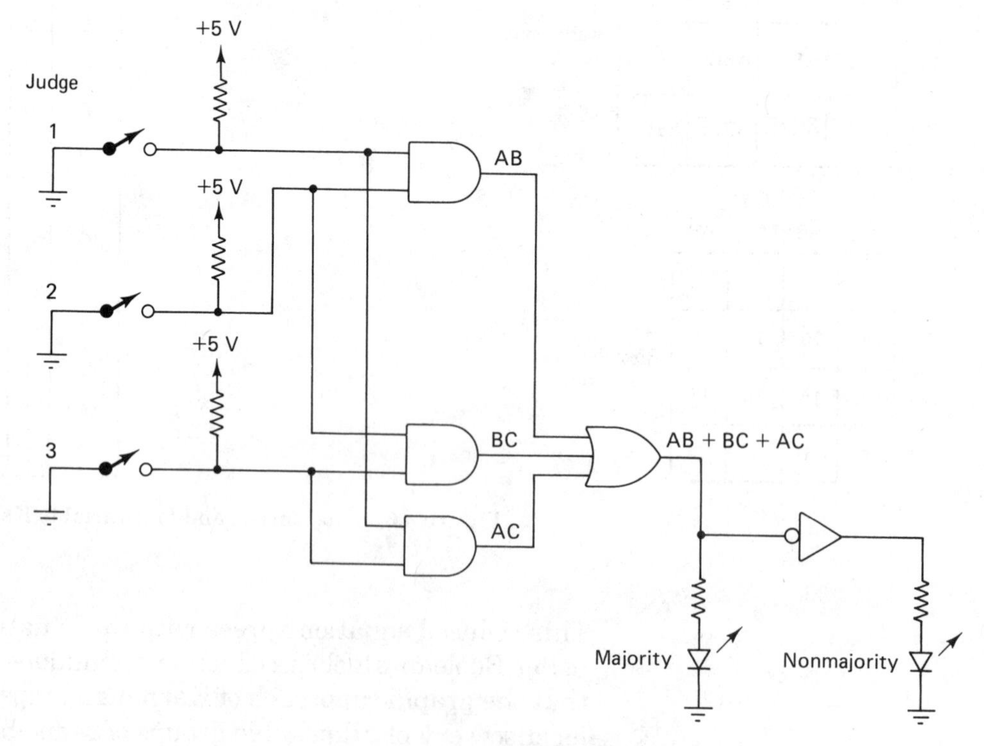

Figure 4.8 Reduced three-judges circuit.

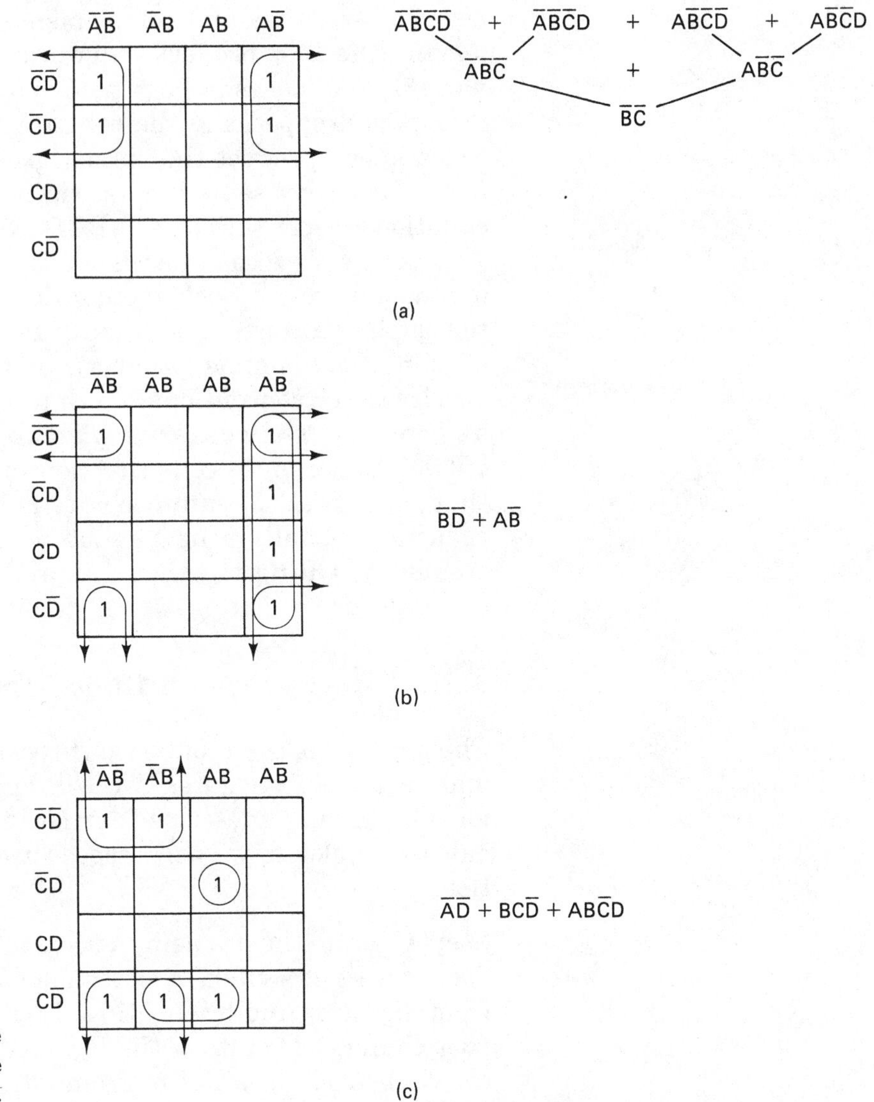

Figure 4.9 Reducible products within the Karnaugh map.

Now refer to Figure 4.9b. In this example there are two sets of quads. This means that the reduced equation will be a sum of two products each of which will have two variables. Remembering to think of the Karnaugh map as a sphere, we can see that the logic 1's in each of the four corners form a quad. The variables A and C appear in both true and complemented forms within this corner quad; therefore, they will be dropped from the first product. The second quad appears in the right column. Do not be concerned that there is a two-cell overlap in the two quads; the rule A + A = A implies that we can use any cell in any number of groups. Within the last column the variables C and D appear in both true and complemented form. They will be dropped from the second product. The reduced equation is simply: $\overline{B}\overline{D} + A\overline{B}$.

Finally, refer to Figure 4.9c. This example illustrates one quad, one pair, and one single variable. Let's examine each group individually. Within the quad the variables C and B appear in both true and complemented form. They are dropped, leaving the first product reduced as: $\overline{A}\,\overline{D}$. Within the pair the variable A appears in both true and complemented form; therefore, it is dropped and the remaining product is $BC\overline{D}$. Finally, there is one product that is not grouped. It cannot be reduced and will be represented in its original form: $AB\overline{C}D$. The reduced equation is: $\overline{A}\,\overline{D} + BC\overline{D} + AB\overline{C}D$.

Let's review the process of using a Karnaugh map to reduce a sum-of-products Boolean equation. First the equation is used to complete a conventional truth table describing each possible combination of inputs (product) and their output logic level. The entries in the conventional truth table are then transferred into an appropriate-sized Karnaugh map. You then must find groups of two, four or eight cells that contain logic 1 levels. Remember that the larger the number of cells in a group, the greater the reduction. Finally a new, reduced sum of products equation is created by ORing together the products that describe each reduced group.

4.2.3 The Room with Three Doors

The problem of the room with three doors is considered a classic application of logic gates. We will approach this problem in the four-step manner described earlier in this chapter. After the circuit is completed, we will inspect it for any possible gate reductions.

Step 1 State the problem. A room has three doors. Next to each door is a light switch. A person should be able to enter or exit from any door and by toggling the light switch adjacent to the door change the state of the light. (By "toggle" we mean to flip a switch from on to off or from off to on.) As an example: We assume the light is originally off; a person enters the room via door A. The switch next to door A is toggled and the light should go on. The person later exits via door C. As the switch next to door C is toggled, the light should go out.

The inputs will be from each of the light switches. The output will turn on an LED that symbolizes the light. (*Note:* In a practical circuit the output would drive an optical isolated solid-state relay.)

Step 2 Complete the truth table. We need to make an initial assumption about the state of the light. We will assume that when all three switches are in their logic 0 state, the light will be out. Whenever the variables from one row to the next change an odd number of times, the light will change states. Whenever the variables change an even number of times, the light will not change states. The preceding two sentences may seem confusing. Picture a situation where two people enter the room at the same time from two different doors. At the same instant they each toggle the light switch next to their door. By toggling two switches they cancel each other out, and the condition of the light does not change. This is what is meant by an even number of variables changing. On the other hand, if three people enter the room at the same time, each by a separate door, the first two toggles will cancel, but the third toggle will change the condition of the light. If one or three variables change from one row to another, we will have an odd number of toggles and the light will change states.

Doors				
C	*B*	*A*	*Light*	*Comment*
0	0	0	Off	Assume that the light is initially off.
0	0	1	On	One variable (A) changed states, so the light changed.
0	1	0	On	Two variables (A and B) changed, so the light didn't change.
0	1	1	Off	One variable (A) changed, so the light changed.
1	0	0	On	Three variables changed, so the light changed.
1	0	1	Off	One variable (A) changed, so the light changed.
1	1	0	Off	Two variables (A and B) changed, so the light did not change.
1	1	1	On	One variable (A) changed, so the light changed.

Step 3 Create a Boolean equation. Each row in the truth table that has a light condition of ON will be a product in the equation

$$\text{light on} = A\overline{B}\,\overline{C} + \overline{A}B\overline{C} + \overline{A}\,\overline{B}C + ABC$$

Step 4 Construct the logic circuit. The circuit shown in Figure 4.10 was realized directly from the Boolean equation in step 3. The Karnaugh map shown in Figure 4.11 indicates that this circuit cannot be reduced. Notice that none of the products are adjacent in the horizontal or vertical cells.

4.2.4 A 2-Line-to-4-Line Decoder

Consider the following problem.

Step 1 State the problem. There are four LEDs on a panel. We would like to be able to turn on any one of the four. The LED that will be turned on will be indicated by a 2-bit input. Two binary digits can represent four unique numbers, 0 through 3. Each LED will be assigned one of these numbers. In digital electronics, we will always start counting with the number 0. The first LED will illuminate whenever the code of 00 is input; the second LED will illuminate with the code 01; the third with 10; and the fourth with 11. This circuit should have an active-low output. That is, the output indicated by the 2-bit select code will go to logic zero, while the other outputs will all go to logic 1. Active-low output circuits are just as common as active-high output circuits.

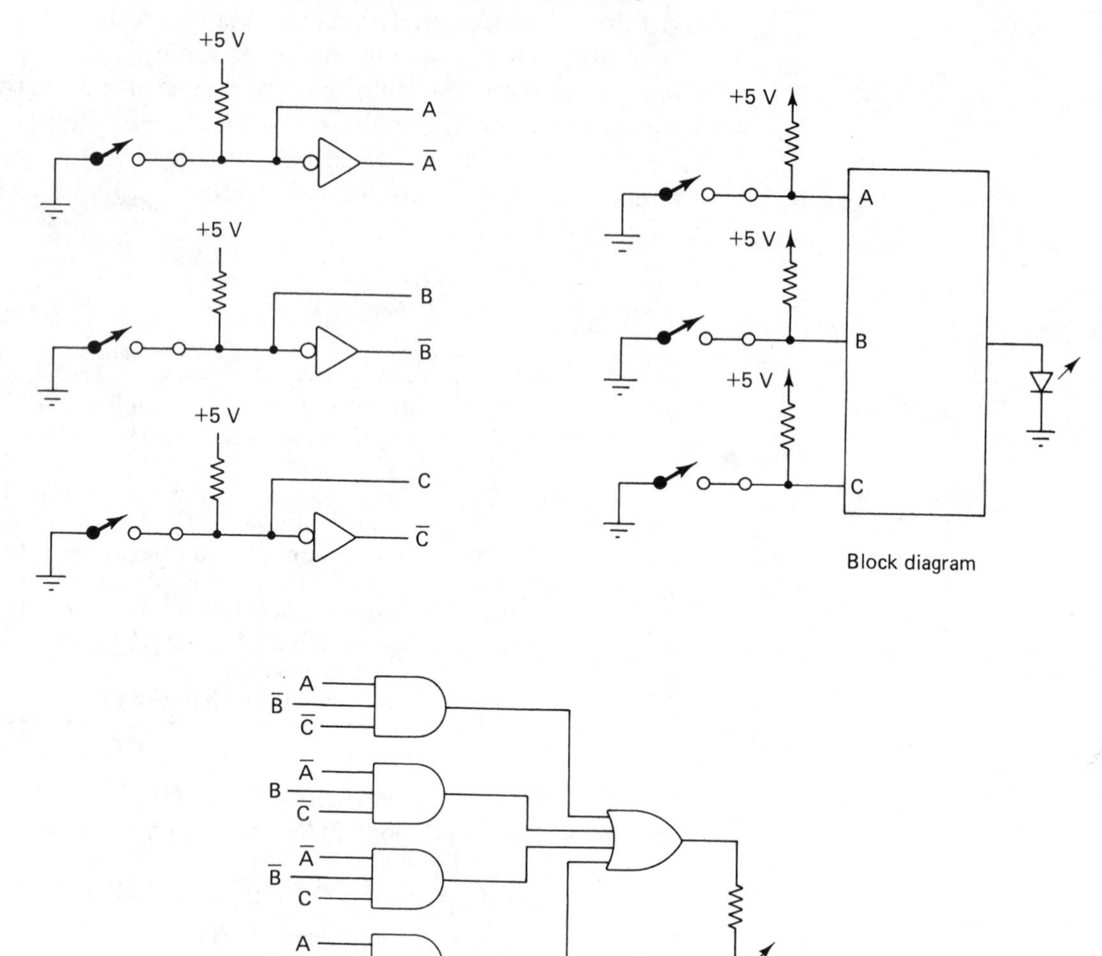

Figure 4.10 The room with three doors.

Step 2 Complete the truth table. This truth table will be different from the previous truth tables because it will have four outputs, one for each LED.

Input		*Output*			
B	A	1	2	3	4
0	0	0	1	1	1
0	1	1	0	1	1
1	0	1	1	0	1
1	1	1	1	1	0

	$\overline{A}\overline{B}$	$\overline{A}B$	AB	$A\overline{B}$
$\overline{C}$	0	1	0	1
C	1	0	1	0

Figure 4.11 Karnaugh map for the room with three doors.

Step 3 Create a Boolean equation. We will create four separate Boolean equations, one for each output:

output 1 = $\overline{A}\overline{B}$ output 2 = $A\overline{B}$ output 3 = $\overline{A}B$
output 4 = AB

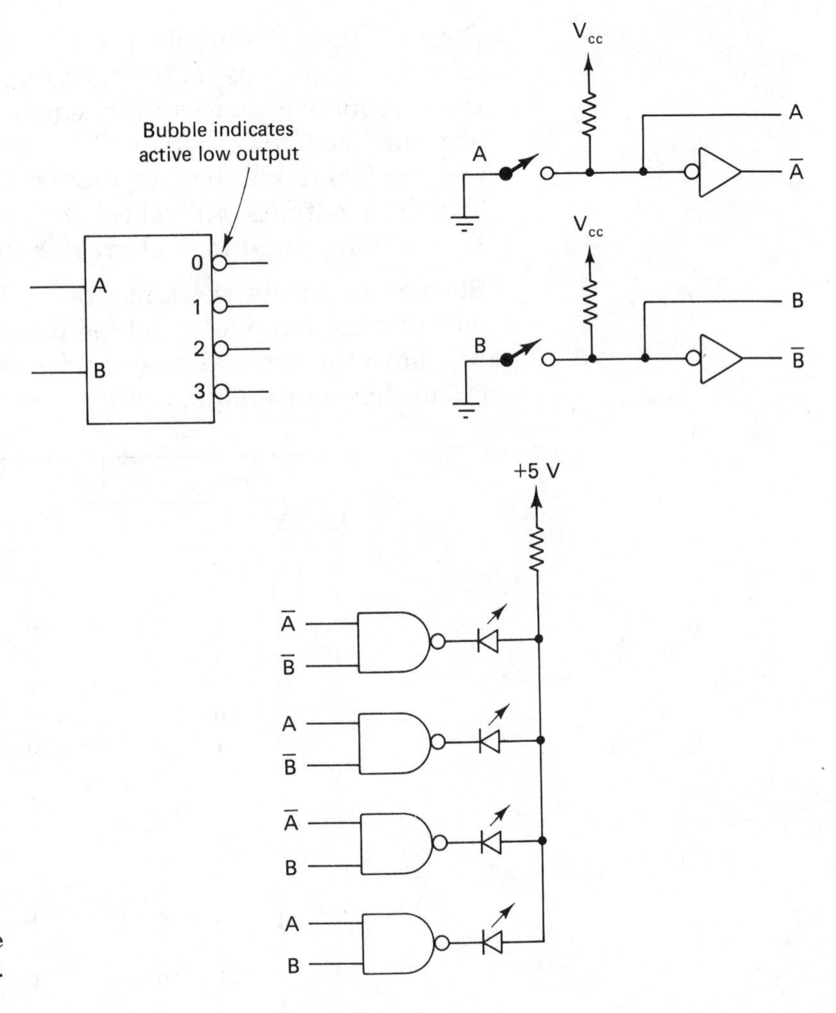

Figure 4.12 2-line to 4-line decoder.

Step 4 Construct the logic circuit. Considering that step 3 contains four equations, one may be inclined to think tht we must construct four separate circuits. We must bear in mind that the only thing common to all four outputs are the two input variables.

Decoders are useful digital circuits. It is rare to find a decoder constructed from discrete gates, as in Figure 4.12. Using medium-scale-integration (MSI) techniques, manufacturers integrate one or more decoders in a single integrated circuit. Chapter 7 will introduce you to these MSI decoders. Notice the name of this circuit: a 2-line-to-4-line decoder. This circuit has two inputs and four outputs, and it provides a decoding function. The names of digital circuits reflect their function. Only one output will ever be low at the same time. That is why all the LEDs can share the same current-limiting resistor.

4.2.5 A 2-Bit Magnitude Comparator

Consider the next procedure, which illustrates a useful digital function.

Step 1 State the problem. A circuit is required that can compare the magnitudes of two 2-bit numbers, and indicate whether the first number is less than, equal to, or greater than the second number. This circuit will be called a 2-bit magnitude comparator. There will be four inputs: 2 bits for A and 2 bits for B. The three outputs will reflect the relative magnitudes of A and B: less than, equal to, and greater than.

Step 2 Complete the truth table. Remember that the inputs will be considered as weighted binary numbers and the inputs will have the decimal magnitudes of 0 through 3. Because this circuit has four inputs, there are 16 possible combinations.

Input				*Output*		
B2	*B1*	*A2*	*A1*	*A < B*	*A = B*	*A > B*
0	0	0	0	0	1	0
0	0	0	1	0	0	1
0	0	1	0	0	0	1
0	0	1	1	0	0	1
0	1	0	0	1	0	0
0	1	0	1	0	1	0
0	1	1	0	0	0	1
0	1	1	1	0	0	1
1	0	0	0	1	0	0
1	0	0	1	1	0	0
1	0	1	0	0	1	0
1	0	1	1	0	0	1
1	1	0	0	1	0	0
1	1	0	1	1	0	0
1	1	1	0	1	0	0
1	1	1	1	0	1	0

Take a moment to examine carefully the outputs of the truth table. Do you agree with the value of each output?

Step 3 Create a Boolean equation. We will create one equation for each output.

$$A < B = \overline{A1}\,\overline{A2}B1\overline{B2} + \overline{A1}\,\overline{A2}\,\overline{B1}B2 + A1\overline{A2}\,\overline{B1}B2 + \overline{A1}\,\overline{A2}B1B2 + A1\overline{A2}B1B2 + \overline{A1}A2B1B2$$

$$A = B = \overline{A1}\,\overline{A2}\,\overline{B1}\,\overline{B2} + A1\overline{A2}B1\overline{B2} + \overline{A1}A2\overline{B1}B2 + A1A2B1B2$$

$$A > B = A1\overline{A2}\,\overline{B1}\,\overline{B2} + \overline{A1}A2\overline{B1}\,\overline{B2} + A1A2\overline{B1}\,\overline{B2} + \overline{A1}A2B1\overline{B2} + A1A2B1\overline{B2} + A1A2\overline{B1}B2$$

Step 4 Construct the logic circuit (see Figure 4.13).

4.3 THE GATE AS AN ELECTRONIC SWITCH

Gates are also used as switches that either pass or block input signals. Many digital circuits have enable inputs. Unless this enable bit is at the proper digital level, the circuit will not function: it is disabled. These enable inputs employ nothing more than simple gates.

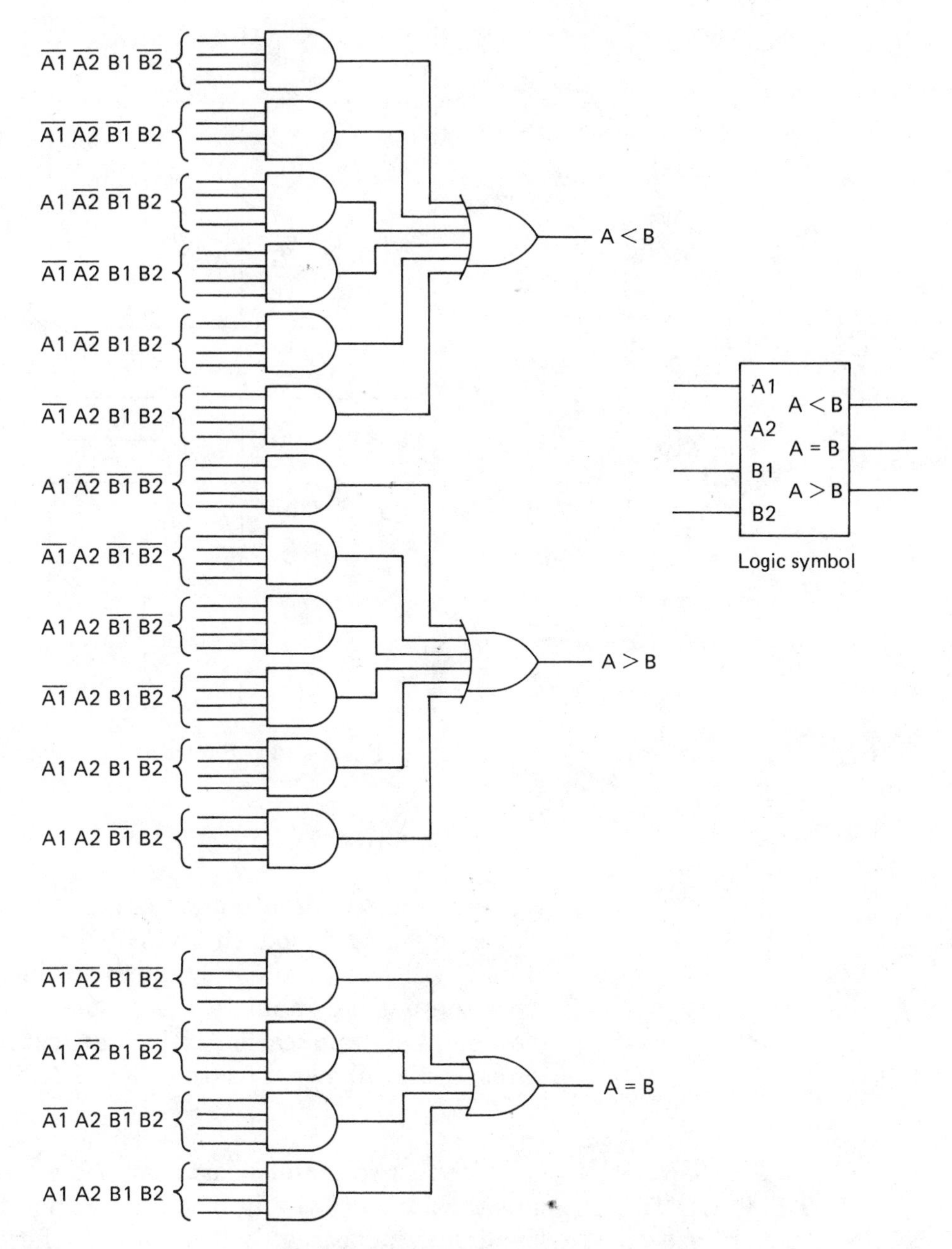

Figure 4.13 2-bit magnitude comparator.

4.3.1 Two-Input Gates as Simple Switches

The gates can be used as electronic switches that either pass data or output a constant value (Figure 4.14). The key to using gates as switches is dynamic input levels. Gates with two inputs are used; one input will have the data applied to it and the other will have a control level applied to it. If the control level is the dynamic input level, the output will be forced to a particular value, regardless of the data input. Here the data input will have a "don't-care" value. A *don't-care value* is denoted by an × in the truth table. It is called a don't-care value because it does not affect the output value.

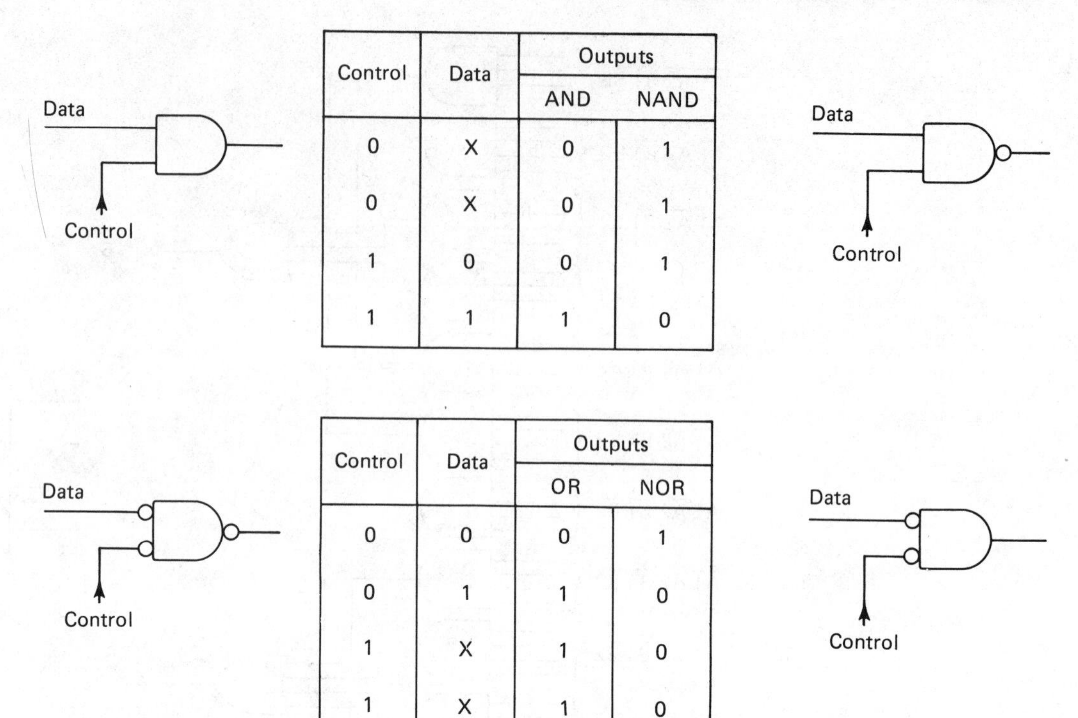

Control	Data	Outputs	
		AND	NAND
0	X	0	1
0	X	0	1
1	0	0	1
1	1	1	0

Control	Data	Outputs	
		OR	NOR
0	0	0	1
0	1	1	0
1	X	1	0
1	X	1	0

Figure 4.14 Simple gates as digital switches.

4.3.2 A 1-of-2 Data Selector

We would like to create a circuit that is the digital equivalent of a single-pole double-throw (SPDT) mechanical switch. This circuit will have two data inputs, A and B. The select bit will steer the desired input to the output. The other input will be blocked. This application is just an extension of using simple gates as digital switches (see Figure 4.15).

4.4 AN IGNITION ENABLE CONTROL CIRCUIT

Consider a two-seater sports car. We would like to design a circuit that enables the ignition only when both doors are closed, the driver has a seat belt on, and if there is a passenger, that person also has a seat belt on. We will approach this problem in much the same manner as the previous applications. Step 2, completing the truth table, will be omitted. There are too many variables in this circuit to create a workable truth table. The Boolean equation can be arrived at without the aid of a truth table.

Step 1 The following table contains the five inputs and the variable names that we will use to represent them in the Boolean equation.

Input	*Door 1*	*Door 2*	*Driver seat belt*	*Passenger seat sensor*	*Passenger seat belt*
Variable name	D1	D2	DSB	PSS	PSB

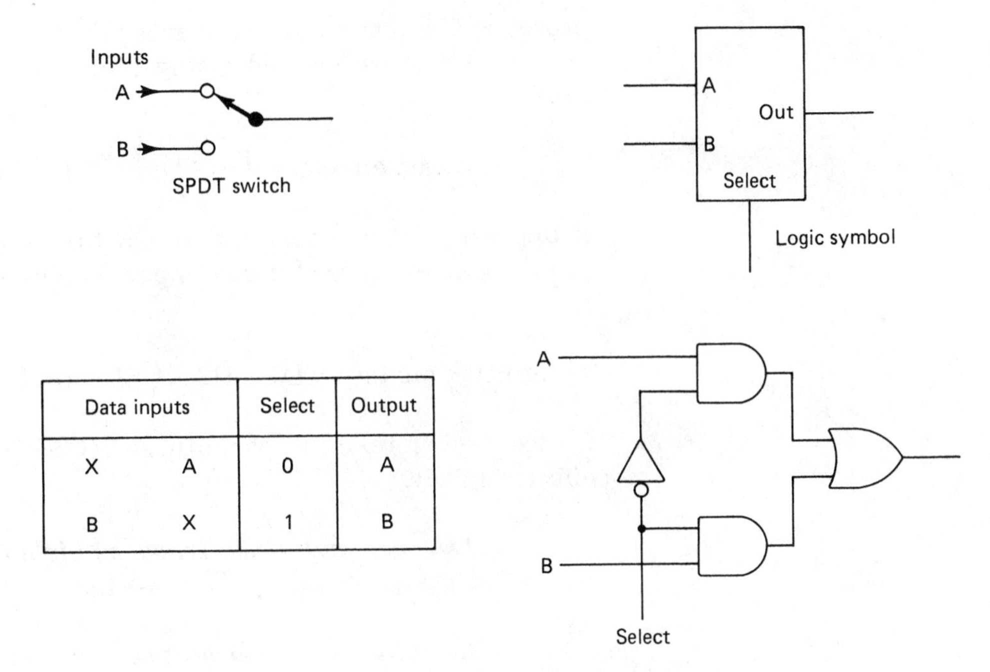

Data inputs		Select	Output
X	A	0	A
B	X	1	B

Figure 4.15 A 1-of-2 data selector.

A sensor switch that acts like a momentary SPST switch will be placed in the doors, seat-belt buckles, and underneath the passenger's seat. This last switch will indicate whether a passenger is present. The output will be a single active-high logic level that indicates when the ignition is enabled.

Step 3 We will skip step 2 because the Boolean equation will be easier to derive from an intuitive examination of the inputs. This is a straightforward problem except for the passenger's seat belt. If there is no passenger, the state of the passenger seat belt (PSB) switch is a don't-care condition. If the passenger seat sensor (PSS) indicates that there is a passenger, the passenger seat belt (PSB) input must indicate that the passenger's seat belt is buckled. We will use a pull-up resistor with each switch; if a switch is open, its associated logic level will be high; if the switch is closed, it will output a logic 0.

Whether or not a passenger is present, the door 1 (D1), door 2 (D2), and driver seat belt (DSB) inputs must all be low. The first part of the Boolean equation should therefore be

$$\text{ignition enable} = \overline{\text{D1}} \cdot \overline{\text{D2}} \cdot \overline{\text{DSB}} \cdots$$

Now let's consider how to introduce the passenger seat sensor and seat belt buckle into the equation. If the passenger seat sensor (PSS) is high, indicating a no-passenger condition, the previous Boolean equation should be true:

$$\text{ignition enable} = \overline{\text{D1}} \cdot \overline{\text{D2}} \cdot \overline{\text{DSB}} \cdot \text{PSS} \cdots$$

What if the passenger seat sensor (PSS) is low, indicating the presence of a passenger? If that is the case, the passenger seat belt must also be low, indicating that the passenger is buckled up:

$$\text{ignition enable} = \overline{\text{D1}} \cdot \overline{\text{D2}} \cdot \overline{\text{DSB}} \cdot \overline{\text{PSS}} \cdot \overline{\text{PSB}} \cdots$$

It appears that we must use an OR function to provide for the presence or absence of a passenger. The final Boolean equation is

$$\text{ignition enable} = \overline{\text{D1}} \cdot \overline{\text{D2}} \cdot \overline{\text{DSB}} \cdot (\text{PSS} + (\overline{\text{PSS}} \cdot \overline{\text{PSB}}))$$

Notice the last terms in parentheses, $(\text{PSS} + (\overline{\text{PSS}} \cdot \overline{\text{PSB}}))$. This effectively says:

If the passenger seat sensor is high, then ignore the status of the passenger seat belt.

But if the passenger seat sensor is low, then the passenger seat belt must also be low.

Figure 4.16 shows the gate realization of this Boolean equation.

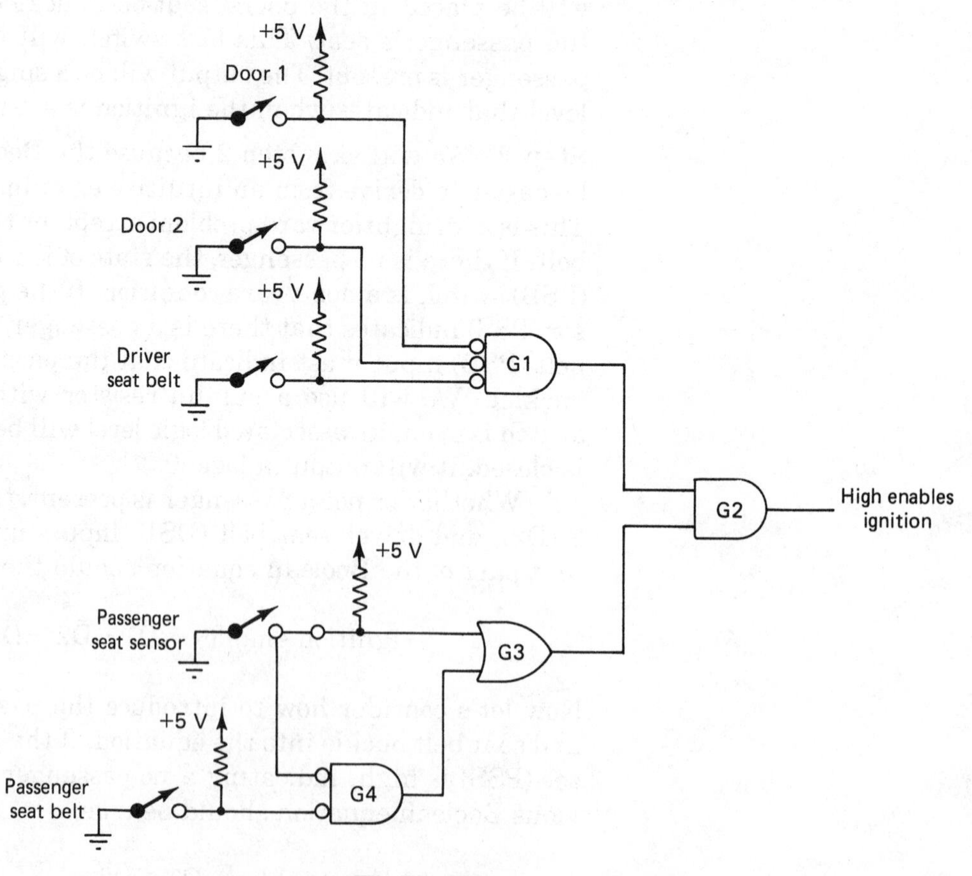

Figure 4.16 Ignition enable circuit.

Gate G1 is a three-input NOR gate. The negative-logic symbol was used to illustrate that when D1 is low AND D2 is low AND DSB is low, the output of G1 will go to its active-high output level. G4 is a two-input NOR gate that also uses a negative-logic symbol. When PSS is low and PSB is low, G1 will go to its active-high output level. G3 is a two-input OR gate that is illustrated with a positive-logic symbol. If PSS is high OR the output of G4 is high, the output of G3 will go to its active high output level. Finally, G2 is a positive-logic two-input AND gate. When the output of G1 is high AND the output of G3 is high, the ignition will be enabled. It is important that you understand the use of positive- and negative-logic symbols. Remember that a properly drawn schematic is an aid to understanding the manner in which the circuit operates. Step through Figure 4.16 several times until you feel that you have achieved a thorough understanding of the circuit operation.

QUESTIONS AND PROBLEMS

4.1. Derive the Boolean equation and complete a truth table for the circuits shown.

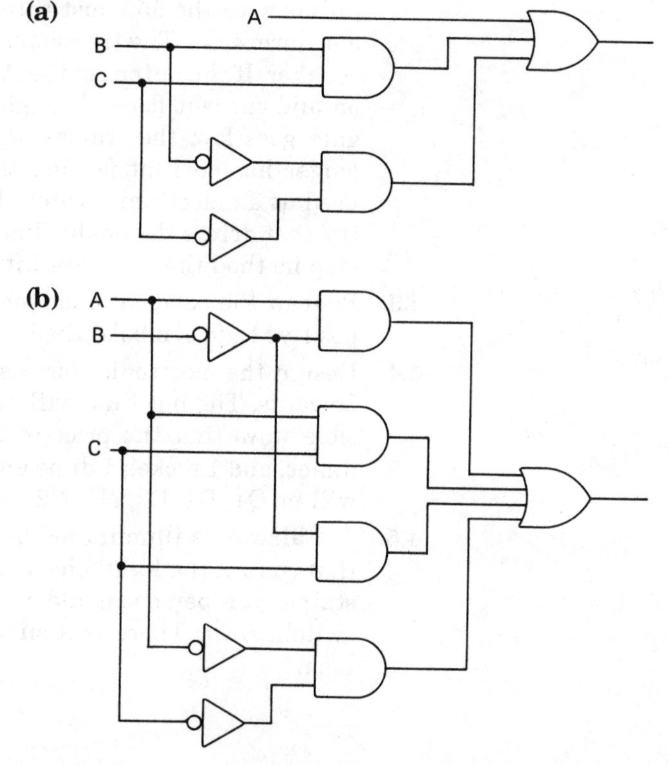

4.2. The circuit shown below should provide the useful function of warning a driver with an audible tone whenever the headlights of a car are on and the ignition switch is off. This circuit will prevent many dead batteries. The circuit will have two logic inputs: the headlight status and the ignition status. A status level of logic 0 indicates tht the headlights or ignition is on. The output will be an audio tone. This tone will be created with the use of a 555 timer, a driving transistor, and a speaker. The 555 timer is configured as a 1-kHz square-wave oscillator. If you are not fa-

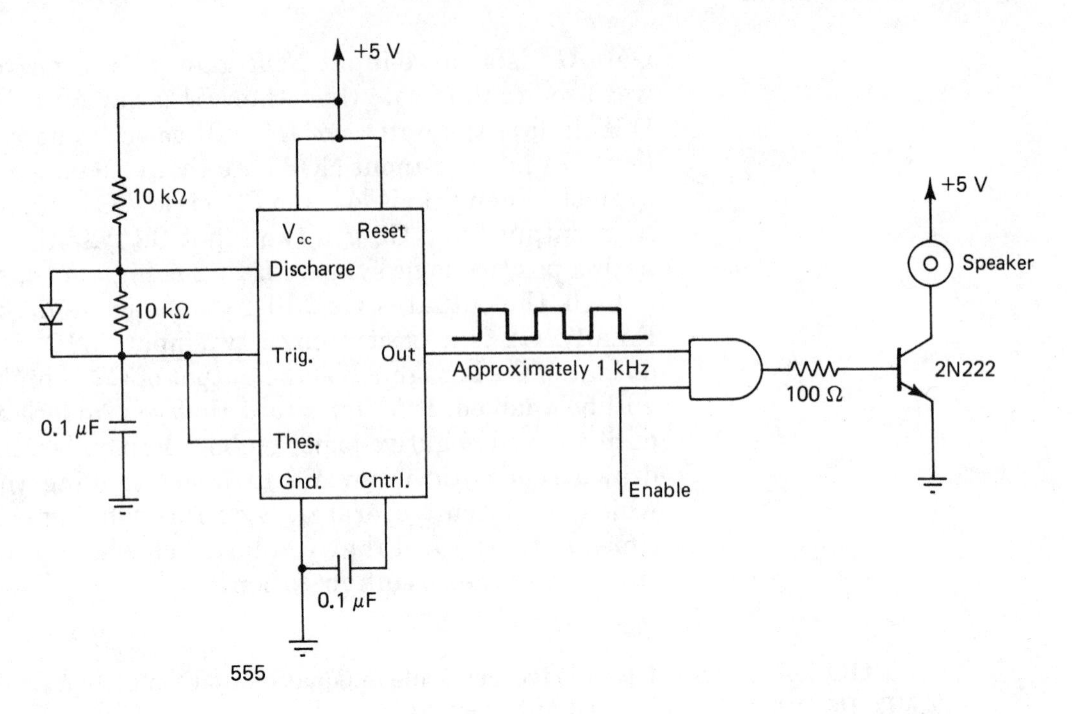

miliar with the 555, just think of it as a black box that outputs a square wave. The transistor provides current gain to drive the speaker. If the output of the AND goes high, the transistor turns on and current flows through the speaker coil. When the AND gate goes low, the transistor turns off and the speaker coil no longer has current flowing through it. The AND gate is to be used as an electronic switch. It is your task to design the circuitry that drives the enable input of the AND gate. Use the three-step method that we have introduced in this chapter.

4.3. Redraw the schematic for the ignition enable circuit using only positive-logic symbols. Explain which schematic is best and why.

4.4. Design the electronics for a soda-pop machine. Each soda costs 25 cents. The machine will not give change. There are four possible ways that the price of 25 cents can be paid: 1 quarter, 2 dimes, and 1 nickel; 1 dime and 3 nickels; or 5 nickels. The inputs will be Q1, D1, D2, N1, N2, N3, N4, and N5.

4.5. A stairway is illuminated by one light. There are two switches that control the light: one at the top and one at the bottom of the stairway. A person should be able to control the light from either switch. *Hint:* There is a simple function that can perform this task.

5

TTL AND CMOS LOGIC FAMILIES

The gates we have been studying can be realized in many integrated-circuit technologies. The two most popular are TTL and CMOS. TTL stands for "transistor-to-transistor logic." The transistors used in TTL are bipolar. CMOS stands for "complementary metal-oxide semiconductor." CMOS employs both PMOS and NMOS transistors (FETs) in a complementary configuration. Many students panic at the mention of bipolar and MOS transistors. Don't panic! Little knowledge of transistor theory is required to understand either TTL or CMOS. Our investigation of these transistors will be at an elementary level.

Integrated circuits can be classified by the complexity of functions that they perform:

Small-scale integration (SSI) Simple logic gates are the types of devices contained in SSI integrated circuits. SSI integrated circuits were widely used until the advent of more complex circuit technologies. Simple SSI devices are used as the "glue" to hold together more complicated circuit technologies. SSI devices come in 14- and 16-pin packages.

Medium-scale integration (MSI) MSI integrated circuits contain subsystems that use the equivalent of 12 or more simple gates. Chapter 7 provides a survey of MSI devices. MSI devices come in 16-, 18-, 20-, 24-, and 28-pin packages.

Large-scale integration (LSI) LSI devices are whole systems on a chip that contain the equivalent of 100 or more gates. The microprocessor is an example of an LSI device. LSI devices come in 28- and 40-pin packages.

Very large scale integration (VLSI) VLSI integrated circuits contain the equivalent of 1000 or more gates. Highly complex systems can be contained in a single integrated circuit. VLSI devices come in 40-, 48-, and 64-pin packages.

SSI is the oldest IC manufacturing process. It was developed in the early 1960s. VLSI is the most advanced circuit technology. Designers want to get more and more functions in a single integrated circuit. The reasons for this approach are simple: less printed circuit board space is required, less power is consumed, much less design time is required, and most important from a technician's point of view, with fewer devices in a circuit there should be fewer problems.

Digital systems are designed around LSI devices. As the complexity of integrated circuits increase, the complexity of new systems will increase. Even though newer designs use fewer ICs, the technician's job of troubleshooting is not going to be easier. On the contrary, as circuit complexities increase, so do the skills required to troubleshoot those circuits.

5.1 THE TTL FAMILY

NAND is the natural function of TTL. We will examine the internal operation of the NAND gate. This will give us insight into the input and output characteristics of TTL. Some important questions that we must consider are:

1. How much current is required to drive a TTL gate with a logic 0? With a logic 1?
2. What is the range of voltages that constitute a valid logic 0? A valid logic 1?

5.1.1 The 7400 Series

The 7400 series of TTL is the most widely used group of integrated circuits in the world. Almost every electronic device in any application contains 7400 series TTL components. When we analyze the 7410 NAND gate (Figure 5.1) you must remember that a transistor in a digital circuit is either in cutoff or saturation. All transistors will be digital switches. We are analyzing this circuit to gain knowledge of the input and output characteristics. We are interested only in the input and output circuitry. The transistors that make up the majority of the IC are of no interest to us. Consider these ICs as "black boxes" that provide certain logic functions. When we look into the input of the black box, all we will see is the input circuitry. When we look into the output of the black box, we will only see the output

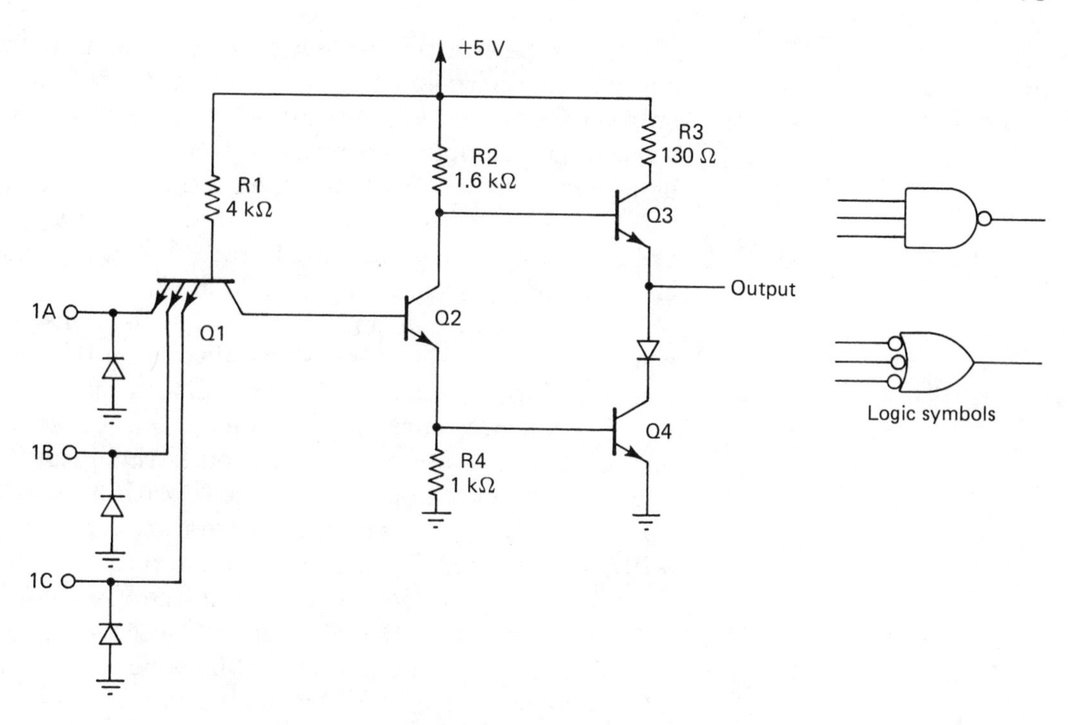

Figure 5.1 7410 three-input NAND gate.

circuitry. The rest of the IC is transparent to the user. There is no reason to concern ourselves with the complex operations of all the internal transistors that make up a digital integrated circuit.

The first thing that you may notice in Figure 5-1 is that Q1 has more than one emitter. Multi-emitter transistors are used to provide a gate with multiple inputs. There is virtually no limit on the number of emitters that can be attached to a transistor. If any of the emitters has a logic 0 applied to it, Q1 will conduct and go into saturation. The emitters can be thought of as being in parallel, and any one of them can provide the path for collector and base current.

The diodes on the emitters are negative-voltage protection diodes. If a negative voltage is applied to the gate, Q1 could be damaged. These diodes clamp any negative voltage to one diode drop.

A logic 0 will forward bias the base–emitter junction of Q1. Q1 will be driven into saturation. In saturation, the voltage on the emitter and collector will be approximately equal. This means that the voltage on the collector of Q1 and the base of Q2 will be 0 V. This 0 V will reverse bias the base–emitter junction of Q2. The voltage on the base of Q1 will be one diode drop, approximately 0.6 V. This leaves 4.4 V to be dropped across R1. This calculates to approximately 1.1 mA of current through R1 and the base–emitter junction of Q1. This 1.1 mA flows into the source of the logic 0. Remember that the source of the logic 0 will usually be another gate. The gate that provides the logic 0 must sink this 1.1 mA of current.

If all the emitters have logic 1's applied to them, Q1 will be cut off. Each base–emitter junction will be reversed biased. The only current that will flow in Q1 will be the base–emitter reverse diode leakage current. In a worst-case situation, this will be a maximum of 40 μA. It appears that the gate that provides the logic 1 input will not be required to source or sink any appreciable amount of current. Figure 5.2 summarizes the input circuitry of the 7400 series of TTL devices.

The standard output of the 7400 series consists of two transistors. One transistor pulls the output up toward +5 V when the gate outputs a logic 1. On a logic 1 output, the pull-up transistor will source current; that means that current will flow out of the gate. The other transistor pulls the output down toward 0 V when the gate outputs a logic 0. With a logic 0 output, the pull-down transistor will sink current; this means that current will enter the gate. This type of active pull-up/pull-down configuration is called a *totem-pole output* (see Figure 5.3). The gate's internal circuitry will assure that Q3 and Q4 will never conduct at the same time. Either Q4 will be conducting and the output will be logic 1, or Q3 will be conducting and the output will be logic 0. The diode, D1, is used by the gate's internal circuitry to make sure that Q4 does not turn on at the wrong time. D1 can be ignored during this analysis of the totem pole.

Let's consider the current capabilities of the totem pole. The pull-up transistor has as a 130-Ω resistor in series with the collector. If the output of the gate is shorted to ground, via another gate or a solder bridge, this collector resistor will limit the output current to a safe value. The trade-off in this design is that the current-limiting resistor will reduce the gate's ability to

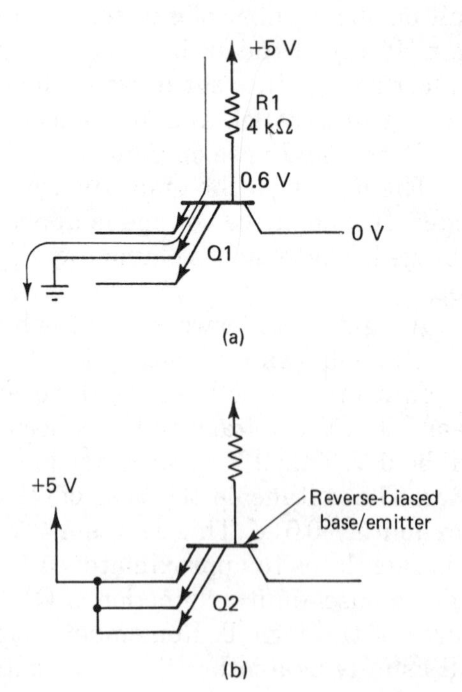

Figure 5.2 Input circuitry of 7410 three-input NAND gate.

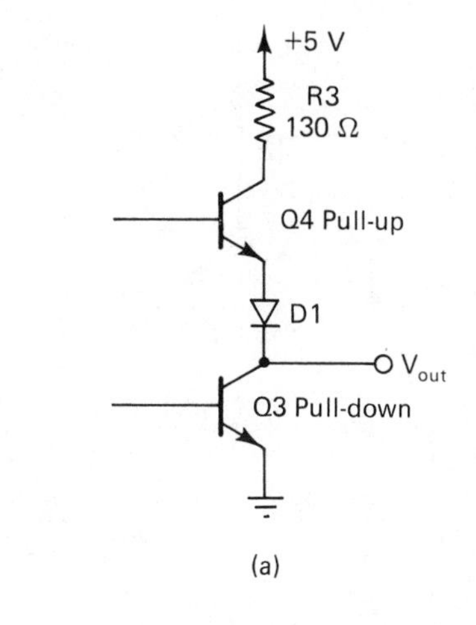

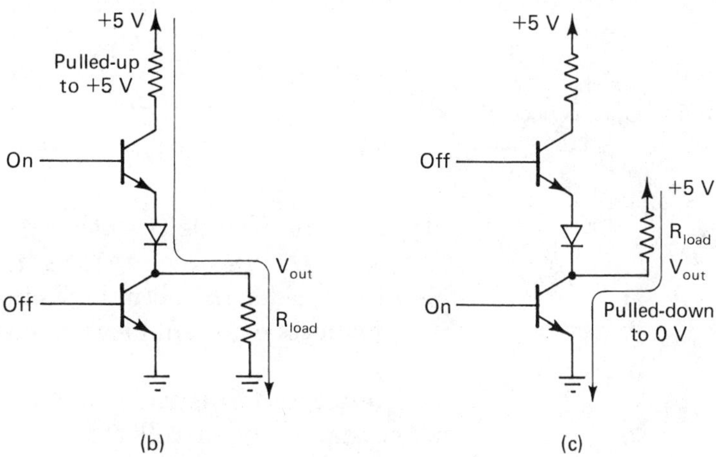

Figure 5.3 Totem-pole outputs.

source current on a logic 1 output. TTL does not source current well. This is an important fact to remember.

The pull-down transistor does not have a current-limiting resistor. It can sink about 40 times more current than the pull-up transistor can source. TTL sinks current well. You will discover that the majority of advanced TTL devices will have active-low outputs. This means that the output of a device will go to a logic 0 level to indicate a true condition. An active-low output takes advantage of TTL's superior ability to sink current.

5.1.2 The Open-Collector Output

The totem pole is the most widely used output structure in TTL. There is another type of output structure called an *open collector* that is used in special applications. To understand open-collector output, we must first consider one limitation of the totem pole.

Totem pole outputs can never be tied together.

If totem-pole outputs are tied together, and one gate is trying to

Output
Pull-down
(a)

O.C.
*
(b)

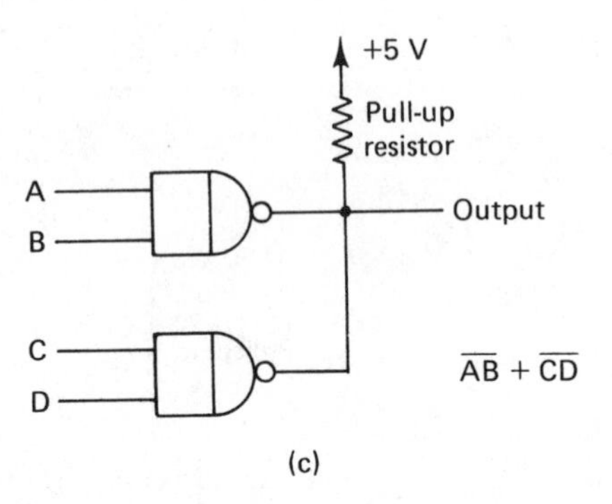

Figure 5.4 Open-collector output.

output a logic 1 while the other gate is trying to output a logic 0, a conflict occurs which results in an indeterminate logic level. (We will consider the actual effect of this occurrence in Chapter 6.) An open collector enables two or more outputs to share a common line.

Figure 5.4a illustrates an open-collector output. The reason for the name "open collector" should be obvious. The output structure is simply a pull-down transistor. If the output of the gate is logic 0, the pull-down transistor is conducting. If the output of the gate is logic 1, the pull-down transistor is cut off. When the pull-down transistor is cut off, the output will float to an indeterminate level. The most important thing to remember about open-collector devices is:

Open-collector outputs require an external pull-up resistor.

This external pull-up resistor will have a value between 1 and 10 kΩ. The pull-up resistor will provide the same function as the pull-up transistor that is contained in the totem pole. The advantage of an open collector is that the output of many gates can share the same pull-up resistor. Consider Figure 5.4b. (*Note:* In schematics, the initials O.C. or an * or a vertical slash across the front of the gate are used to indicate an open-collector output.)

If gate 1 OR gate 2 outputs a low, then the common node will go low; otherwise, the common node will be high.

We can tie together an infinite number of open-collector gates sharing a common pull-up resistor; it takes only a low output on one of the gates to pull the common node low. An output that connects two or more open-collector gates together is called a *wired-OR output*. Think about the name: wired-OR. The pull-up resistor provides an OR function.

If the first gate OR the second gate OR the third gate . . . OR the Nth gate goes low, then the output will go low.

It is not yet an appropriate time to provide applications of wired-OR connections. This task will be deferred until Chapter 11.

5.1.3 Valid TTL Logic Levels

We have made the ideal assumption that a logic 1 is equal to +5 V and a logic 0 is equal to 0 V. Due to factors such as loading effects and nonideal transistor characteristics, there is an acceptable range of voltages for both logic 1's and logic 0's. Consider Figure 5.5. TTL is specified to accept an input voltage in the range of 2.0 to 5.0 V as a logic 1. It is also specified to accept a voltage between 0 and 0.8 V as a logic 0. All the circuits supplying inputs to TTL gates are required to meet these specifications. An important question concerns the range of voltages between 0.8 and 2.0 V. Any voltage that lies within this range is neither a logic 1 nor a logic 0. These voltages are called *indeterminate levels* and always indicate a circuit malfunction.

If you find a voltage in a TTL circuit that is greater than 0.8 V and less than 2.0 V, this indicates a circuit malfunction.

There are many circuit faults that can cause a voltage to fall into this indeterminate range. In Chapter 6 we will address the problem of troubleshooting logic levels. You should memorize the acceptable ranges for logic 1 and logic 0 input for TTL circuits.

Figure 5.5b illustrates the specified output levels of TTL devices. These levels are guaranteed for standard TTL only if the gate is sourcing less than 0.4 mA on a logic 1 or sinking less than 16 mA on a logic 0. These current limitations can be easily understood by analyzing Figure 5.3.

Assume that the pull-up transistor is saturated and the totem pole is pulling the output up to logic 1. Any source current flowing out of the gate must pass through R3. If the load current

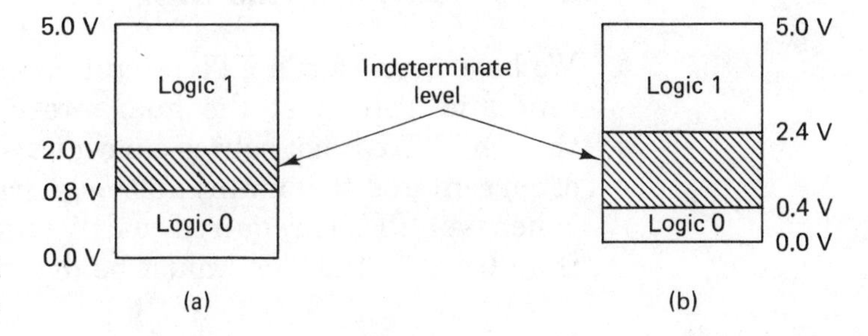

Figure 5.5 Valid TTL acceptable input (a) and guaranteed output (b) levels.

increases, the voltage drop over R3 will increase. As the voltage drop over R3 increases, the output voltage will decrease. R3 and the load make up a simple voltage divider. It is specified that the gate cannot source more than 0.4 mA and still maintain an output greater than 2.4 V. A TTL logic 1 output that is less than 2.4 V usually indicates that the gate is being excessively loaded. Most often this is due to a short in the device that this output is driving: a solder bridge or an unetched trace on the printed circuit board.

When the pull-down transistor is conducting, the gate is outputting a logic 0. A saturated transistor can act just like a resistor. The saturated pull-down transistor has a bulk resistance of approximately 25 Ω. As the quantity of current it is sinking increases, the voltage drop over it increases. It is specified to be able to sink 16 mA and still maintain an output voltage that is less than 0.4 V.

Designers take all the TTL specifications into consideration when they design a circuit. Some circuits you will troubleshoot may have design errors that reveal themselves only in certain situations. When you troubleshoot a circuit you must assume that it is designed properly, unless you have some solid evidence to the contrary.

5.1.4 Noise Immunity

You may have noticed that the TTL output levels illustrated in Figure 5.3b are 0.4 V closer to the ideal level of +5 V and 0 V than the acceptable input levels in Figure 5.5a. That means that a TTL gate can output a logic 1 of 2.4 V and it can drop down to 2.0 V and still be recognized by the next TTL input as a logic 1. On the other hand, a gate can output a logic 0 of 0.4 V and it can increase to 0.8 V and still be recognized by the next TTL input as a logic 0. What is the importance of this 0.4-V voltage difference between guaranteed outputs and acceptable inputs? This 0.4 V is the level of noise that a TTL circuit can tolerate. All circuits have a certain amount of noise. This noise is an ac quantity that amplitude-modulates the signals in a circuit. If a TTL logic 1 output is 2.4 V, it can have −0.4 V of noise added to it and still be an acceptable logic 1 input. A logic 0 output of 0.4 V can have +0.4 V of noise added to it and still be accepted as a logic 0 input.

5.1.5 Driving Loads with TTL

We have seen that a TTL circuit can source only 0.4 mA of current and still meet the guaranteed logic 1 output level. Although TTL cannot source current well, it can sink up to 16 mA of current and still maintain the guaranteed logic 0 output level. Whenever TTL is required to drive any load that requires more than 0.4 mA, that load must be driven with a logic 0. Consider

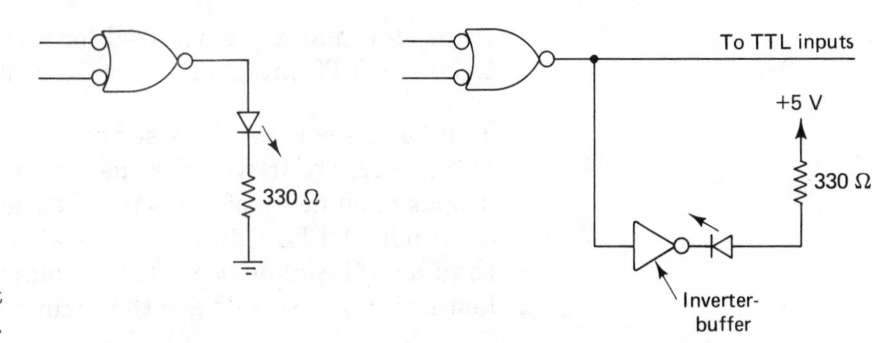

Figure 5.6 Driving current loads with TTL.

Figure 5.6. Let's use an LED as an example of a load requiring an appreciable amount of current. This LED requires 8 mA of drive current. If we try to illuminate this LED with a logic 1 output, we will have a problem. The output of the gate will fall below an acceptable TTL logic 1 level and any TTL inputs that this gate may also be driving will not recognize this output level as a TTL high. The solution is to buffer the LED with an inverter. The logic 1 will be inverted into a logic 0 and easily drive the LED. The valid logic 1 output of this gate may then be passed onto other gates.

5.1.6 The Subfamilies of TTL: LS, S, L, ALS, and AS

The 7400 series of TTL is a relatively old circuit technology. It has two major drawbacks: it requires appreciable amounts of bias current from the +5 V power supply and it has certain speed limitations. The subfamilies of TTL were developed to help overcome these two problems. We will overview each of these subfamilies in their order of popularity.

In an ideal gate the output switches the instant that the input changes. As an example: If a logic 0 is placed on the input of an inverter, the output should instantaneously go to logic 1. In real circuits there is a switching time called *propagation delay.* The phenomenon of propagation delay is due to the parasitic base–collector capacitance of transistors. A transistor cannot be instantaneously saturated or cut off. This base–collector capacitance must be charged each time the transistor is driven into saturation, and discharged each time the transistor is cut off. The propagation delay for standard TTL gates is approximately 15 ns. This is such a short period of time that you may never notice this input-to-output delay. The specification of propagation delay is an important parameter that circuit designers must consider carefully.

The 74LS00 series This is the most popular form of TTL. The initials LS stand for "low-power Schottky." The 74LS00 series was designed to supersede the older standard TTL family. The LS familiy requires much less bias current from the power supply and it is also faster than standard TTL. It employs Schottky transistors. Schottky transistors are a special form of bipolar transistor that are optimized for fast switching speeds. Eighty to 90% of TTL integrated circuits are of the 74LS00 series.

The 74S00 series This series also uses Schottky transistors. Until recently it was the fastest of all TTL subfamilies. The propagation delay of Schottky TTL is about 50% less than that of standard TTL. 74S00 devices also require much more power than do 74LS00 devices. This subfamily will be used wherever fast switching speeds are the primary concern.

The 74L00 Series The purpose of this series is low power. The 74L00 has the lowest power requirement, but it is also the slowest of all the TTL subfamilies. 74L00 TTL has been displaced by CMOS devices.

The 74ALS00 and 74AS00 Series The advanced low-power Schottky and advanced Schottky are the most recently developed TTL subfamilies. These subfamilies offer much greater speeds at lower power levels. The 74AS series is twice as fast as the 74S family. The devices in this family are pin-for-pin compatible with all other TTL subfamilies, but their major emphasis will be as LSI integrated circuits.

5.1.7 Some Common TTL Gates

All TTL ICs are available in each of the subfamilies except the ALS and AS. These two subfamilies are not yet widely employed. The following illustrations of common TTL gates will use the LS family because it is the most popular.

You will notice that all the pinouts indicate another series that starts with the numbers 54 instead of 74. The 54 series meets some more demanding specifications involving power supply voltages and temperature operating ranges. The 54 series is a "Mil Spec" series. The term "Mil Spec" is short for "military specifications." Because the 54 series meets higher specifications than the 74 series:

A 54 series IC can always be substituted for a 74 series IC

but

A 74 series IC should never *be substituted for a 54 series IC.*

The best rule to follow when replacing components is always to use an exact replacement part; do not experiment with ICs that you feel may be equivalent. IC manufacturers distribute cross-reference charts that list all the valid equivalent part numbers for any given IC. Consider the following part number:

SN74LS00

The first two letters, SN, indicate the manufacturer (SN is used by Texas Instruments). You can ignore these first two letters. The numbers, 74, indicate a 74 series integrated circuit. LS is the low-power Schottky subfamily. 00 indicates the part number within the series. In TTL this intraseries part number will be two or three digits long. It will take a bit of on-the-job exposure before you will feel confident about properly identifying ICs. You must also be careful not to mistake the manufacturer's date code for the part number. If you have any doubts, don't be embarrassed to ask for assistance. This is much preferred to replacing an IC with the wrong device.

On data sheets you will often see a letter in parentheses following the part number. This letter indicates the type of package in which the actual chip is enclosed. Common packages are: J, ceramic and N, plastic. The package type is usually not critical. For most applications ceramic and plastic packages are interchangeable.

The 74LS00 Quadruple Two-Input Positive NAND Gates (Figure 5.7a) This may be the most popular integrated circuit ever produced. "Quadruple," usually shortened to "quad," indicates that this IC contains four NAND gates. Each of the NAND gates has two inputs. Data sheets always show the positive logic symbol and Boolean equation representing the gate. This does not imply that the positive representations are any more important than the negative representations. It would be too troublesome to show both positive and negative representations of every gate. The data sheet assumes that you know and understand the negative symbols and equations of each gate.

The 74LS03 Quadruple Two-Input Positive NAND Gates with Open-Collector Outputs (Figure 5.7b) This device is exactly the same as the 74LS00 except that it has an open-collector instead of a totem-pole output. Note that the diagram shown does not indicate that this device is open collector. The O.C., *, or vertical slash will be added to the gate symbol when it appears in a schematic diagram.

The 74LS02 Quadruple Two-Input Positive NOR Gates (Figure 5.7c) This is the standard TTL NOR gate. Notice that the pinouts for the inputs and outputs are different from those in the 74LS00. The only standard pin positions for TTL ICs are ground and V_{cc}. Ground will be at the bottom right-hand corner, pin 7 for 14-pin DIPS and pin 8 for 16-pin DIPS. V_{cc} will be at the top left-hand corner, pin 14 for 14-pin DIPS and pin 16 for 16-pin DIPS. Having the power and ground at standard positions is not only convenient for troubleshooting but also simplifies the power-ground bus layouts on printed circuit boards.

The 74LS04 Hex Inverters (Figure 5.7d) The prefix "hex" stands for the number 6. The 74LS04 contains six inverters in one 14-pin package. The right-hand pinout is a special Mil Spec

QUADRUPLE 2-INPUT POSITIVE-NAND GATES

00

positive logic:
$Y = \overline{AB}$

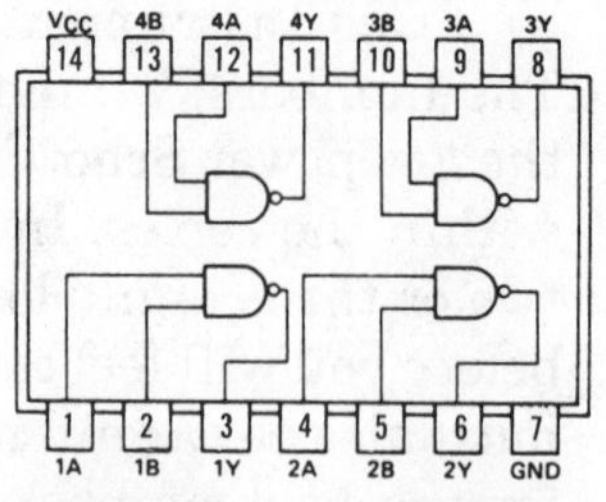

SN5400 (J) **SN7400 (J, N)**
SN54H00 (J) **SN74H00 (J, N)**
SN54L00 (J) **SN74L00 (J, N)**
SN54LS00 (J, W) **SN74LS00 (J, N)**
SN54S00 (J, W) **SN74S00 (J, N)**

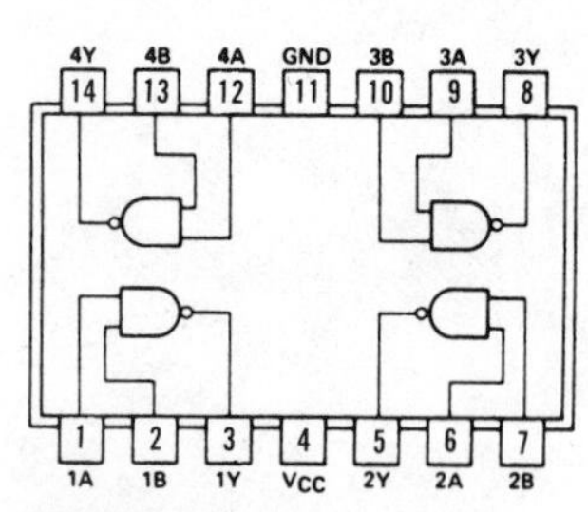

SN5400 (W)
SN54H00 (W)
SN54L00 (T)

(a)

QUADRUPLE 2-INPUT POSITIVE-NAND GATES WITH OPEN-COLLECTOR OUTPUTS

03

positive logic:
$Y = \overline{AB}$

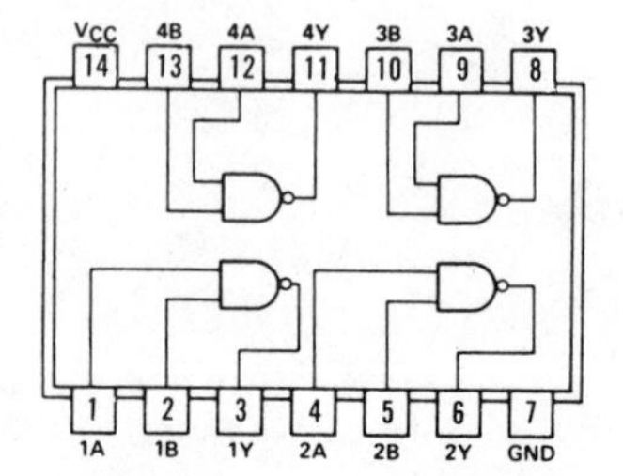

SN5403 (J) **SN7403 (J, N)**
SN54L03 (J) **SN74L03 (J, N)**
SN54LS03 (J, W) **SN74LS03 (J, N)**
SN54S03 (J, W) **SN74S03 (J, N)**

(b)

QUADRUPLE 2-INPUT POSITIVE-NOR GATES

02

positive logic:
$Y = \overline{A+B}$

SN5402 (J) **SN7402 (J, N)**
SN54L02 (J) **SN74L02 (J, N)**
SN54LS02 (J, W) **SN74LS02 (J, N)**
SN54S02 (J, W) **SN74S02 (J, N)**

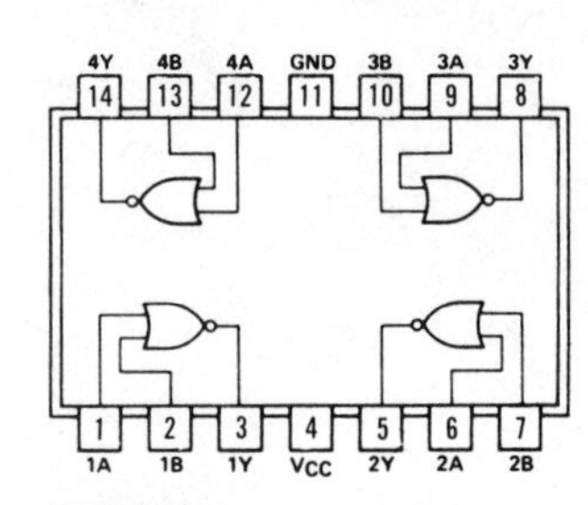

SN5402 (W)
SN54L02 (T)

(c)

HEX INVERTERS

04

positive logic:
$Y = \overline{A}$

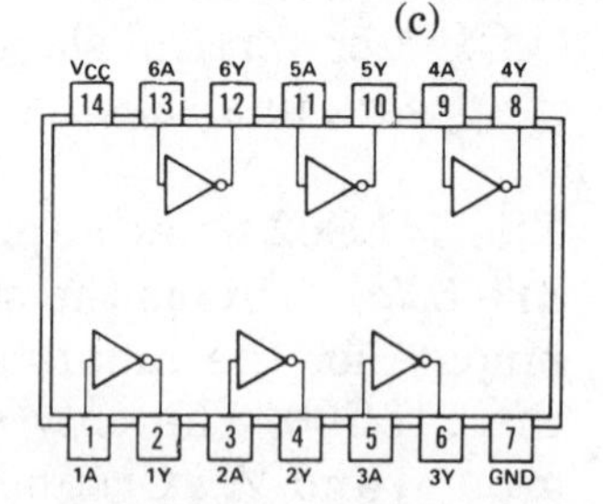

SN5404 (J) **SN7404 (J, N)**
SN54H04 (J) **SN74H04 (J, N)**
SN54L04 (J) **SN74L04 (J, N)**
SN54LS04 (J, W) **SN74LS04 (J, N)**
SN54S04 (J, W) **SN74S04 (J, N)**

SN5404 (W)
SN54H04 (W)
SN54L04 (T)

(d)

Figure 5.7 (a) 74LS00 quad two-input NAND gates; (b) quad two-input NAND gates, open collector; (c) 74LS02 quad two-input NOR gates; (d) 74LS04 hex inverters; (e) 74LS05 hex inverters, open collector; (f) quad two-input AND gates; (g) 74LS10 triple three-input NAND gates; (h) 74LS32 quad two-input OR gates; (i) quad two-input XOR gates.

HEX INVERTERS WITH OPEN-COLLECTOR

05

positive logic:
$Y = \overline{A}$

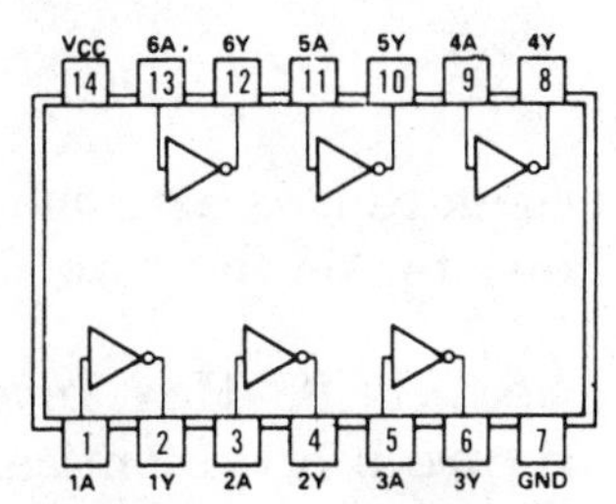

SN5405 (J) SN7405 (J, N)
SN54H05 (J) SN74H05 (J, N)
SN54LS05 (J, W) SN74LS05 (J, N)
SN54S05 (J, W) SN74S05 (J, N)

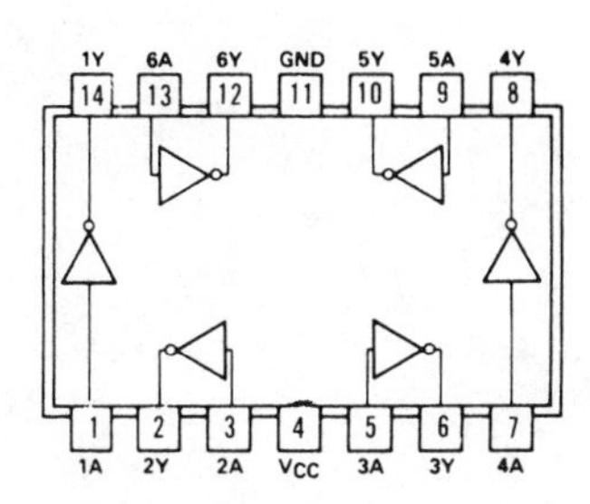

SN5405 (W)
SN54H05 (W)

(e)

QUADRUPLE 2-INPUT POSITIVE-AND GATES

08

positive logic:
$Y = AB$

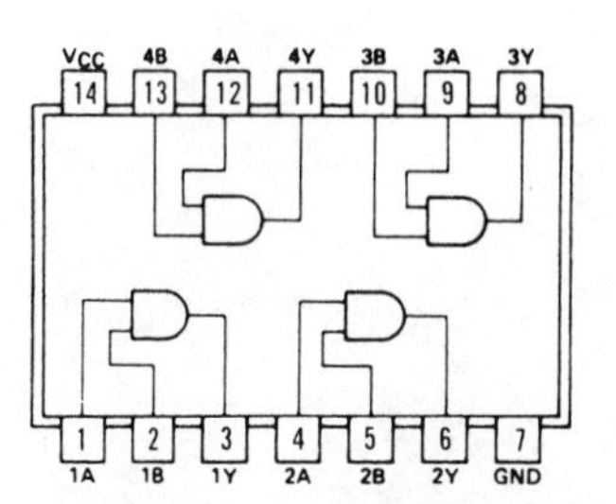

SN5408 (J, W) SN7408 (J, N)
SN54LS08 (J, W) SN74LS08 (J, N)
SN54S08 (J, W) SN74S08 (J, N)

(f)

TRIPLE 3-INPUT POSITIVE-NAND GATES

10

positive logic:
$Y = \overline{ABC}$

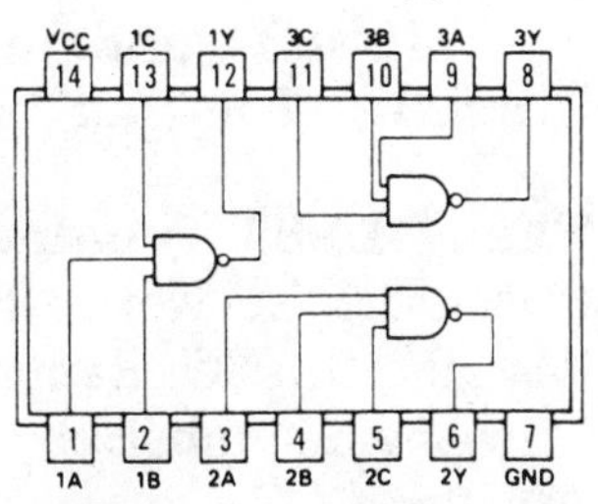

SN5410 (J) SN7410 (J, N)
SN54H10 (J) SN74H10 (J, N)
SN54L10 (J) SN74L10 (J, N)
SN54LS10 (J, W) SN74LS10 (J, N)
SN54S10 (J, W) SN74S10 (J, N)

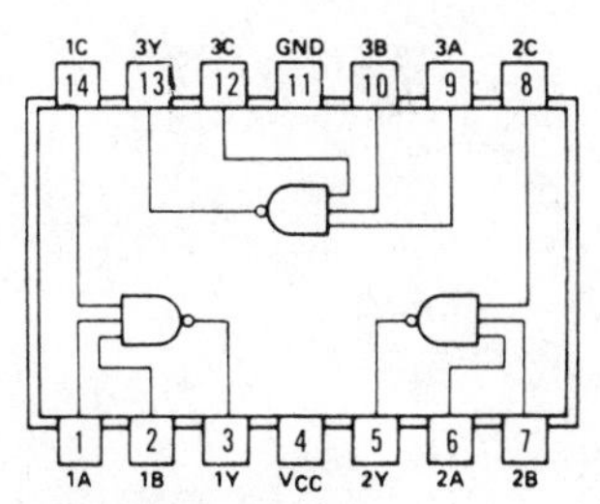

SN5410 (W)
SN54H10 (W)
SN54L10 (T)

(g)

QUADRUPLE 2-INPUT POSITIVE-OR GATES

32

positive logic:
$Y = A+B$

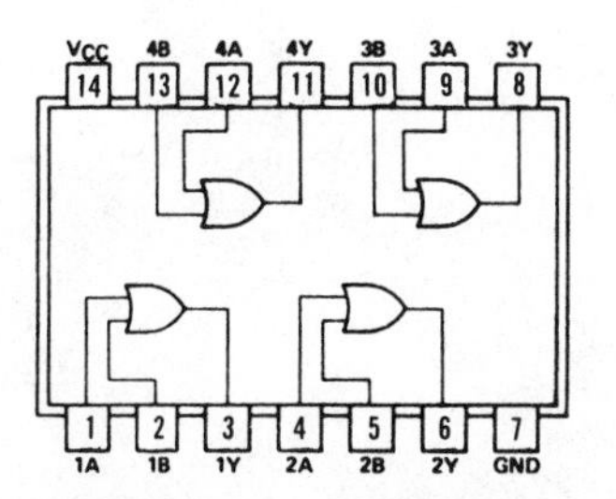

SN5432 (J, W) SN7432 (J, N)
SN54LS32 (J, W) SN74LS32 (J, N)
SN54S32 (J, W) SN74S32 (J, N)

(h)

QUADRUPLE 2-INPUT EXCLUSIVE-OR GATES

86 $Y = A \oplus B = \overline{A}B + A\overline{B}$

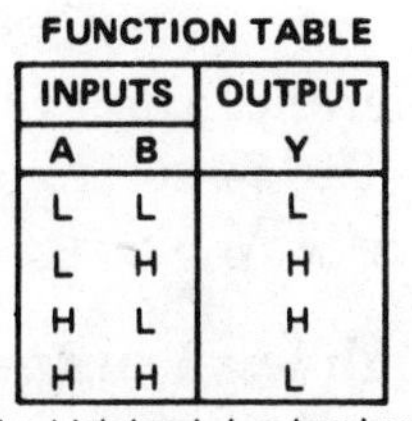

FUNCTION TABLE

INPUTS		OUTPUT
A	B	Y
L	L	L
L	H	H
H	L	H
H	H	L

H = high level, L = low level

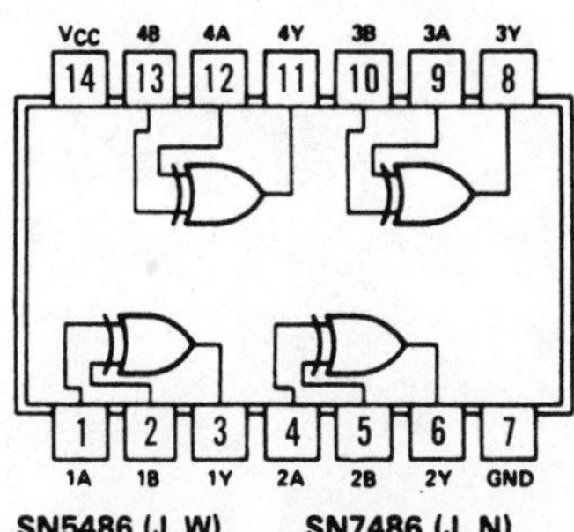

SN5486 (J, W) SN7486 (J, N)
SN54LS86 (J, W) SN74LS86 (J, N)
SN54S86 (J, W) SN74S86 (J, N)

(i)

design that is extremely uncommon. Most of the extra pinouts indicated on the data sheets are obscure and can be ignored.

The 74LS05 Hex Inverters with Open-Collector Outputs (Figure 5.7e) Identical to the 74LS04 except that the 74LS05 has open-collector outputs.

The 74LS08 Quadruple Two-Input Positive AND Gates (Figure 5.7f) This gate performs the most basic logic function. Notice that the pinout for the 74LS08 is identical to the pinout for the 74LS00. It is helpful to memorize pinouts as quickly as possible. If you memorize the pinouts of these simple devices, you will have a much easier time wiring and troubleshooting circuits in your digital lab.

The 74LS10 Triple Three-Input Positive NAND Gates (Figure 5.7g) SSI devices are limited to 14- and 16-pin packages. The IC manufacturers can only fit three three-input NAND gates into one package.

The 74LS32 Quadruple Two-Input Positive OR Gates (Figure 5.7h) The 74LS32 shares the same pinout as the 74LS00.

The 74LS86 Quadruple Two-Input Exclusive-OR Gates (Figure 5.7i) Notice that the word "positive" was not used to describe this IC. Remember that the XOR does not have a negative-logic symbol because it is really a composite function of AND and OR.

5.1.8 A TTL Data Sheet

Technicians do not need to concern themselves with most TTL specifications. That is the job of the circuit designer. Nonetheless, there are certain specifications that you must be aware of to test and troubleshoot digital circuits efficiently. Figure 5.8 shows a TTL data sheet. We will examine each of the parameters.

Note: The 54H and 74H series comprise a little used subfamily that we have not discussed. We will focus on the column that describes the 54LS and 74LS series.

The following specifications are the recommended operating conditions. If these conditions are met, the electrical characteristics in the second table will be guaranteed.

Supply voltage, V_{cc} The 54 series have a wider power supply tolerance than the 74 series. The 54 series V_{cc} range is 5 $\pm$ 0.5 V. The 74 series is 5 $\pm$ -0.25 V. You will find that V_{cc} is usually 5 ± 0.1 V.

High-level output current, I_{OH} The minus sign symbolizes that the current is flowing out of the IC. The maximum amout of current that the IC can source and still maintain the guaranteed minimum logic 1 output level of 2.4 V is 0.4 mA.

recommended operating conditions

	54 FAMILY 74 FAMILY	SERIES 54 SERIES 74 '00, '04, '10, '20, '30			SERIES 54H SERIES 74H 'H00, 'H04, 'H10, 'H20, 'H30			SERIES 54L SERIES 74L 'L00, 'L04, 'L10, 'L20, 'L30			SERIES 54LS SERIES 74LS 'LS00, 'LS04, 'LS10, 'LS20, 'LS30			SERIES 54S SERIES 74S 'S00, 'S04, 'S10, 'S20, 'S30, 'S133			UNIT
		MIN	NOM	MAX	MIN	NOM	MAX	MIN	NOM	MAX	MIN	NOM	MAX	MIN	NOM	MAX	
Supply voltage, V_{CC}	54 Family	4.5	5	5.5	4.5	5	5.5	4.5	5	5.5	4.5	5	5.5	4.5	5	5.5	V
	74 Family	4.75	5	5.25	4.75	5	5.25	4.75	5	5.25	4.75	5	5.25	4.75	5	5.25	
High-level output current, I_{OH}	54 Family			−400			−500			−100			−400			−1000	μA
	74 Family			−400			−500			−200			−400			−1000	
Low-level output current, I_{OL}	54 Family			16			20			2			4			20	mA
	74 Family			16			20			3.6			8			20	
Operating free-air temperature, T_A	54 Family	−55		125	−55		125	−55		125	−55		125	−55		125	°C
	74 Family	0		70	0		70	0		70	0		70	0		70	

electrical characteristics over recommended operating free-air temperature range (unless otherwise noted)

PARAMETER	TEST FIGURE	TEST CONDITIONS†			SERIES 54 SERIES 74 '00, '04, '10, '20, '30			SERIES 54H SERIES 74H 'H00, 'H04, 'H10, 'H20, 'H30			SERIES 54L SERIES 74L 'L00, 'L04, 'L10, 'L20, 'L30			SERIES 54LS SERIES 74LS 'LS00, 'LS04, 'LS10, 'LS20, 'LS30			SERIES 54S SERIES 74S 'S00, 'S04, 'S10, 'S20, 'S30, 'S133			UNIT
					MIN	TYP‡	MAX	MIN	TYP‡	MAX	MIN	TYP‡	MAX	MIN	TYP‡	MAX	MIN	TYP‡	MAX	
V_{IH} High-level input voltage	1, 2				2			2			2			2			2			V
V_{IL} Low-level input voltage	1, 2			54 Family			0.8			0.8			0.7			0.7			0.8	V
				74 Family			0.8			0.8			0.7			0.8			0.8	
V_{IK} Input clamp voltage	3	V_{CC} = MIN,	I_I = §				−1.5			−1.5						−1.5			−1.2	V
V_{OH} High-level output voltage	1	V_{CC} = MIN, I_{OH} = MAX	V_{IL} = V_{IL} max,	54 Family	2.4	3.4		2.4	3.5		2.4	3.3		2.5	3.4		2.5	3.4		V
				74 Family	2.4	3.4		2.4	3.5		2.4	3.2		2.7	3.4		2.7	3.4		
V_{OL} Low-level output voltage	2	V_{CC} = MIN, V_{IH} = 2 V	I_{OL} = MAX	54 Family		0.2	0.4		0.2	0.4		0.15	0.3		0.25	0.4			0.5	V
				74 Family		0.2	0.4		0.2	0.4		0.2	0.4		0.25	0.5			0.5	
			I_{OL} = 4 mA	Series 74LS												0.4				
I_I Input current at maximum input voltage	4	V_{CC} = MAX		V_I = 5.5 V			1			1			0.1						1	mA
				V_I = 7 V												0.1				
I_{IH} High-level input current	4	V_{CC} = MAX		V_{IH} = 2.4 V			40			50			10							μA
				V_{IH} = 2.7 V												20			50	
I_{IL} Low-level input current	5	V_{CC} = MAX		V_{IL} = 0.3 V									−0.18							mA
				V_{IL} = 0.4 V			−1.6			−2						−0.4				
				V_{IL} = 0.5 V															−2	
I_{OS} Short-circuit output current♦	6	V_{CC} = MAX		54 Family	−20		−55	−40		−100	−3		−15	−20		−100	−40		−100	mA
				74 Family	−18		−55	−40		−100	−3		−15	−20		−100	−40		−100	
I_{CC} Supply current	7	V_{CC} = MAX			See table on next page															mA

†For conditions shown as MIN or MAX, use the appropriate value specified under recommended operating conditions.
‡All typical values are at V_{CC} = 5 V, T_A = 25°C.
§I_I = −12 mA for SN54'/SN74', −8 mA for SN54H'/SN74H', and −18 mA for SN54LS'/SN74LS' and SN54S'/SN74S'.
♦Not more than one output should be shorted at a time, and for SN54H'/SN74H', SN54LS'/SN74LS', and SN54S'/SN74S', duration of short-circuit should not exceed 1 second.

Figure 5.8 TTL data sheet.

Low-level output current, I_{OL} LS TTL cannot sink current as well as standard TTL, 4 mA for 54LS and 8 mA for 74LS. This is the maximum amount of current that the device can sink and still maintain a logic 0 output level of less than 0.4 V.

Operating free temperature range, T_A This is an important specification for Mil Spec devices. The 54LS series operate from −55 to 125°C; the 74LS series has a much smaller temperature range, 0 to 70°C.

The following electrical characteristics are guaranteed only if the recommended operating conditions are met.

High-level input voltage, V_{IH} This 2-V level is the minimum voltage that will be accepted as a logic 1.

Low-level input voltage, V_{IL} The maximum voltage that will be recognized as a logic 0 input is 0.7 V for 54LS and 0.8 V for 74LS.

Input clamp voltage, V_{IK} This is the voltage at which the input protection diodes will clamp a negative input signal.

High-level output voltage, V_{OH} 54LS 2.5V and 74LS 2.7V. This is the minimum logic 1 output level if the recommended operating conditions are met. Notice that this is slightly different than the 2.4 V that we have previously cited for standard TTL. The typical value is 3.4 V. TTL is overspecified; by "overspecified" we mean that the specifications always depict the worst-case conditions and the IC is likely to perform much better than the specifications indicate. The typical values indicate the levels measured for the average TTL device. Despite the fact that these typical values are not guaranteed, these are the levels you are likely to see when testing and troubleshooting TTL.

Low-level output voltage, V_{OL} We have seen in standard TTL that this output is guaranteed not to exceed 0.4 V. In the 54LS series it is also 0.4 V; in 74LS it is 0.5 V. The typical value is 0.25 V.

High-level input current, I_{IH} The device that is providing a high logic level, typically the output of another TTL gate, must be able to source 20 μA for each input it is driving. Remember that the maximum high-level output current, I_{OH}, was specified as -400 μA. This means that one LS gate can drive the inputs of 20 LS gates with a logic 1 input.

$$20\ \mu\text{A per gate} \times 20\ \text{gates} = 400\ \mu\text{A of current}$$

The number of outputs a particular gate is capable of driving is referred to the *fanout* of the gate. When an LS device is driving other LS devices, it has a fanout of 20. The different TTL subfamilies all have different current output and input requirements. If one subfamily is driving another, these differences must be taken into account.

Low-level input current, I_{IL} The device providing a logic 0 input to an LS device must be capable of sinking -0.4 mA. The minus sign indicates that the current will be flowing out of the gate that is being driven. Notice that standard TTL has a I_{IL} of -1.6 mA. This is four times greater than the LS spec! That is a major reason why LS is the most popular form of TTL. The 74LS family has a maximum low-level output current, I_{OL}, of 8 mA. If an LS gate can sink a maximum of 8 mA and each gate it is driving requires 0.4 mA of sink current, one LS gate can drive 20 other LS gates with a logic 0 output. This agrees with our previous fanout calculation of 20.

Short-circuit output current, I_{OS} The totem-pole pull-up transistor is protected by a collector resistor. This specification describes what will happen if the output of a gate is shorted to ground while the gate is attempting to output a logic 1. It is specified that only one gate in any package can be shorted at a time, and the maximum duration of the short is 1 second. Many specifications are given a range of values. Here the range of short-circuit output current is 20 to 100 mA.

The most important specifications deal with acceptable and guaranteed input and output voltage levels. The other specifications are presented to provide you with insight into the other electrical parameters of TTL. You will not use these specs on a day-to-day basis, but it is important that you know where they can be found and how to interpret them.

5.2 CMOS: THE OTHER LOGIC FAMILY

The ideal digital device should have:

1. No power dissipation
2. An infinitely fast switching speed
3. A noise immunity equal to 50% of the power supply voltage

CMOS approaches these ideal characteristics. To many people in the electronics industry TTL is the only logic family that provides SSI and MSI devices. Recent years have seen a great surge in the use of CMOS logic devices.

CMOS devices have many advantages over their TTL counterparts:

1. Ultra-low-power supply requirements
2. Excellent noise immunity
3. Wide power supply voltage range

CMOS logic gates are constructed from *N*-channel and *P*-channel enhancement-mode MOSFETs. (MOSFETs are also known as IGFETs.) Many transistor classes spend little or no time studying FETs (field-effect transistors). A quick course on FETs is in order. We will not try to make you an instant authority on FETs, but just give you enough information to appreciate the input and output characteristics of CMOS logic devices.

5.2.1 A Comparison of Bipolar and Field-Effect Transistors

When most people mention the term "transistor" they are referring to the bipolar junction transistor (BJT). Field-effect transistors are known as FETs. BJTs and FETs are both three-terminal devices. The input that controls the operation of the BJT is called the *base*. If we source some current into the base of an *NPN* transistor, the collector current will be equal to the base current times a gain factor called *beta*. The BJT is called a *cur-*

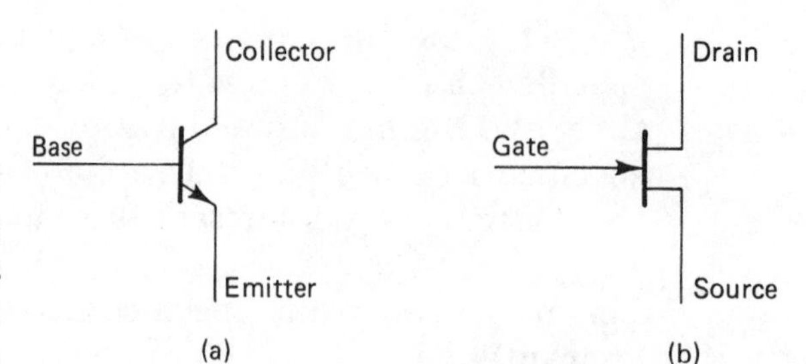

Figure 5.9 BJT and FET: (a) current driven, low input impedance; (b) voltage driven, high input impedance.

rent-driven device. It depends on base current to control the collector current. The analogous control input of the FET is called the *gate*. The collector lead of the FET is called the *drain* and the emitter lead the *source*. Unlike the BJT the FET is a *voltage-driven device.* This means that little current is required from the device that provides the input signal to drive the FET. That is the reason for the extremely low power requirements of CMOS. Figure 5.9 illustrates the schematic symbols of the BJT and FET.

There are two major types of FETs: junction FETs, (JFETs) and metal-oxide semiconductor FETs (MOSFETs). Because CMOS logic devices are constructed from MOSFETs we will focus our attention on them. BJTs require two voltages for bias: V_{cc} (the voltage to the collector) and V_{ee} (the voltage to the emitter). In TTL devices $V_{cc} = 5$ V and $V_{ee} = 0$ V. V_{ee} is most often designated as the ground pin. CMOS also requires two voltages: V_{dd} (the voltage to the drain) and V_{ss} (the substrate voltage). V_{ss} is most often referenced to ground. As we have mentioned, CMOS gates are constructed from *P*-channel and *N*-channel enhancement-mode MOSFETs. Enhancement-mode FETs are normally in the cutoff state. They will conduct only if an appropriate gate voltage is applied. The channel refers to whether the gate is constructed from *P* or *N* material. Figure 5.10 shows the schematic symbols for these MOSFETs. The imaginary diode indicates whether the channel is *P* or *N*. If the anode of the diode is pointing at the channel, the MOSFET is a *P*-type. If the diode's cathode is pointing at the channel, the MOSFET is *N*-type. The FET will conduct when this diode is reversed biased!

A logic 0 will cause the P-channel MOSFET to saturate

and

A logic 1 will cause the N-channel MOSFET to saturate.

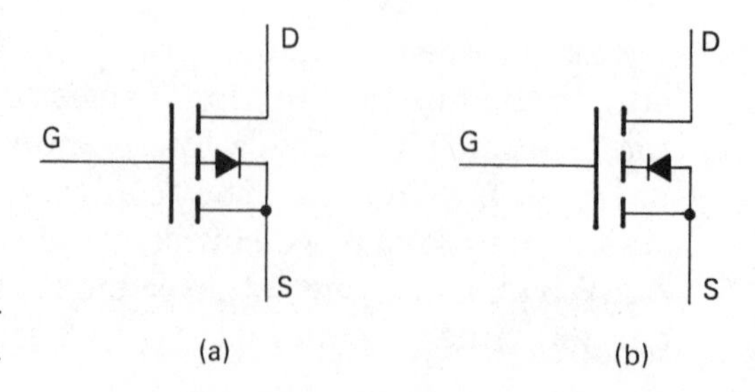

Figure 5.10 MOSFETs: (a) *P*-channel; (b) *N*-channel.

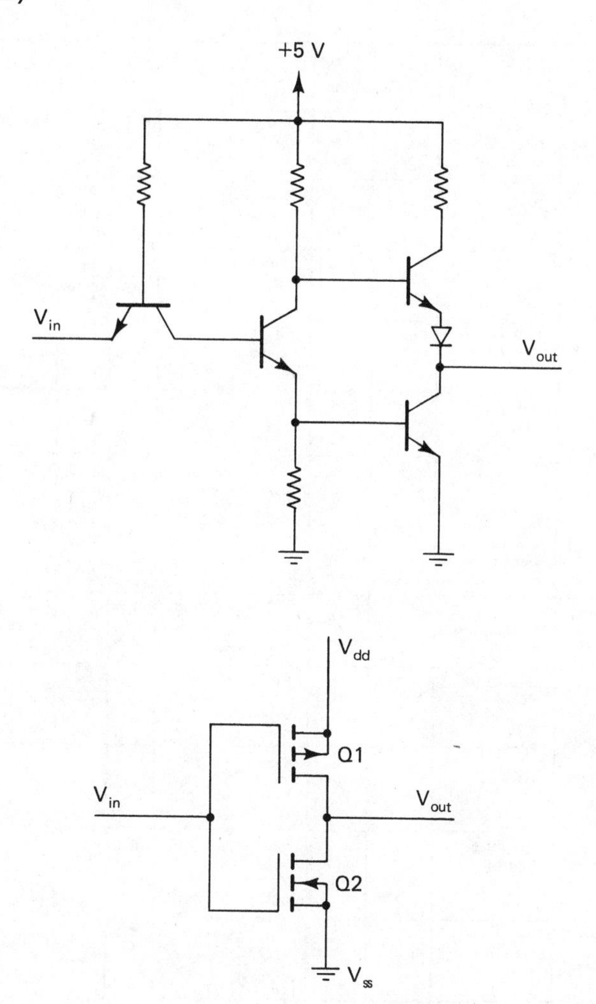

Figure 5.11 TTL and CMOS inverters.

This is all that you have to remember to understand the internal operation of a CMOS gate. Figure 5.11 illustrates a TTL inverter and a CMOS inverter. The first impression that you should get from Figure 5.11 is the simplicity of the CMOS inverter. It requires only two MOSFETs, one *P*-channel and the other *N*-channel. Compare this with the TTL inverter, which requires: four BJTs, four resistors, and one diode.

Let's analyze the operation of the CMOS inverter. We only need to remember one rule:

To drive a MOSFET into saturation we must reverse-bias the channel diode.

A logic 0 will drive a *P*-channel MOSFET into saturation and a logic 1 will drive an *N*-channel MOSFET into saturation. If a logic 0 is applied to the input of the CMOS inverter, Q1 will be driven into saturation and Q2 will be cut off. The output will be pulled up to V_{dd}, a logic 1. If a logic 1 is applied to the input, Q1 will be cut off and Q2 will be driven into saturation. The output will be pulled down to V_{ss}: a logic 0. Notice that Q1 and Q2 will never both be in the same state. One is always saturated while the other is cut off.

PARAMETER		TEMP. RANGE	V_{DD} (Vdc)	CONDITIONS	T_{LOW}* Min	T_{LOW}* Max	+25°C Min	+25°C Typ	+25°C Max	T_{HIGH}* Min	T_{HIGH}* Max	Units
I_{DD}	Quiescent Device Current GATES	Mil	5	$V_{IN} = V_{SS}$ or V_{DD}		0.25			0.25		7.5	uAdc
			10			0.5			0.5		15	
			15			1.0			1.0		30	
		Comm	5	All valid input combinations		1.0			1.0		7.5	
			10			2.0			2.0		15	
			15			4.0			4.0		30	
	BUFFERS, FLIP-FLOPS	Mil	5	$V_{IN} = V_{SS}$ or V_{DD}		1.0			1.0		30	uAdc
			10			2.0			2.0		60	
			15			4.0			4.0		120	
		Comm	5	All valid input combinations		4			4		30	
			10			8			8		60	
			15			16			16		120	
	MSI	Mil	5	$V_{IN} = V_{SS}$ or V_{DD}		5			5		150	uAdc
			10			10			10		300	
			15			20			20		600	
		Comm	5	All valid input combinations		20			20		150	
			10			40			40		300	
			15			80			80		600	
V_{OL}	Low-Level Output Voltage	All	5	$V_{IN} = V_{SS}$ or V_{DD} $\lvert I_O \rvert < 1uA$		0.05			0.05		0.05	Vdc
			10			0.05			0.05		0.05	
			15			0.05			0.05		0.05	
V_{OH}	High-Level Output Voltage	All	5	$V_{IN} = V_{SS}$ or V_{DD} $\lvert I_O \rvert < 1uA$	4.95		4.95			4.95		Vdc
			10		9.95		9.95			9.95		
			15		14.95		14.95			14.95		
V_{IL}	Input Low Voltage B Types	All	5	V_O = 0.5V or 4.5V		1.5			1.5		1.5	Vdc
			10	V_O = 1.0V or 9.0V		3.0			3.0		3.0	
			15	V_O = 1.5V or 13.5V $\lvert I_O \rvert < 1uA$		4.0			4.0		4.0	
	UB Types	All	5	V_O = 0.5V or 4.5V		1.0			1.0		1.0	Vdc
			10	V_O = 1.0V or 9.0V		2.0			2.0		2.0	
			15	V_O = 1.5V or 13.5V $\lvert I_O \rvert < 1uA$		2.5			2.5		2.5	
V_{IH}	Input High Voltage B Types	All	5	V_O = 0.5V or 4.5V	3.5		3.5			3.5		Vdc
			10	V_O = 1.0V or 9.0V	7.0		7.0			7.0		
			15	V_O = 1.5V or 13.5V $\lvert I_O \rvert < 1uA$	11.0		11.0			11.0		
	UB Types	All	5	V_O = 0.5V or 4.5V	4.0		4.0			4.0		Vdc
			10	V_O = 1.0V or 9.0V	8.0		8.0			8.0		
			15	V_O = 1.5V or 13.5V $\lvert I_O \rvert < 1uA$	12.5		12.5			12.5		

Absolute Maximum Ratings (Voltages referenced to V_{SS}):

DC Supply Voltage	V_{DD}	−0.5 to +18	Vdc
Input Voltage	V_{IN}	−0.5 to V_{DD} +0.5	Vdc
DC Input Current (any one input)	I_{IN}	±10	mAdc
Storage-Temperature Range	T_S	−65 to +150	°C

Recommended Operating Conditions:

DC Supply Voltage	V_{DD}	+3 to +15	Vdc
Operating-Temperature Range, T_A			
Military-Range Devices		−55 to +125	°C
Commercial-Range Devices		−40 to +85	°C

Figure 5.12 4000B CMOS data sheet.

5.2.2 The 4000B Series of CMOS Devices

The "B" means that this series is buffered. The buffer consists of a CMOS driver circuit. It gives all the members of this family similar output characteristics. The 40 is often designated as 140 or 340 by various manufacturers. Thus a 4001B and a 14001B and a 34001B are all the same device. You will also discover 4000UB CMOS devices. The UB suffix denotes unbuffered CMOS. The major difference between the 4000B and 4000UB series is drive ability. The UB series is employed when CMOS digital devices are used in a special linear mode as oscillators

PARAMETER		TEMP. RANGE	V_{DD} (Vdc)	CONDITIONS	LIMITS							Units
					T_{LOW}*		+25° C			T_{HIGH}*		
					Min	Max	Min	Typ	Max	Min	Max	
I_{OL}	Output Low (Sink) Current	Mil	5	V_O = 0.4V, V_{IN} = 0 or 5V	0.64		0.51			0.36		mAdc
			10	V_O = 0.5V, V_{IN} = 0 or 10V	1.6		1.3			0.9		
			15	V_O = 1.5V, V_{IN} = 0 or 15V	4.2		3.4			2.4		
		Comm	5	V_O = 0.4V, V_{IN} = 0 or 5V	0.52		0.44			0.36		
			10	V_O = 0.5V, V_{IN} = 0 or 10V	1.3		1.1			0.9		
			15	V_O = 1.5V, V_{IN} = 0 or 15V	3.6		3.0			2.4		
I_{OH}	Output High (Source) Current	Mil	5	V_O = 4.6V, V_{IN} = 0 or 5V	-0.25		-0.2			-0.14		mAdc
			10	V_O = 9.5V, V_{IN} = 0 or 10V	-0.62		-0.5			-0.35		
			15	V_O = 13.5V, V_{IN} = 0 or 15V	-1.8		-1.5			-1.1		
		Comm	5	V_O = 4.6V V_{IN} = 0 or 5V	-0.2		-0.16			-0.12		
			10	V_O = 9.5V, V_{IN} = 0 or 10V	-0.5		-0.4			-0.3		
			15	V_O = 13.5V, V_{IN} = 0 or 15V	-1.4		-1.2			-1.0		
I_{IN}	Input Current	Mil	15	V_{IN} = 0 or 15V		±0.1			±0.1		±1.0	uAdc
		Comm	15	V_{IN} = 0 or 15V		±0.3			±0.3		±1.0	uAdc
I_{OUTmax}	3-State Output Leakage Current	Mil	15	V_{IN} = 0 or 15V		±0.4			±0.4		±12	uAdc
		Comm	15	V_{IN} = 0 or 15V		±1.6			±1.6		±12	uAdc
C_{IN}	Input Capacitance per Unit Load	All	—	Any Input					7.5			pF

*T_{LOW} = −55° C for Military Temp. Range device, −40° C for Commercial Temp. Range device
*T_{HIGH} = +125° C for Military Temp. Range device, +85° C for Commercial Temp. Range device
▲ Reprinted from JEDEC Tentative Standard No. 13-B, "JEDEC Standard Specification for Description of B-series CMOS Devices."

Figure 5.12 *(cont.)*

or amplifiers. UB devices are slightly faster than B devices because they do not have the extra buffer stage. We will examine a CMOS data sheet (Figure 5.12) in much the same manner that we examined a TTL data sheet. It is important that a technician understand the strengths and weaknesses of both TTL and CMOS integrated-circuit technologies.

Recommended operating range, dc supply voltage: +3 to +15 V TTL required +5 V. CMOS has a wide range for its V_{dd} requirement. Often CMOS will use a V_{dd} of +5 V or +12 V. The higher V_{dd} values are required for faster speeds and higher noise margins. Typical propagation delays for CMOS are:

V_{dd} (V)	*Propagation delay (ns)*
5	125
10	60
15	45

As V_{dd} is increased, the propagation delay decreases proportionally, but the power consumption also increases proportionally.

All the specification will be illustrated for three V_{dd} values: 5 V, 10 V, and 15 V.

Low-level output voltage (0.05 V), V_{OL} This is extremely close to the ideal logic output of 0 V! CMOS dc parameters are all this close to ideal. This is one reason that CMOS has gained such popularity.

High-level output voltage, V_{OH} A CMOS logic 1 is within 0.05 V of the respective V_{dd}. Again, this is almost an ideal logic level.

Input low voltage, V_{IL} CMOS will accept any input level that is up to, approximately, 30% of V_{dd}. For a V_{dd} of +5 V, a voltage up to 1.5 V will be accepted as a logic 0. Considering that the guaranteed low-level output voltage is 0.05 V, this gives 5-V CMOS a noise immunity of 1.5 V, 30% of V_{dd}. Compare this with the TTL noise immunity of 0.4 V, which is only 8% of V_{cc}. As CMOS's V_{dd} increases, so does its noise immunity. This is one of CMOS's most useful parameters.

Input high voltage, V_{IH} This is the lowest voltage that will be accepted as a logic 1. We again find that this voltage is approximately 30% less than V_{dd}. This means that we have the same noise margin for both logic 0 and logic 1 levels. 0 V to 30% of V_{dd} will be accepted as a logic 0. V_{dd} to V_{dd} − (30% of V_{dd}) will be accepted as a logic 1. The area between these two extremes is considered to be indeterminate. If you ever find a voltage level between 30 and 60% of V_{dd}, this indicates a circuit malfunction.

Output low-sink current, I_{OL} With LS TTL this spec was 8 mA. For a V_{dd} of 5 V, the CMOS spec is 0.5 mA. It is obvious that CMOS is not good at driving loads. This is not a serious limitation, as you will see shortly.

Output high-source current, I_{OH} TTL has a source current spec of 0.4 mA. A CMOS device with a V_{dd} of +5 V will source 0.20 mA.

Input current, I_{IN} Unlike TTL, the input to a CMOS gate is perfectly symmetrical. This means that the current requirements for a logic 0 and a logic 1 are of the same magnitude with the opposite polarity. The spec is ±0.3 μA. This is an incredibly low input current. You can think of CMOS as having an infinite input impedance. Even though CMOS is so poor at sinking and sourcing current, its input current requirements are so low that one CMOS output can drive approximately 50 CMOS inputs. CMOS has a fanout of 50, LS TTL has a fanout of only 20. A CMOS output can drive two TTL LS inputs. This is an important factor in systems that employ both LS TTL and CMOS.

5.2.3 Special Handling Requirements for CMOS Devices

If a device has a high impedance, it is susceptible to static discharge damage. A person walking across a common waxed floor or carpet generates 20 to 30 kV of static charge. Even though CMOS has input protection diodes, it can still be easily damaged or destroyed by static discharge. It is always good practice not to touch the pins of an integrated circuit. When working with CMOS, a person should be grounded before any device is handled. The grounding is usually accomplished with a wrist strap. Another precaution is never to place the legs of a CMOS device in plastic trays, on tabletops, or in Styrofoam. Black conductive foam or pink conductive bags should be used to store CMOS ICs and CMOS circuit boards. A popular trick is to line plastic trays with aluminum foil to make them conductive.

5.2.4 Some Common CMOS Gates

You can find all the common TTL logic functions in CMOS. There are also advanced functions in CMOS that are not found in TTL. Some common CMOS gates are shown in Figure 5.13.

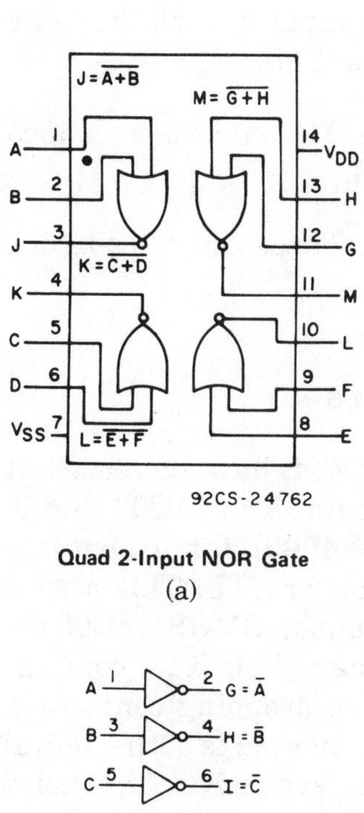

Quad 2-Input NOR Gate
(a)

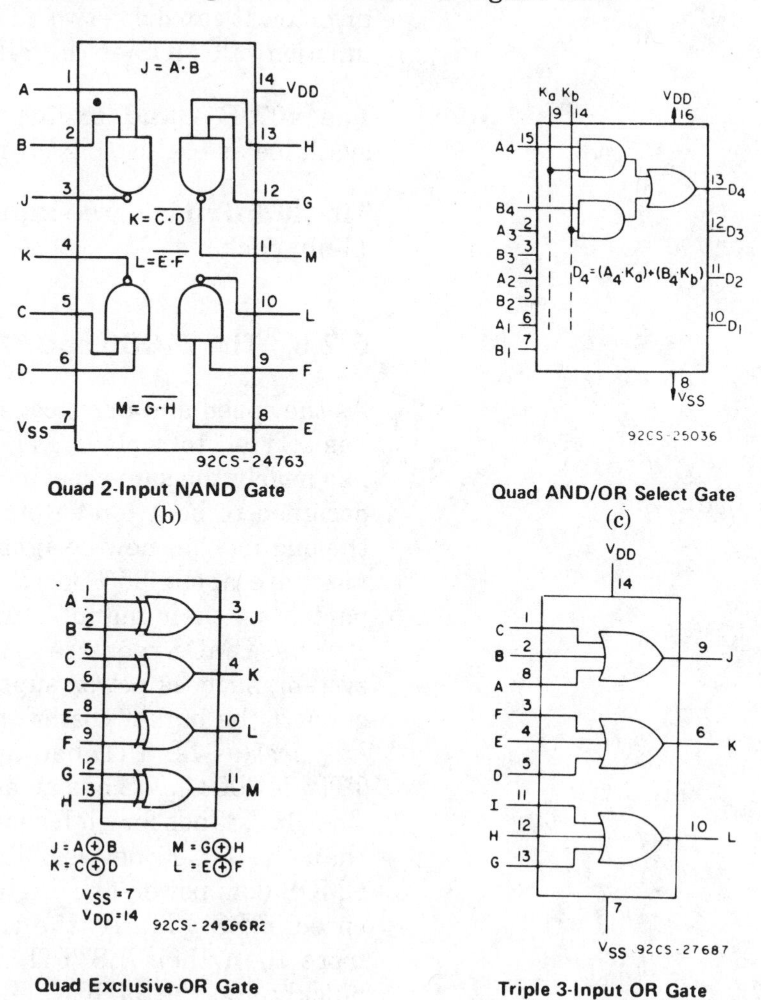

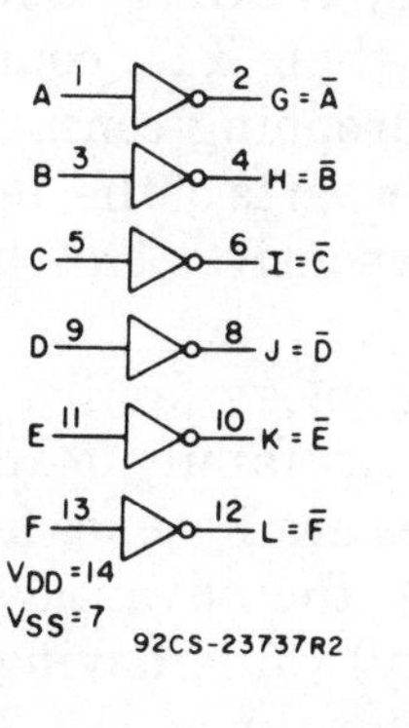

Hex Inverter
(d)

Figure 5.13 (a) 4001B; (b) 4011B; (c) 4019B; (d) 4069B; (e) 4070B; (f) 4075B.

The 4001B Quad two-input NOR gate NOR is the natural function of CMOS. Notice that the power pins, V_{dd} and V_{ss}, are in the corners of the DIP just as they were in TTL. The equivalent TTL gate is the 74LS02.

The 4011B Quad two-input NAND gate Notice that although this gate is the logical equivalent of the 74LS00, it is not pin-for-pin compatible. The inputs and outputs use different pins; only the power pins are the same. There is a pin-for-pin compatible CMOS series, 74C00. We will examine this series in the next section.

The 4019B Quad AND/OR select gate The inputs K_A and K_B are control inputs. They are common to all four AND/OR gates. If K_A is high, then the A inputs, A1 through A4, are enabled. If K_B is high, the B inputs are enabled. You can think of this device as four two-input AND/OR gates with common control lines stuffed into one integrated circuit.

The 4069B Hex inverters This is the general-purpose CMOS inverter. It can drive two TTL LS inputs. It plays much the same function in CMOS as the 74LS04 does for TTL.

The 4070B Quad exclusive-OR The XOR function is also available as exclusive NOR in the 4077B.

The 4075 Triple three-input OR gate Three OR gates in one 14-pin package.

5.2.5 The 74C00 and 74HC00 Series

As the speed and drive capabilities of CMOS have increased, it has started to replace TTL in new designs. The 74C00 family has nearly the same specifications as the 4000B family, but it is designed to be a pin-for-pin replacement for TTL. This assists the engineer in new designs that will employ CMOS. 74C00 devices are about 25% more expensive than their TTL counterparts, but their manufacturing costs are dropping constantly. Because CMOS requires so little power other parts of the digital system, such as power supplies, are less expensive. This helps support the higher cost of the 74C00 series.

Just as LS TTL had an advanced technology family, ALS TTL, 74C00 CMOS has an advanced technology family, 74HC00. The "H" stands for high speed. These devices are 10 times faster than the 74C00 devices. They maintain all the advantages of CMOS (low power and high noise immunity) while having the speed of LS TTL. At the time this is written, they cost 300% more than their LS TTL counterparts. This price will drop quickly and the 74HC family is sure to become an active competitor in the LS TTL market. The day of TTL dominance may be coming to an end!

5.3 AREAS OF APPLICATION FOR TTL AND CMOS

Let's review the strong and weak points of LS TTL and the 4000B CMOS families.

LS-TTL

Advantages	*Disadvantages*
1. Short propagation delay	1. Requires high power
2. Cost	2. Low noise immunity
3. Years of market dominance	

4000B CMOS

Advantages	*Disadvantages*
1. Ultra-low power	1. Speed
2. High noise immunity	2. Static sensitive
3. Wide range of power supply voltages	3. Market resistance to new technology

TTL has been the standard technology for SSI and MSI logic devices during the last 10 years. CMOS has greatly improved in recent years. Even though CMOS is challenging TTL for some applications, each has clear advantages in particular fields. You should expect to find TTL devices in these following applications:

1. General-purpose SSI and MSI logic circuits
2. High-speed digital circuitry
3. Circuitry that is required to sink and source current at medium to high levels

The following applications lend themselves to CMOS devices:

1. Portable battery-powered equipment, or any application where low power consumption is required.
2. Hostile environments where radio-frequency interference (RFI) or electromechanical interference (EMI) is a concern. Modern machine shops have drills and lathes that are controlled by digital electronics. These logic devices must have high immunity to the electromechanical interference that ac motors create.

5.4 INTERFACING TTL AND CMOS

Consider the following problem. A system is using TTL with a +5-V power supply where valid logic 1 is between +5 and 2.4 V and a valid logic 0 is between 0 and 0.4 V. This same system also employs CMOS devices that use a V_{dd} of +12 V. A valid logic 1 for the CMOS devices is between +12 and 8.4 V and the valid logic 0 is between 0 and 3.6 V. If we drive a CMOS gate with a TTL gate, both TTL logic 0's and logic 1's may appear to be CMOS logic 0's. On the other hand, if we drive a TTL gate with the output of a CMOS gate, the logic 1 output of the CMOS gate could destroy the TTL input transistor.

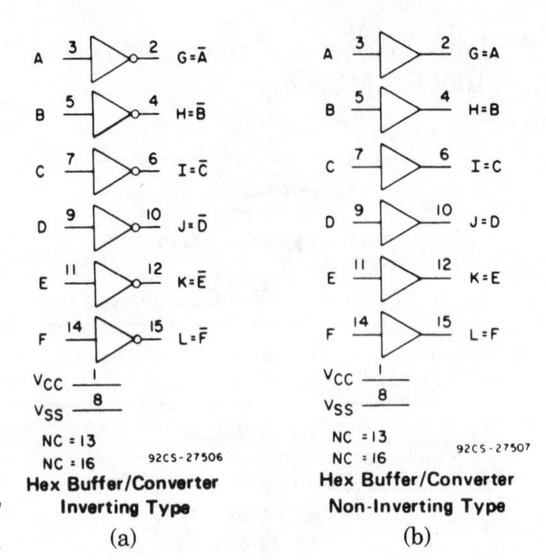

Figure 5.14 Hex buffers/converters.

The electronics term *interface* defines the place where two independent systems are connected. The problem we are considering is: How can we interface the different logic levels of CMOS and TTL? There are many devices designed to be used as TTL/CMOS interfaces.

5.4.1 The CMOS 4049/4050 Buffers-Converters

These two CMOS devices are used to convert CMOS logic levels to TTL logic levels (Figure 5.14). Notice that the power pin is designated as V_{cc}—not the usual designation of V_{dd} for CMOS devices. The 4049 converts logic levels and inverts the logic sig-

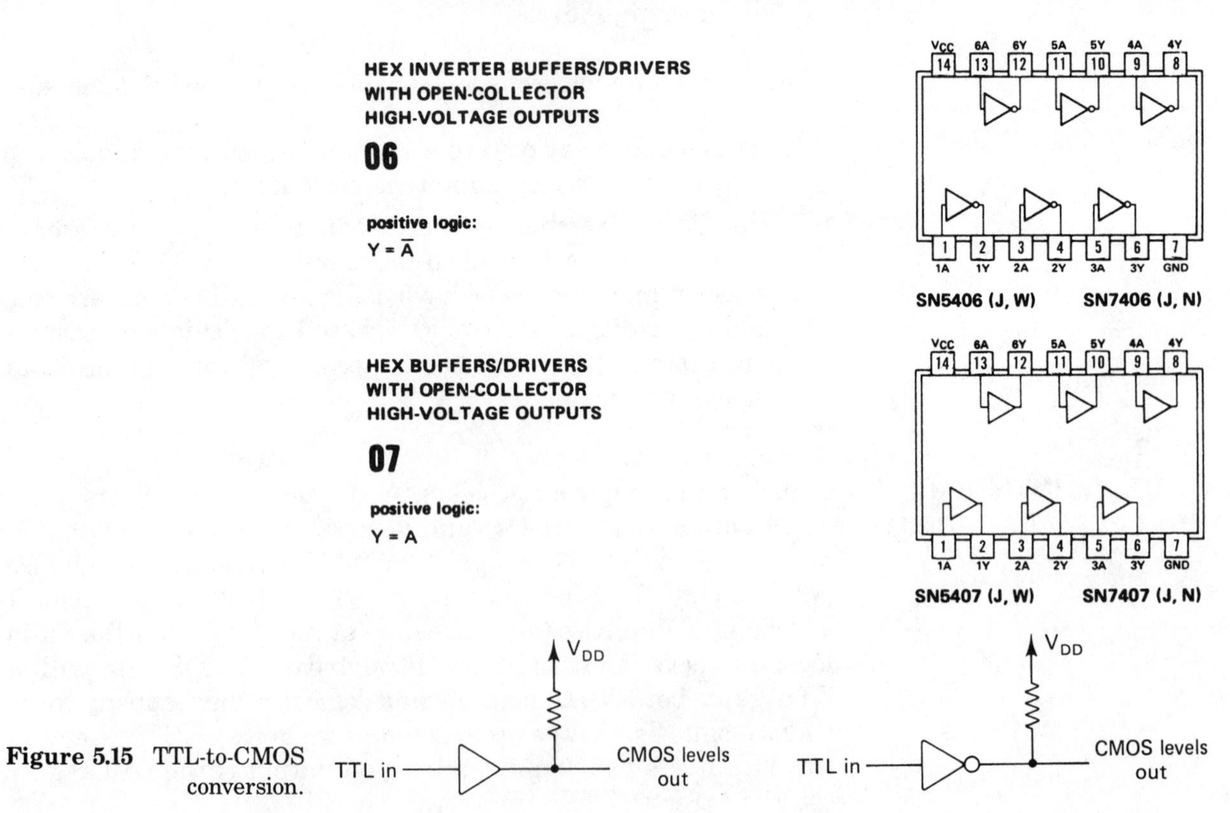

Figure 5.15 TTL-to-CMOS conversion.

nal. The 4050 only converts logic levels. The output of these converters can drive two standard TTL inputs.

5.4.2 The 7406-7407 Open-Collector Buffer/Interface Gates

The 7406/7407 devices convert TTL logic levels to CMOS logic levels (Figure 5.15). This is accomplished by connecting the external pull-up resistor to V_{dd} of the CMOS devices. A logic 0 output will be equal to 0 V and a logic 1 output will be pulled up to V_{dd} via the external pull-up resistor. The maximum voltage the pull-up resistor can be connected to is +30 V. This is well beyond the range of CMOS V_{dd}. These higher pull-up voltages are used to drive other devices (such as incandescent displays) that require high voltages.

QUESTIONS AND PROBLEMS

Use the pinouts of the TTL and CMOS devices contained in this chapter to answer the following questions.

5.1. Redraw the reduced three-judges circuit. Make a list of all the TTL ICs that would be required to build the circuit. Indicate the physical pin number of each IC on its appropriate lead in the schematic.

5.2. Design a 1-of-2 data selector employing the 4019B.

5.3. Repeat Problem 5.1 with the ignition enable circuit.

5.4. Repeat Problem 5.3 with CMOS 4000B gates.

5.5. What is the typical logic 1 output of TTL? CMOS with a +12-V V_{dd}?

5.6. What is the typical logic 0 output of TTL? CMOS with a + 12-V V_{dd}?

5.7. Consider the schematic shown. Under what input conditions will the LED illuminate? Redesign this circuit using gates with totem-pole outputs. What is the advantage of using open-collector gates in applications such as this?

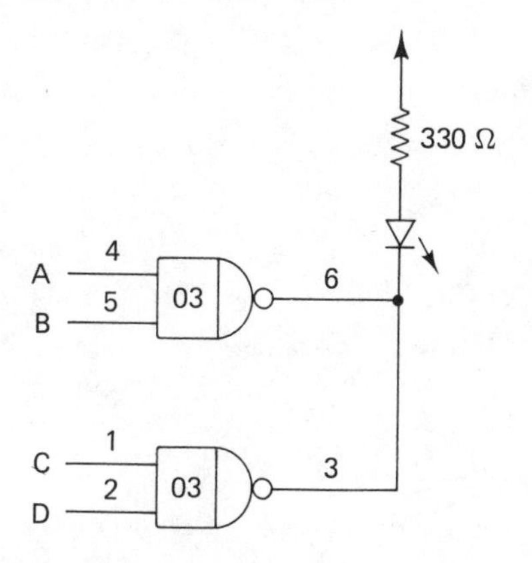

5.8. What is the typical propagation delay of:

(a) An LS TTL gate?

(b) An S TTL gate?

(c) A CMOS gate with a V_{dd} of 5 V? 10 V? 15 V?

5.9. A lumber mill wants to install digital circuits to increase the efficiency of the cutting saws. Considering the environment, should they use TTL or CMOS circuitry? Why?

5.10. The circuit shown will output a valid logic 0, but the logic 1 output falls into the indeterminate range. What is the problem?

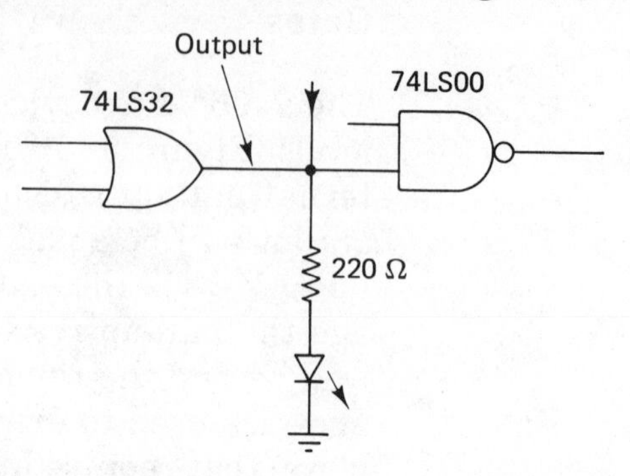

5.11. What range of voltages are indeterminate for TTL? +12-V CMOS? +5-V CMOS?

5.12. Why is LS the most popular form of TTL?

5.13. What is the greatest limiting factor of 74HC CMOS devices?

5.14. Explain all the elements of the following part number:

DM74AS00(J)

5.15. What precautions should be taken when a technician is handling CMOS devices?

5.16. A buzzer requires a voltage of +25 V. What TTL device can be used to drive this buzzer?

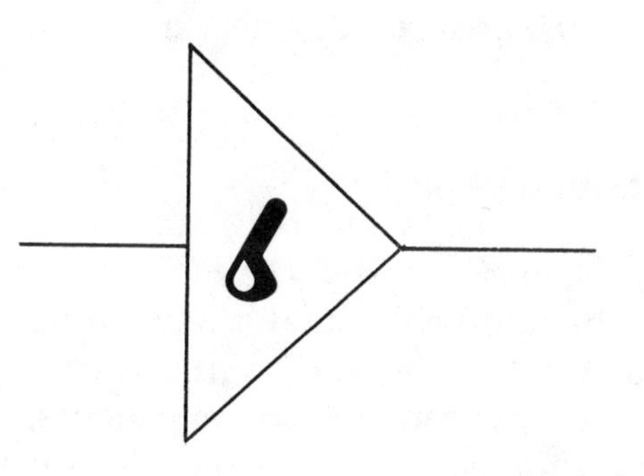

TROUBLESHOOTING LOGIC LEVELS

The electronics technician's primary responsibility is the repair of malfunctioning circuits. To repair these circuits efficiently, you must understand how digital circuits function and have the ability to use basic test equipment. The oscilloscope and digital multimeter are the two most widely used test instruments. Although the scope and digital multimeter (DMM) are useful, there are other pieces of test equipment designed exclusively for the test and repair of digital circuits. The logic analyzer, logic probe, logic driver, logic clip, and logic comparator offer many advantages in specialized situations. These test instruments will be introduced in the latter part of this chapter.

6.1 THE TECHNICIAN'S WORKING ENVIRONMENTS

Technicians work in many different environments. Circuit boards must be tested, and if malfunctioning, repaired. These functional circuit boards are then integrated into systems. Typically, a small computer may consist of three to five circuit boards. A technician must test and repair these systems. Other technicians work in research and development (R&D) areas. Here they assist engineers in the design, prototyping, and debugging of new circuits. Still other technicians work out in the field at customer sites. They perform preventive maintenance and repair customers' machines. Here is a short description of each environment.

6.1.1 The Board Test Environment

Commonly known as *subassembly test* (SAT) or *PCB test,* this is the place where most technicians will start their careers. In many ways it is the most technically challenging area. After printed circuit boards (PCBs) are stuffed with components, they must be tested and, if malfunctioning, repaired. Technicians work with detailed test procedures written by test engineers. Often there will be special test systems and other apparatus to assist in the testing and troubleshooting of PCBs. Board test techs must troubleshoot down to the component level, unlike other environments which only require the technician to find the bad PCB. In many system environments, technicians need only troubleshoot down to the bad PCB. The board test environment is challenging because these units have never worked before and they can have a wide variety of problems. This chapter is written specifically for the board test environment.

6.1.2 The Final Systems Test Environment

After PCBs are tested and repaired, they are integrated into a system. The systems test tech must check out the complete system. Systems test techs also work from detailed test procedures. They must understand how all the PCBs in the system interact. They are required to understand the system wiring and interconnect diagrams. Most systems test techs troubleshoot only down to the board level. This means that after the bad PCB is found, they replace it with a good unit and continue with the system test. The bad PCB is returned to board-test to be repaired and later integrated into a new system. Systems test techs often come from a board test environment. In board test they learn each individual board. In systems they learn how all the PCBs interact to perform a systems function.

6.1.3 The Field Service Environment

Field service technicians perform a wide variety of tasks. They troubleshoot many different systems to a board level, offer the customer technical assistance, and often perform a sales function. The field service tech must be capable of working with little or no supervision. People skills are often as important as technical skills to the field service tech. The ability to soothe an irate customer should not be underestimated. These techs usually drive company cars and are often required to dress in suits and ties. Travel, both local and out of town, may be required as well as long working hours.

6.1.4 The Customer Service Environment

When electronic systems malfunction, customers will often send the equipment back to the manufacturer to be repaired. Customer service techs work on a wide variety of PCBs and systems. Sometimes they troubleshoot down to the component level; other times they replace the bad PCB with a known good one from stock. The customer service tech is a combination of the board test tech and the systems tech.

6.1.5 The R&D Environment

Research and development is the glamor field of electronics. Often, the glamor is more image than substance. R&D techs must work closely with high-powered, potentially temperamental design engineers. Besides the normal electronic skills, R&D techs must have good mathematical and mechanical ability. They build circuits and mechanical test fixtures. Long, tedious hours can be spent checking the wiring of a prototyped circuit. Weeks and months can be spent recording and analyzing data samples. The greatest attraction of the R&D environment is that R&D techs generally advance quickly into the engineering ranks.

6.2 THE PRINTED CIRCUIT BOARD

Most production digital electronic designs are implemented on a PCB. Many of the problems that you will experience as a board test tech will be related directly to the characteristics of the PCB.

6.2.1 The Evolution of a Design

After a design engineer has finished a preliminary schematic, the circuit must be physically realized. This is usually accomplished by a technique called *wire wrapping*. Integrated-circuit sockets are inserted into perforated fiberboard, called *perf board*. These sockets have long, four-sided posts. A wire-wrapping tool is used to wind single-strand 30-gauge wire, called *Kynar wire*, around these four-sided posts. In this manner, the point-to-point wiring is accomplished. The integrated circuits are then inserted into the sockets, and the design can be tested.

Wire wrapping has many advantages in prototyping and limited production applications. New designs must go through many levels of debugging before they are ready for full-scale production. A wire-wrapped design can easily be modified. Wires can be removed, added, or moved to a different location. PCBs have a long development time and are not easily modified. R&D techs spend a good deal of time wire wrapping and debugging wire-wrapped prototyped designs.

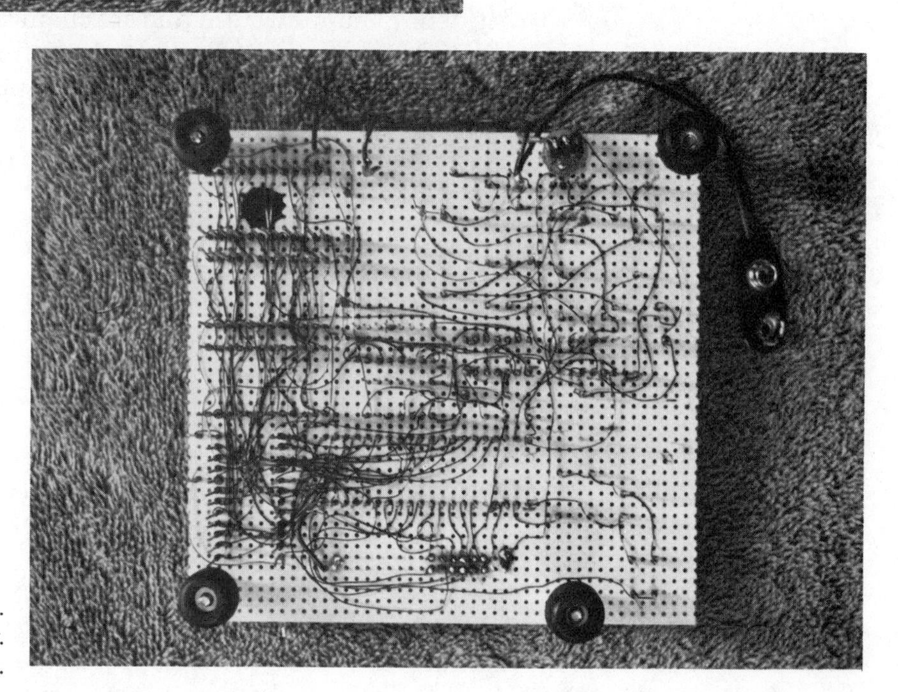

Wire-wrapped board.
(a) top view; (b) bottom view.
Photograph by Peter Ballinger.

After the prototype design has been fully debugged, updated schematics must be produced. These schematics are then given to a PCB designer, who will design the PCB layout. The PCB layout will be accomplished by attaching black tape onto a Mylar sheet. The black tape will represent circuit traces that will connect templates of DIP packages and other components that will reside on the PCB. The process of translating a schematic diagram into a finished tape-up is time consuming and prone to error.

The finished tape-up is given to a photographer. A picture is taken of the taped-up sheet of Mylar. The negative or positive of this picture is then given to a printed circuit board manufacturer. During a chemical process, the copper on the nontaped areas of the PCB is etched away. After the etching is complete, the PCB is thoroughly cleaned. Then a greenish insulating liquid called *solder mask* is applied. Finally, the PCB must have holes drilled in all the places where ICs or components will be inserted.

The bare board must be populated with ICs and all the other components. The act of populating a PCB with components is called *stuffing*. Stuffing PCBs can be done by machine, but is often done by hand. This populated board is then ready to be soldered. The machines that solder PCBs are called *wave solder machines*. The populated PCBs are run through the wave solder machine, where all the solder connections are performed automatically. Certain plastic components such as LEDs and switches cannot be wave soldered because they would melt under the intense heat. These components must be hand soldered after the rest of the PCB is complete. The PCB is now ready to be tested in a board test environment.

The traces on the PCB are called, collectively, the *artwork*. If any mistakes are found in the artwork, the original tape-up must be modified and the whole process repeated. This is why it is important that the original wire-wrapped circuit be thoroughly debugged and a new, correctly updated schematic be available to the PCB designer.

6.2.2 Elements of a PCB

The bare PCB can have many manufacturing defects that will require troubleshooting. It is important that the technician understand all the physical elements of a PCB. PCBs can have artwork on one side, both sides, and even many layers of artwork. The majority of digital PCBs are double-sided. Multilayer PCBs are used in complicated lightweight military applications. They can be extremely difficult to troubleshoot. We will concentrate our efforts on double-sided PCBs. One side of the PCB contains the electrical components and is called the *component side*. The other side of the PCB will contain the solder connections and is called the *solder side*. Refer to the following illustrations: The ICs are arranged in rows and columns. When you are troubleshooting a PCB, you will constantly be referring back and forth between the schematic and the PCB. Each gate on the schematic will have an identifying number. This number will relate to a physical IC on the PCB. The rows on the pictured PCB are identified with the letters A through E. The columns are numbered 1 through 13. A typical gate on the schematic may be labeled C11. This indicates that the physical IC can be found in row C, column 11.

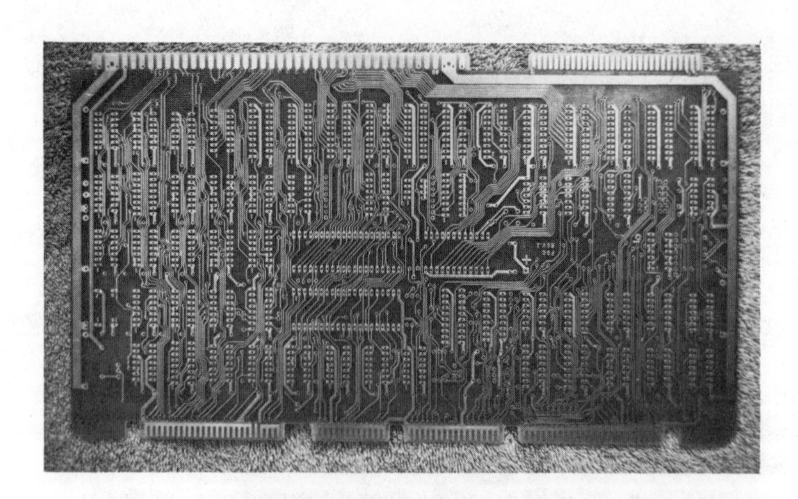

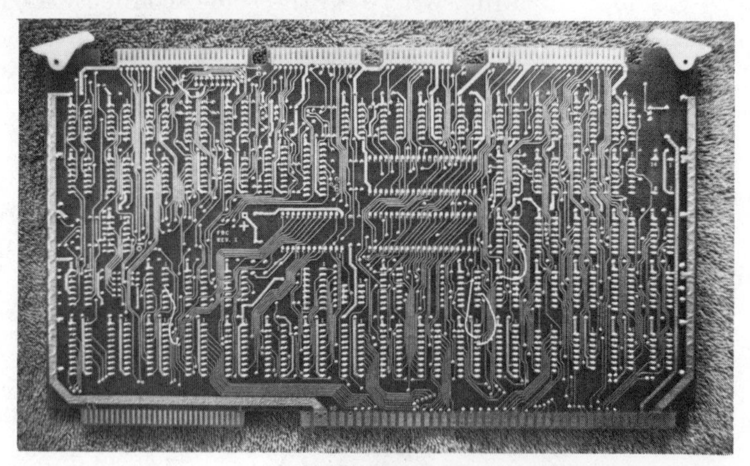

(a) Solder side of an unstuffed board; (b) Solder side of a stuffed board; (c) Component side of a stuffed board. *Photograph by Peter Ballinger.*

Consider the following terms describing various components that constitute a PCB:

Trace A trace is the conducting medium on a PCB that connects pins of IC and other components together.

Artwork As stated previously, artwork refers to the PCB layout. If a PCB is said to have an artwork problem, this means that the PCB layout must be modified to correct an electrical problem. Typically, this problem can be a trace running to the wrong pin of an IC, or traces that are laid out too closely and have a high probability of shorting together.

Pad ICs are plugged into holes in the PCB. Pads are the oblong conductors that encircle these holes on both the component and solder side of the PCB. When the leads of the devices are soldered, the pads are the surface area on which the solder flows to make a solid conductive bond. A square pad will designate pin 1 of an IC, the cathode of a diode or LED, the emitter of a transistor, or the positive lead of an electrolytic capacitor. This is a useful fact to remember. When you have the PCB turned upside down and are checking signals, it is often difficult to locate pin 1 of an IC. Just look for the square pad.

Feedthrough On double-sided PCBs there must be a way to route traces from one side of the board to the other; that is the function of feedthroughs. Feedthroughs are plated holes: the trace on one side travels through this plated hole to the other side of the PCB.

Fingers There must be a physical interface that allows PCBs to be inserted into a system. This interface usually takes the form of an edge connector. An *edge connector* is a rectangular plastic receptor that contains an upper and lower row of spring tension conductors. Notice the parallel groups of conductors perpendicular to and near the edge of the PCB. One of these conductors is called a *finger*. A group of fingers will plug into an edge connector to be interfaced into the system. Edge connectors come in many sizes, from 26 to 100 pins.

Power and ground buses You will also notice traces that are much wider than the rest. These traces are the power and ground buses. These buses carry considerable amounts of current, so the resistance of the trace becomes an important factor. You will notice many identical capacitors on the component side of the PCB. These capacitors are called *bypass caps* and usually have a value of 0.1 μF. They are placed between the power bus and the ground bus every 1 to $1\frac{1}{2}$ inches. The function of these caps is to short any noise on the power bus to ground. When many gates are turning on and off simultaneously, switching spikes may be induced on the V_{cc} bus. This switching noise does not occur in CMOS systems because of the low current requirements of CMOS. If you examine the picture of the bare PCB closely, you should be able to see the V_{cc} bus and ground bus as they run from IC to IC. Near the outside edges of the PCB you will notice large, electrolytic capacitors, located between V_{cc} and ground. These caps range in value from 22 to 100 μF. They also act as bypass caps.

Title and reversion number Each PCB should have a title or a number that identifies the unit. Many PCBs even have serial numbers. No matter how well the initial design is debugged, the

PCB will still go through many design and artwork changes. These changes are usually denoted by a revision number or letter. The original unit will be revision 0. After an artwork or component change is required, the new unit will be revision 1. In this manner old PCBs can be identified and updated. When a modification to existing design is required, the engineer will release an engineering change order (ECO). Each PCB should have a history of ECOs that describe the various component and artwork modifications.

IC sockets Most ICs will be soldered directly into the PCB. There are times when it is not desirable to solder an IC directly into the PCB; in these cases IC DIP sockets are used. Sockets are plastic devices which are wave soldered into the PCB. The ICs are then hand stuffed into these sockets. The advantage of sockets is obvious: ICs can be removed quickly and easily and replaced without soldering. Sockets are often used for LSI devices. The process of desoldering and resoldering a 28- to 64-pin device can be frustrating and time consuming. The traces and pads of the PCB can also be torn off if the device is not desoldered properly. If the IC is found not to be defective, it can be reused (which is not often the case with hard-soldered ICs). CMOS devices are often socketed due to their static sensitivity; because of this sensitivity a higher percentage of CMOS devices malfunction during the initial checkout in board test.

However, sockets are not perfect and can introduce problems into the system. When an IC is originally stuffed into a new socket, it may not make good electrical contact, due to the tightness of the socket. It is a good habit to check all socketed devices quickly to assure that they are stuffed correctly. In electronic equipment that is subject to vibrations, ICs may work themselves loose from the sockets. Most important of all, they often cost more than the ICs that are placed into them. For these reasons the use of IC sockets is limited. In the PCB pictured, all the devices are placed in sockets, because the unit is a prototype.

6.3 POSSIBLE DEFECTS IN NEWLY STUFFED PCBS

As a board test technician you will test PCBs right after they have been stuffed and wave soldered. Large companies have quality assurance (QA) people who inspect incoming PCBs. Smaller companies cannot afford this luxury. A new PCB can have many possible problems other than just malfunctioning ICs. The following is a list of common problems that you will find as a production board test tech.

Bad components ICs, transistors, and other active devices can malfunction. They can be bad straight from the manufacturer or be destroyed in the wave-soldering process. Capacitors can be open, shorted, or out of tolerance.

Solder bridges A solder bridge is "a splash of solder that connects two or more independent conductors resulting in a short." Solder bridges occur during the wave-soldering operation. The solder bridge can be between traces, between pads, or between a trace and a pad. Wherever traces run close together, there is a possibility that a solder bridge can occur. Usually, they are visible to the unaided eye, but sometimes solder bridges are so narrow that they cannot be seen without the aid of a strong light and a magnifying glass. To remove a solder bridge, a tech can cut it with a razor knife, or heat the solder bridge and remove it with a vacuum desoldering tool or solder wick.

Unetched traces Sometimes the chemical etching process will fail to etch away the copper between adjacent conductors. Most often, the unetched traces are between a pad and an adjacent trace. If you notice an unetched trace that occurs chronically, report it to the test engineer. The artwork may need to be modified.

If the PCB manufacturer does not take great care in blowing clean the PCBs before the solder mask is applied, small slivers of etched away copper may short across adjacent conductors. These shiny copper slivers will have the appearance of fine solder bridges. If you discover shorts of this type, report them to the test engineer. The PCB manufacturer will be cautioned to clean the boards before applying solder mask. To simplify matters, we will refer to these sliver-like shorts as a subset of unetched traces.

ICs stuffed backwards It is common to find ICs stuffed with pin 1 in the pin 8 or 9 position. When this happens the IC is almost always destroyed. Sometimes an IC inserted backwards will short out V_{cc}. It is always a good habit to perform a quick visual check before applying power to a new PCB.

Transistors and other polarity-sensitive devices stuffed backwards A square pad indicates the emitter lead for a transistor, the cathode of a diode or LED, or the positive lead of an electrolytic capacitor. If electrolytic caps are plugged in backwards they can explode with a loud bang. LEDs and diodes will not be damaged. Transistors may or may not be damaged, depending on the configuration.

The wrong devices stuffed Another common stuffing problem is inserting the wrong IC, transistor, or passive component. The best way to assure that all the proper components are stuffed is to compare the unit under test to a known good unit. Lay these PCBs side by side, and make sure that all the components match. Check the color code of the resistors and the values of the capacitors.

Open/cut traces Sometimes a trace may have a small cut or break in it. This can occur at the point where the trace meets a pad, a feedthrough, or a finger. An ohmmeter should be used to establish if two points in question are continuous. Another possible cause of an open happens when an IC is being changed. Often when you are resoldering an IC into place, the trace will fold back and create a small gap.

Open feedthroughs Feedthroughs are plated. During wave soldering, solder usually wicks through feedthroughs, creating a good connection between the top and bottom traces. If you have an open between two points that should be continuous, be sure to check any feedthroughs for continuity.

Bent leads under ICs When an IC is stuffed into the PCB or socket, a lead can miss the hole and be bent under the IC. It is often difficult to see these bent under leads. If you suspect this problem, turn the PCB over to look at the soldered side. A small amount of lead should stick through each pad. With ICs that are in sockets, simply pull out the device, visually inspect it, and then reinsert it into the socket.

Cold solder joints A good solder connection should have just enough solder to cover the middle of the pad. The solder should be bright and shiny. A connection with too much or too little solder, or a dull gray appearance, may be cold. By "cold" we mean that it is not forming a connective joint between the trace and the pad. To repair a cold solder joint, remove the bad connection with a vacuum desoldering tool and resolder it using good technique and just enough solder to cover the middle of the pad. Check the connection with an ohmmeter to ensure that it is continuous.

Bad socket IC sockets can malfunction. Usually, the malfunction takes the form of an open between the IC lead and the pad. It is important to take all measurements on the lead of the IC, not the pad. This will assure that the trace and pad are connected, the IC or socket lead is connected to the pad with a good solder connection, and the socket is conducting properly.

Dirty fingers If a PCB appears to have a wide variety of problems, it is always good practice to check the fingers. Fingers should be bright and shiny. Sometimes they will corrode or be coated with a foreign insulating substance. If you suspect that a PCB may have dirty fingers, clean them until they are bright and shiny with an ordinary pencil eraser. Be sure to remove any trace of eraser bits before you reinsert the PCB into the edge connecter.

Shorted bypass capacitors A typical PCB may have 40 bypass caps. Visualize 40 capacitors in parallel, between V_{cc} and ground. If one of these caps is shorted, the power supply will current limit, and V_{cc} will measure as a few tenths of a volt. Determining which bypass cap is shorted can be a difficult task. We will discuss techniques for finding a shorted bypass capacitor later in this chapter.

Schematic and artwork errors When a PCB is revised to correct a problem, another problem can be introduced. If you are troubleshooting a newly revised PCB, be alert for possible artwork problems. If your schematic and the PCB artwork do not match, this indicates a problem. Bring this discrepancy to the attention of a test engineer.

Wrong manufacturer of a particular component This is a problem that plagues engineers and technicians. Most ICs are manufactured by many different companies. If one manufacturer cannot meet sales demands, another one will increase production to pick up the slack. If a 74LS00 is required, it can be purchased from at least 10 different manufacturers. Each of these 74LS00s should be 100% compatible. There will be isolated cases when a particular IC will function correctly only if it is sourced by a particular manufacturer. If you experience this problem, be sure to bring it to the attention of a test engineer. The circuit will have to be modified, or the purchasing department should be instructed to purchase this IC from a specific manufacturer.

6.4 TROUBLESHOOTING OPENS

The following table summarizes the most common causes of opens and the appropriate action to rectify the problem.

Causes of opens	*Required action*
Open/cut traces	Apply bead of solder or install jumper across broken trace
Cold solder joints	Reheat or remove solder with desoldering pump and resolder
Open feedthroughs	Resolder feedthrough, assuring that solder wicks to both sides of PCB
Bent legs underneath ICs	If IC is in socket, remove, straighten leg, and reinsert If IC is soldered into PCB, remove and replace IC
Internal opens in ICs	Remove and replace IC
Bad IC socket	Remove and replace socket
Dirty fingers	Remove oxidation with eraser
Artwork errors	Contact test engineer

The most important questions that you should be asking yourself are:

1. What is the voltage that one would measure on an open input?
2. What logic level does a gate interpret an open input as: a logic 1 or a logic 0?

6.4.1 The TTL Open Circuit

Most malfunctioning gates will be stuck at either a logic 0 or a logic 1. If you observe the output of gate switching back and forth between valid logic levels, this usually indicates that the gate is good. That does not imply that the inputs driving the gate are good. Refer to Figure 6.1, and consider the following facts.

The output of a 74LS04 inverter is connected to one input of a 74LS00 quad two-input NAND gate. The trace connects these two points by way of a feedthrough on the PCB. During the etching process, the copper lining the feedthrough was etched away. Therefore, this feedthrough is not electrically continuous and acts as an open between the inverter and NAND gates.

The input to the NAND gate is said to be *floating*. Floating means that it is not connected to any output. An open between an output and an input will always result in a floating input. You are troubleshooting this malfunctioning circuit. An oscilloscope probe is placed on the output of the inverter, and it appears to be switching correctly. You now place the scope probe on the output of the NAND gate in question. The output of the NAND gate, pin 3, appears to be switching correctly. We have now established that both gates are probably good. What about the floating input? We place a scope probe on pin 1 of the LS00. We do not observe a valid logic 1 or logic 0; instead, we read a dc voltage of approximately, 1.6 V.

A floating TTL input will measure between 1.4 and 1.8 V dc.

This is the single most important fact that a technician troubleshooting a TTL open can possess. If you are measuring the voltages on TTL inputs and find one between 1.4 and 1.8 V—*Stop*. You have discovered a floating TTL input.

Now that you have discovered a floating input, what is your next action? Obviously, we must find the open. The preceding table lists light common causes of opens.

You will use your eyes and an ohmmeter to find the cause of an open.

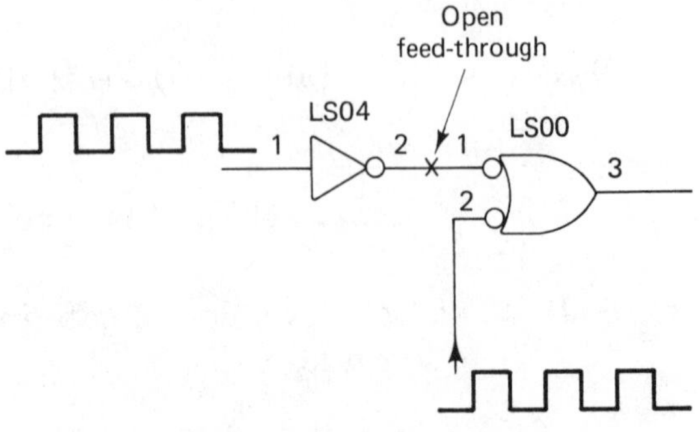

Figure 6.1 Open TTL input.

After you have located a floating input by observing a dc voltage between 1.4 and 1.8 V, you must power-off the circuit. Ohmmeters can be used only on deenergized circuits. The ohmmeter should be placed in the most sensitive range, typically 200 Ω full scale. The ohmmeter will be used as a device to test the electrical continuity between two points. We will place one lead of the ohmmeter on pin 2 of the LS04; the other lead will be placed on pin 1 of the LS00. The ohmmeter will read infinite resistance. This verifies our assumption of an open between the inverter and pin 1 of the NAND gate.

The traces on PCBs are coated with a green insulating material called solder mask. This material helps to prevent solder bridges between traces during the wave-soldering process. You will require a set of ohmmeter probes with needle-sharp points. When measuring continuity between two points in the same trace, you must push your probe through the insulating solder mask and into the trace itself. It will require a bit of practice to become proficient at piercing traces. Use your ohmmeter to assure that the connections of the IC legs and the pads on the PCB are continuous. This will check for the following problems:

1. Cold solder joint between IC leg and pad
2. Bent leg underneath the IC
3. Bad IC socket

Because no edge connector is involved in this circuit, dirty fingers will not be considered as a possible cause of this open. Artwork errors are a rarity, but you should always keep that possibility in the back of your mind. There are no internal opens in either IC; the inverter outputs a valid TTL signal, and the NAND is showing 1.6 V on the floating input, an indication that the device is internally connected.

This leaves us with two possibilities: an open feedthrough or an open/cut trace. The skill to follow traces from point to point on a double-sided PCB requires a good deal of practice to develop. Traces flip from side to side via feedthroughs and run under ICs. Feedthroughs can also emerge underneath ICs. It is extremely helpful to have an unstuffed PCB of the unit that you are troubleshooting. You can use this bare board as an uncluttered model of the traces and feedthroughs. You will usually have to request an unstuffed PCB from your boss, but it is well worth the trouble.

The quickest and most efficient way to find an open/cut trace or open feedthrough is to continually cut the possible area of the open by half. Starting at the output of the inverter, we will find the approximate halfway point between the two ICs. If this measures continuous, the open is on the second half of the trace; if it measures open, the open is on the first half of the trace. Repeat this process until you locate the open. With an etched away feedthrough, you may have to insert a piece of Kynar wire to assure that the solder will completely wick through to both sides. If you find a wide cut in a trace, you may also have to add a small jumper wire to connect the two broken ends. These are manual skills that must be learned on the job. If you can find some junk PCBs, they are excellent devices for learning how to repair the various problems that can occur, without running the risk of mutilating a good unit.

We now know that a floating TTL input will measure in the indeterminate voltage range, between a valid logic 0 and a valid logic 1. Consider the following case. The input of a 74LS04 inverter is floating: What will be the output of the inverter, a logic 0 or logic 1? The real question here is: Does a floating input look more like a logic 0 input or a logic 1 input? A logic 0 input must be able to sink about 1 mA of current; a floating input cannot sink any current. On the other hand, a logic 1 input does nothing more than reverse bias the base–emitter junction of the input transistor; it must deliver only a minute amount of reverse diode leakage current. It is obvious that a floating input and a logic 1 input have exceedingly similar characteristics. Therefore:

A floating TTL input appears to be a logic 1.

Before the open feedthrough in Figure 6.1 was repaired, the output of the NAND would have been the inverted input on pin 2. The floating input behaved as a logic 1, causing the NAND gate to act as an inverter. If the input to a particular gate is required to be a permanent logic 1 (such as using NAND gates as inverters or reducing the number of inputs of an AND gate), the input should be tied to V_{cc} via a 1- to 10-kΩ resistor, not left floating. The reason for this precaution is simple: If an input is left floating, it acts as an antenna for noise in the system. This noise can cause the voltage level on a floating input to randomly glitch between logic 1 and logic 0 input levels—thus introducing random output levels.

Technicians will often use the floating input characteristic of TTL inputs as a convenient means of forcing a logic 1 input at a particular gate. Refer to Figure 6.2. You can think of this circuit as consisting of three major parts: input circuitry to the AND gate, enable circuitry, and output circuitry. It may be much easier to test and troubleshoot this circuit if the AND gate was always enabled. This can easily be accomplished. If the LS08 is in a socket: we would remove it, bend pin 2 out with the aid of needle-noise pliers, and reinsert the IC with pin 2 sticking out of the socket. This floating input would look like a logic 1, and the LS08 would be enabled. If the LS08 was hard-soldered onto the PCB, we would cut the trace next to the pad for pin 2. This would simulate a broken trace condition and cause pin 2 to float. After the test is over, you must remember to reinsert pin 2 into the IC socket or resolder the broken trace.

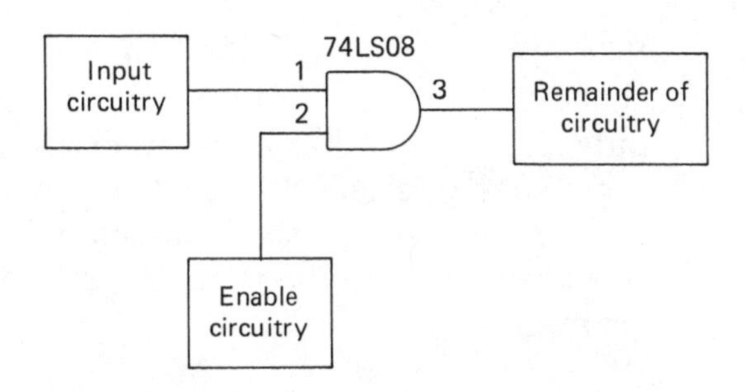

Figure 6.2 AND gate as a digital switch.

6.4.2 The CMOS Open Circuit

If a CMOS input is left floating, it will eventually destroy itself. The inputs of unused gates of TTL ICs can be left floating and no harm will come to the IC. (In high-speed TTL systems, all unused inputs are usually tied to V_{cc} via a 1- to 10-kΩ resistor to improve switching speeds.) All the unused inputs in a CMOS IC must be tied to V_{dd} via a resistor or ground. Remember: In a CMOS device only one input MOSFET can be conducting, while the other must be cutoff. The cutoff MOSFET acts as a load resistor for the saturated MOSFET. If a CMOS input is left floating (through a design error or any of the other possible causes of opens), noise can cause both MOSFETs to conduct simultaneously. If this happens, neither MOSFET will have a load to current limit it and the CMOS IC will be destroyed. We have stated that a floating TTL input will act as an antenna and pickup noise. Because the input impedance of CMOS inputs is in the range 10 to 50 MΩ, they will be affected much more adversely by noise than the low-input-impedance TTL devices. A floating CMOS input will not exhibit a specific voltage, as did the floating TTL input. An oscilloscope will usually reveal random noise transients on a floating CMOS input. Thus a floating CMOS input does not appear to be either a logic 1, as did TTL, or a logic 0—it switches with the noise transients.

A consideration is whether a CMOS device must be replaced when it is found to have had a floating input, due to any of the reasons discussed previously. This is a difficult question to answer in absolute terms. It is our opinion that if, after the cause of the open is repaired, the gate performs as claimed, it is probably good and need not be replaced. This will be the case in most situations. There is a possibility that the gate may have suffered enough damage to degrade its performance in the future and ultimately shorten its effective life. The most conservative action is always to replace the device, but experience has proven that this practice is overkill, in addition to being expensive and time consuming.

6.5 TROUBLESHOOTING SHORTS

Within a digital PCB, it is a simple truth that all inputs must be connected to some output. The only "pure" inputs (i.e., inputs that are not driven by the output of another gate) are introduced onto the PCB via the fingers and edge connector. We are going to concern ourselves with the most common types of shorts: shorted outputs to ground, shorted outputs to V_{cc} or V_{dd}, outputs shorted to other outputs, and V_{cc} shorted to ground. Finding shorts in a board test environment may be the single most difficult task facing the technician. This is due to the nature of shorts and the way they manifest themselves. Consider the following table.

Causes of shorts	*Required action*
Solder bridges	Heat and remove with desoldering tool or cut with razor knife
Unetched traces	Cut with razor knife
Internal shorts in ICs	Replace IC
Shorted bypass caps	Find and replace
Artwork errors	Contact test engineer

6.5.1 Isolating Outputs

The output of one gate may drive the inputs of many other gates. This increases the complexity of finding the cause of a short. Consider Figure 6.3. Assume that the output of G1 appears to be stuck at a logic 0 level. A partial list of possible causes may be:

1. G1 is stuck internally low.
2. G2, G3, or G4 may be internally shorted, loading down the output of G1.
3. A solder bridge somewhere on the trace between the output of G1 and the inputs of G2, G3, or G4 may be shorting the output of G1 to another source that is loading it down.
4. An unetched trace may be causing the same problem as the possible solder bridge.

How do we systematically isolate the output of G1 and the inputs of G2, G3, and G4 so that we can locate the circuit fault? The first step will be to isolate the output of G1. By isolating G1

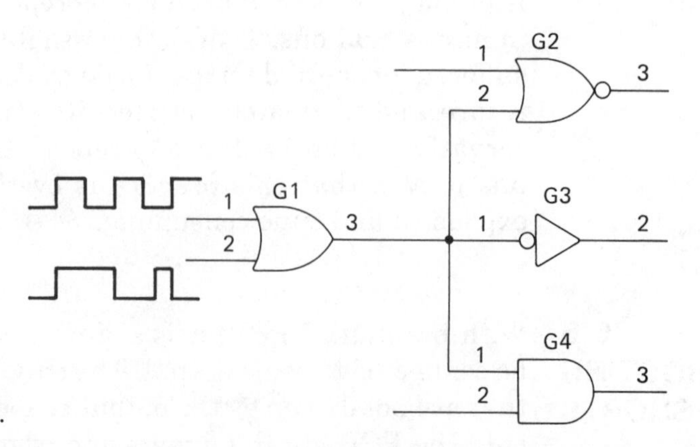

Figure 6.3 Multinode circuit.

we will determine if G1 is internally stuck low. If G1 is in a socket, our task is simple: Remove G1 from the socket, bend out pin 3 of G1 with needle-noise pliers, and reinsert G1. If G1 is not in a socket, we have two different choices: Cut the trace from G1 pin 3, just after the pad, or cut pin 3 of G1 with a pair of diagonal cutters, just above the pad on the component side of the PCB. (The latter method of isolating a pin is difficult, and not preferred.) Now the output of G1 cannot possibly be loaded down from external sources. If G1 is still stuck low (assuming that either pin 1 or pin 2 of G1 is high), replace G1. It is internally stuck low.

If G1 will now switch properly, we know that G1 is good and the problem lies elsewhere. We will now reconnect pin 3 of G1 by reinserting the leg or soldering the cut trace. Now we must isolate pin 2 of G2, using the same technique that was employed to isolate the output of G1. If output of G1 will now go high, replace G2; it is internally shorted and loading down the output of G1.

If G2 tests as good, you must use the same technique on G3 and G4. If G3 and G4 test as good, the problem must be an unetched trace or a solder bridge. Take a moment to make a good visual inspection of the traces and other artwork that is associated with the output in question. In most cases, you can visually locate the problem. If a good visual inspection of the PCB in a well-lit area does not reveal the short, it is time to use an ohmmeter. As always when using an ohmmeter, the first step is to deenergize the circuit. Then check for shorts between adjacent legs of the devices involved:

1. Is pin 1 or pin 3 of G2 shorted to pin 2 of G2?
2. Is pin 2 of G3 shorted to pin 1 of G3?
3. Is pin 2 of G4 shorted to pin 1 of G4?

Follow the trace from G1—pin 3 as it runs toward G2, G3, and G4. Use the needle-sharp points of the ohmmeter probes to check for any possible microscopic shorts or unetched traces between any trace that runs close to the one in question. Be patient and methodical and you will find the problem. Shorts of this type can be time consuming to find. The authors have personally spent more than two hours finding one microscopic short on an extremely dense PCB. Even after you have found another trace that is shorted to the one in question, you still may not be able to locate the actual cause of the short! When this happens, you will be required to drag the razor knife between the two traces, hoping to cut the invisible short. Granted, this will deface the PCB somewhat, but there are times when it is the only solution.

Once again, be sure to refer to an unstuffed PCB. Without the clutter of all the components, you will be able to see the points that are the most likely to be shorted to the trace in question. The denser the artwork, the more difficult it will be to find the shorts. Use your eyes, a strong light, a magnifying glass, and an ohmmeter to assist you in pinpointing the short.

6.5.2 Shorts in TTL Circuitry

When a technician is quickly scanning a circuit with an oscilloscope, it is easy to miss an obvious symptom of a short. A typical TTL low is 0.1 to 0.2 V. If you ever find a TTL logic 0 that looks like 0 V, it is probably shorted to ground. When probing TTL circuitry with an oscilloscope it is best to keep the vertical sensitivity at 2 V per division; 5 V per division is too course and you will miss a logic 0 that is really a short to ground; 1 V per division is too fine, and you will not be able to display both channels of the scope simultaneously. Remember to look for a slight displacement whenever you see a logic 0 displayed on the scope's CRT.

A typical TTL logic 1 is between 3.2 and 3.6 V. If you measure a TTL logic 1 that is 5 V, the gate must be under no-load conditions (an open must exist between the output of the gate and the input to the gate that it is driving) or the output must be shorted to V_{cc}.

On a 14-pin IC, pin 6 can be shorted to pin 7 (ground) with a solder bridge. Also, pin 13 can be shorted to pin 14 (V_{cc}) with a solder bridge. Be sure to check the obvious before going into detailed troubleshooting procedures.

Consider the output of a TTL gate that is shorted to V_{cc}. When the gate outputs a logic 0, the pull-down transistor will have 5 V placed across it with no current limit! This can easily burn out the pull-down transistor of the totem pole. Be aware that after finding the short, you may also be required to replace an IC whose destruction was a direct result of the short!

On the other hand, consider a TTL output that is shorted to ground. Recall that the pull-up resistor is current-limited by an internal, 130-Ω collector resistor. When the output of the gate tries to go high, the pull-up transistor will be current limited, even with the output shorted to ground. The TTL specifications only guaranteed the gate for a short of 1-second duration. Experience has shown that in the majority of cases, the gate will not be harmed from an output shorted to ground. After you find and remove the cause of the short, the gate should function correctly.

6.5.3 The CMOS Short

CMOS logic levels approach the ideal of ground and V_{dd}. Therefore, a CMOS output short to ground or V_{dd} is more difficult to see than the same short in TTL. Just remember that if a CMOS gate is stuck high or low, it could be shorted to V_{dd} or ground, respectively.

CMOS outputs can be shorted to ground or V_{dd} for a limited time and not be damaged. After the short is found and removed, test the gate's ability to switch high and low. If it appears to behave normally, it is probably not damaged and need not be replaced. CMOS devices employing higher V_{dd} values are more likely to sustain damage because they are capable of delivering higher amounts of short-circuit current. It is the internal dissipation of heat that will damage an IC output structure.

6.5.4 Shorted TTL Totem-Pole Outputs

What happens when two totem-pole outputs are shorted to each? Consider Figure 6.4. The input to G1 is a 1-kHz, 50% duty cycle, TTL level square wave. The input to G2 is exactly the same signal except that it has a frequency of 500 Hz. The timing diagram in Figure 6.4b is divided into four parts: T1, T2, T3, and T4. We will analyze the output of the shorted totem poles for each time period.

T1 The inputs to both G1 and G2 are high; therefore, both active pull-down transistors in G1 and G2 saturate and the output will go to logic 0. Notice that both outputs are pulling in the same direction and actually assist each other.

T2 The input to G1 has gone low, but the input to G2 is still

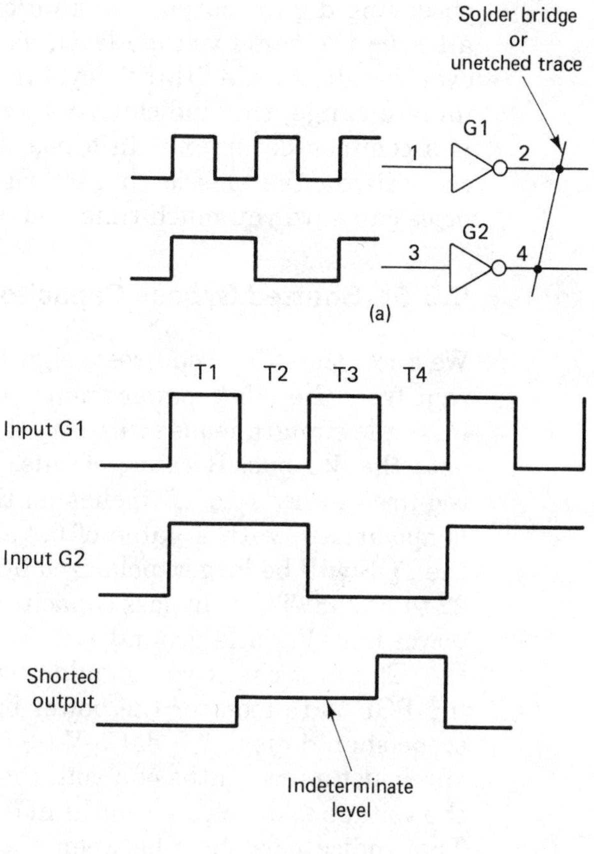

Figure 6.4 Shorted totem-pole outputs.

high. The output of G1 wants to go high; the active pull-up transistor in G1 turns on and saturates. The output of G2 wants to go low; the active pull-down transistor in G2 stays saturated. It appears that because of the short between the two outputs, there is a conflict. G1 wants to go high, and G2 wants to go low. We know that TTL is much better at sinking current than it is at sourcing current. The pull-up transistor in G1 will source between 20 and 60 mA of short-circuit current. The pull-down transistor in G2 will be trying to sink all this current. The result will be a voltage level at the shorted output in the indeterminate range, somewhere between a valid logic 0 and a valid logic 1. With LS TTL this voltage level will be approximately 0.8 to 1.2 V.

If you ever measure a TTL output voltage between 0.8 and 1.2 V, this indicates shorted totem-pole outputs.

T3 During this time period the input to G1 goes high while the input to G2 goes low. Once again, there will be a conflict at the shorted outputs that results in an indeterminate voltage between 0.8 and 1.2 V.

T4 Both inputs are low, so both outputs go high. There is no conflict and the shorted node will pull up to a valid logic 1.

Review the output timing diagram in Figure 6.4b. When you are observing digital outputs on an oscilloscope, the logic lows will all have the same voltage level, as will the logic highs. Whenever you discover a "third" level in the low part of the indeterminate range, this indicates a shorted totem-pole output. This is a common occurrence in a board test environment. Quickly recognizing the classic characteristic signal of shorted totem poles can save you much time and effort.

6.5.5 Shorted Bypass Capacitors

We know that TTL requires a significant amount of biasing current from the +5-V power supply. On a moderate-to-large TTL PCB, the simultaneous switching of many gates can induce noise onto the V_{cc} bus. Because of this, noise bypass capacitors are required every 1 to 1.5 inches on the PCB. These caps will be nonpolarized, with a value of 0.1 μF. Also, along the border of the PCB will be larger, polarized bypass capacitors in the range 22 to 100 μF. These bypass capacitors will be placed between the power bus (V_{cc}) and ground.

The first check you should always make on a malfunctioning PCB is to monitor the power bus with an oscilloscope. The scope should display a flat 5-V ($\pm$ 0.25 V) dc signal. If V_{cc} measures just a few tenths of a volt, the power supply is probably in the voltage foldback current limit (commonly called "crow-bar"). That indicates a short between V_{cc} and ground.

There are many possible causes of a short between V_{cc} and ground:

1. Solder bridge
2. Unetched trace
3. Internal short in an IC
4. An IC or other active device inserted backwards
5. An electrolytic bypass cap inserted backwards
6. A shorted bypass cap

After you have discovered what appears to be a V_{cc} short to ground, your first step should be to isolate the power supply from the PCB. Measure the +5-V output of the isolated power supply to assure that it is functioning correctly. (*Important note:* Most digital systems use high-efficiency switching power supplies. Switching power supplies, unlike analog power supplies, need to be loaded before they will output a proper voltage level. If you have any questions about the difference between an analog and a switching power supply, consult your instructor or test engineer.)

If the isolated power supply checks, you should then take a long moment for a complete visual inspection of the PCB. When ICs are internally shorted they will generate a significant amount of heat. Touch all the ICs, checking for one that seems to be running hotter than the others. (Be careful; it is easy to burn your finger.) If the visual inspection and touch tests do not reveal the problem, it is time to resort to the use of test equipment.

A shorted bypass capacitor or internally shorted IC will measure from 5 to 15 Ω, whereas a solder bridge or unetched trace short will measure a dead short of 0 Ω. Measure the resistance between the power and ground bus; this will further narrow the problem down to a component or PCB fault.

The greatest problem in finding V_{cc}-to-ground shorts is the difficulty of isolating the components. As we stated previously, a moderate-size TTL PCB will have more than 40 bypass caps, each one in parallel. If you are using a typical digital multimeter (DMM), the most sensitive resistance scale is probably 200 Ω, full scale. This is not sensitive enough to track down the short through resistance readings. You must start cutting traces and lifting components. Power buses are wide traces. They are difficult to cut and correctly resolder. The most efficient way of approaching this problem, as we have previously done, is to cut the PCB in halves until you find the short. Remember that the ICs and bypass caps will be organized in rows; you should continue to isolate the short down to the bad row. At this point, you can start to desolder one leg of capacitors or isolate the V_{cc} pin on ICs until you find the bad component. Once again, this is a difficult and time-consuming process. Only effort and patience will help you locate the cause of the short.

If you are fortunate enough to have a high-resolution ohmmeter, the task of finding the short will be much faster and easier. A typical high-resolution ohmmeter will have a full-scale reading of 2 Ω. The high-sensitivity scale is typically labeled "Low Ohms." There are high-resolution ohmmeters that measure resistances as small as 10 mΩ. Use the high-resolution ohmmeter with a set of needle-pointed probes. Move the probes between the power and ground buses until you find the least resistance; this will be the point of the short!

6.6 SPECIALIZED DIGITAL TEST EQUIPMENT

Wherever you may work, small company or large, there will always be an oscilloscope and a DMM. Most college electronics labs will rely solely on these two pieces of test equipment. One of your highest priorities should be to become an expert in the use of the scope and DMM.

Because of digital's two-state nature, there exist many types of specialized digital test equipment. Instead of a CRT or a digital display, these specialized test instruments will use an LED for an indicator. The six most common digital test instruments are:

1. The logic analyzer
2. The logic probe
3. The logic pulser
4. The logic clip
5. The logic comparator
6. The digital current tracer

Except for the logic analyzer, these instruments are easily mastered and can greatly speed up the process of troubleshooting digital circuits. Their use is not as widespread as that of the oscilloscope or DMM, but they are increasing in popularity at a fast rate. The logic analyzer is the most complex and powerful digital test instrument. The introduction to logic analyzers is usually reserved for your first class in microprocessors.

6.6.1 The Logic Probe

Logic probes use an LED to indicate the logic state of the node that they are monitoring. Like a good DMM probe, the logic probe ends in a needle-sharp point. The logic probe derives its power from V_{cc} and ground of the circuit under test. The LED will glow brightly to indicate a logic 1. The LED will be unlit to indicate a logic 0. This would seem to cover the two possible conditions of a logic circuit—but we know better! An important question is: How will the logic probe react to an indeterminate logic level? It will glow dimly. Thus we can distinguish between a logic 1, logic 0, and an indeterminate level. As we stated previously, most gate failures are stuck levels: logic 1 or logic 0. The LED will blink to indicate a node that is active.

Most logic probes have the ability to detect TTL or CMOS thresholds by the setting of a mode switch. They can also catch fast, single pulses with a memory mechanism. The advantage of logic probes is the speed at which they can be used. When you are using a scope or DMM you must look up and take your eyes off the circuit to see the reading or display; with a logic probe the display (an LED) is at the circuit board level and easily seen without removing your eyes from the PCB. Because of this feature, you can scan the PCB much faster, looking for stuck nodes or indeterminate levels. A logic probe is not the appropriate test instrument for all digital circuits, but for many applications its ease of use and speed is hard to beat.

6.6.2 The Logic Pulser

The logic pulser is the ideal complement to a logic probe. The logic pulser generates high or low logic levels or trains of pulses. In this manner you can inject a signal into a gate and observe the output of the gate with a logic probe.

The greatest convenience in using the logic pulser is that no outputs need be isolated to inject the logic pulser signal. The logic pulsers can sink or source more than 500 mA of current. When a logic pulser is touched to a node, this great current sinking/sourcing ability enables it to override any other outputs that may be connected to the node. If you want to exercise a particular node, set the pulser for a high or low logic level, or a train of pulses, and use the needle point to touch an IC pin or pierce a trace. You then monitor these pulses, or level, downstream with a logic probe.

Like a good logic probe, a good logic pulser can also support both TTL and CMOS. They also have many different output options: steady logic level, single pulse, 10 pulses, 100 pulses, or a continuous stream of pulses. Logic pulsers and logic probes make a powerful digital troubleshooting team.

6.6.3 The Logic Clip

There will be many instances when you want to monitor many pins of the same IC simultaneously. A dual-trace scope can display only two outputs simultaneously. Consider the power of monitoring all three inputs and the output of a 74LS10 triple three-input NAND gate simultaneously. Knowing the state of each input, you could easily check the output to assure that the device is functioning correctly. That is the function of the logic clip. The logic clip snaps onto the top of IC dip packages and monitors the activity via 16 LEDs. If the device you are testing has only 14 pins, you will ignore the last two LEDs on the logic clip.

Logic clips contain their own buffers to avoid loading down the device under test. They are usually intelligent enough to find V_{cc} and ground in devices with nonstandard power pins. Like logic probes and logic pulsers, they can usually support both TTL and CMOS logic families.

6.6.4 The Logic Comparator

The logic comparator monitors the inputs and outputs of two separate ICs. The first IC will be the device that you are testing on the PCB; the second IC will be a known-good reference IC and will reside in the logic comparator. A clip will be snapped onto the IC under test. The inputs and outputs of both devices will be compared. Any differences between these two ICs will cause an LED to illuminate, indicating the pin where the mismatch is occurring.

Logic comparators are supplied with reference ICs that are soldered onto a small board that clips into the comparator. Each time you want to test a different type of IC the appropriate logic board containing the known-good IC must be inserted into the logic comparator.

6.6.5 The Digital Current Tracer

A frustrating and time-consuming problem in digital troubleshooting is shorts: internal shorts in ICs, solder bridges, and unetched traces. We have discussed the methods of isolating outputs and inputs until the short is located. A digital current tracer can be an extremely useful device for finding shorts of all types.

Refer to Figure 6.3. We discussed all the possible ways to isolate the outputs and input of this circuit to discover where the short may be. Wherever the short is, it must be sinking and sourcing current from the output of G1. If we had a way of tracing the path of the current, it would lead us directly to the cause of the short. This is the function of the digital current tracer.

The current tracer senses the magnetic field changes that accompany an ac signal. The LED at the tip of the current probe will glow proportionally to the amount of current it senses. To find the short, apply a signal to the input of pins of G1 to cause the output to toggle between a logic 0 and logic 1. Because there is a malfunction in the node, the voltage level will not toggle, but the current will toggle between the output of G1 and the source of the short. It is this ac current that the probe will follow. Placing the probe at the output of G1 should cause its LED to glow brightly. Slowly follow the trace as it leaves pin 3 of G1. If the LED starts to dim, you have gone past the short and must double back. The digital current tracer will work with any logic family because it senses current, not voltage. The digital current probe can greatly improve a board test technician's troubleshooting efficiency.

QUESTIONS AND PROBLEMS

6.1. In what situations do you expect to encounter wire-wrapped circuits? What are the advantages of wire-wrapped circuits? What are their disadvantages?

6.2. What are the advantages of PCBs? Disadvantages of PCBs?

6.3. What problems can occur when a PCB is being stuffed? What is the easiest way to spot these problems?

6.4. What problems can occur when a PCB is being wave soldered?

6.5. What problems can occur when a PCB is being photoetched?

6.6. Name the situations where you expect to encounter ICs in sockets. What are the advantages of sockets? Disadvantages?

6.7. Why is it useful to have an unpopulated PCB on hand?

6.8. Why is it useful to have a known good PCB on hand?

6.9. What is the mechanism that allows PCBs in a multiboard system to communicate with each other?

6.10. How does one distinguish a power or ground trace from other traces on a PCB?

6.11. What does a square pad designate?

6.12. What is the function of a feedthrough?

6.13. What characteristics indicate a cold solder joint? Dirty fingers?

6.14. What places on a PCB are most likely to have solder bridges? Unetched traces?

6.15. What is the easiest method to assure that all the proper components are stuffed and in the correct orientation?

6.16. How can one recognize when a lead on an IC is bent underneath the IC?

6.17. What is the function of bypass capacitors? Why don't CMOS PCBs require them?

6.18. How do you know if you have discovered an artwork error? What action should you take?

6.19. How can one isolate a particular pin on an IC? In what situations would this be an appropriate action?

6.20. Why does a circuit have to be deenergized to use an ohmmeter?

6.21. A TTL input measures 1.5 V. What is the problem?

6.22. An oscilloscope displays noise spikes on the input of a CMOS device. What is the possible problem?

6.23. Why must all CMOS inputs be terminated?

6.24. What logic level does a floating TTL input appear to be? Why?

6.25. What logic level does a floating CMOS input appear to be? Why?

6.26. What test instrument is used to find opens? Shorts?

6.27. Explain what is meant by the phrase: "Continually cut the circuit in half until the fault is located."

6.28. When the input to a TTL gate is required to be a steady logic 1, why is it pulled up to V_{cc} via a resistor instead of just left floating?

6.29. Which is more difficult to troubleshoot: shorts or opens? Why?

6.30. Will a TTL gate whose output is shorted to V_{cc} be damaged? Why?

6.31. Will a TTL gate whose output is shorted to ground be damaged? Why?

6.32. What does a third "logic level" between 0.8 and 1.2 V indicate in TTL circuitry indicate?

6.33. If a circuit fails when you first power-up, what should be your first check?

6.34. What is the characteristic of switching power supplies that is important to remember?

6.35. When can heat be used to help troubleshoot a circuit?

6.36. Define a high-resolution ohmmeter.

6.37. What are the advantages of using a logic probe and logic pulser to troubleshoot a circuit?

6.38. Why can't a digital current tracer be used to troubleshoot a V_{cc}-to-ground short?

6.39. What is the greatest limitation of the logic comparator?

6.40. In what situations is a logic clip useful?

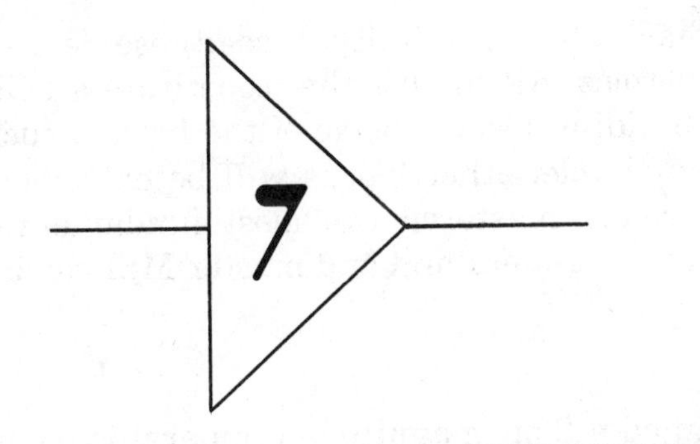

SURVEY OF MSI DEVICES

Any digital device, from a basic logic circuit to the most sophisticated computer, could be designed and constructed solely from simple gates. There are many faults to this approach: the huge number of ICs required would consume extensive amounts of power and PCB space; the process of circuit and PCB design would be long and tedious; debugging and troubleshooting difficulty increases proportionally with the number of gates in the circuit; and the expense of using simple gates would be cost prohibitive. For these reasons, and many others, SSI and MSI devices play only a supportive role in the design of new circuits. LSI and VLSI devices dominate new designs. Where 50 to 100 SSI and MSI ICs were once used to implement a circuit, one LSI device and a handful of SSI/MSI devices will now perform the same function.

MSI devices are used to support microprocessor, memory, input/output circuitry, and many other advanced digital functions. At this point in your digital education it is difficult to illustrate these MSI devices in meaningful circuits. Consider them as the nuts and bolts that hold together advanced LSI devices. We will closely examine their pinouts and function tables, and provide general references to their actual roles in advanced circuits. The last few chapters will bring all the loose ends together, and you will have the opportunity to see these devices in their actual support roles. All the ICs that constitute a PCB can be considered as building blocks. Some of the blocks, such as the basic gates, are simple; other blocks will be extremely complex. You have already mastered the most fundamental blocks. Your next task is to understand and master MSI combinational devices.

7.1 THE MAGNITUDE COMPARATOR

In Chapter 4 we designed a 2-bit magnitude comparator using AND and OR gates. The circuit required a total of 22 gates. We have already discussed the limitations of such a circuit. For practical applications a magnitude comparator must have a width of at least 4 bits. The IC should provide a means of connecting many comparators together so any number of bits can be compared.

7.1.1 The 74LS85 4-Bit Magnitude Comparator

Specifications sheets for MSI devices have three major parts that technicians should be interested in: pinout, function table, and word description. They also include an equivalent gate circuit, detailed specifications, and test information. Your major focus should be on the pinout, function table, and word description. Consider Figure 7.1. The circuit symbol in the pinout is usually the same symbol that is used to represent the device in schematics. The word descriptions provided by the manufacturers are highly technical in nature; they are not intended as tutorials. It is expected that as your technical expertise increases, you will learn to read the compact and technical word descriptions provided with the data sheets. On the job, these word descriptions will often be your sole source of information.

The Pinout The pins on the 74LS85 fall into five groups:

1. V_{cc} and ground
2. Word A
3. Word B
4. Outputs
5. Cascade inputs

V_{cc} and ground All TTL devices will have pins of V_{cc} and ground. As we already know, these pins will most often be located on the corners of the IC. The 74LS85 is a 16-pin DIP and, as you would expect, pin 8 is ground and pin 16 is V_{cc}.

Word A and word B The 74LS85 compares two 4-bit binary quantities. Word A is the first group of 4 bits, while word B is the second group. Here the term "word" is used in its most informal sense; it designates nothing more than a group of binary inputs. The binary value of word A will be compared with the binary value of word B.

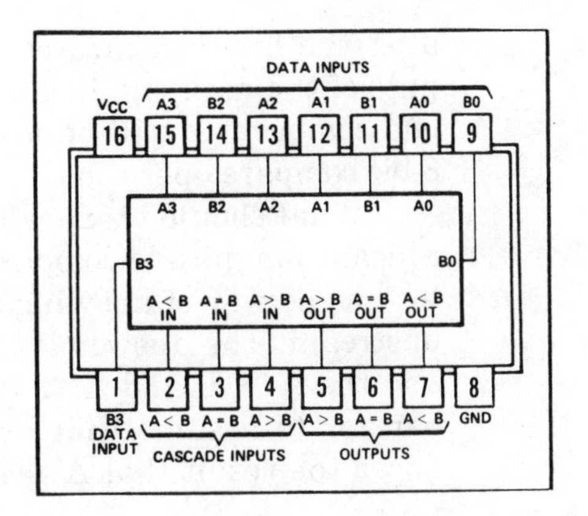

FUNCTION TABLES

COMPARING INPUTS				CASCADING INPUTS			OUTPUTS		
A3, B3	A2, B2	A1, B1	A0, B0	A > B	A < B	A = B	A > B	A < B	A = B
A3 > B3	X	X	X	X	X	X	H	L	L
A3 < B3	X	X	X	X	X	X	L	H	L
A3 = B3	A2 > B2	X	X	X	X	X	H	L	L
A3 = B3	A2 < B2	X	X	X	X	X	L	H	L
A3 = B2	A2 = B2	A1 > B1	X	X	X	X	H	L	L
A3 = B3	A2 = B2	A1 < B1	X	X	X	X	L	H	L
A3 = B3	A2 = B2	A1 = B1	A0 > B0	X	X	X	H	L	L
A3 = B3	A2 = B2	A1 = B1	A0 < B0	X	X	X	L	H	L
A3 = B3	A2 = B2	A1 = B1	A0 = B0	H	L	L	H	L	L
A3 = B3	A2 = B2	A1 = B1	A0 = B0	L	H	L	L	H	L
A3 = B3	A2 = B2	A1 = B1	A0 = B0	L	L	H	L	L	H

'85, 'LS85, 'S85

A3, B3	A2, B2	A1, B1	A0, B0	A > B	A < B	A = B	A > B	A < B	A = B
A3 = B3	A2 = B2	A1 = B1	A0 = B0	X	X	H	L	L	H
A3 = B3	A2 = B2	A1 = B1	A0 = B0	H	H	L	L	L	L
A3 = B3	A2 = B2	A1 = B1	A0 = B0	L	L	L	H	H	L

Figure 7.1 Pinout and function table for the 74LS85.

Outputs The 2-bit magnitude comparator that we designed in Chapter 4 had three outputs: A = B, A > B, and A < B. The 74LS85 also has these same outputs. Consider the comparison as if we are subtracting the value of word B from word A: if the result is equal to 0, A = B; if the result is a positive number, A > B; if the result is a negative number, A < B. Notice that one, and only one, output can ever be true for any given combination of word A and word B.

Cascade inputs You may be familiar with the definition of the term "cascade" as it applies to amplifiers. When amplifiers are cascaded, the output of one amplifier becomes the input of the next amplifier—like a series of waterfalls, where the pool under one waterfall provides the water for the next waterfall.

An important characteristic of an MSI comparator is that it must have built-in provisions for expansion. That is the function of the cascade inputs. The cascade inputs are: A = B, A > B, and A < B. Your first reaction to this information should be to state that the cascade inputs are labeled the same as the outputs. That is the clue that explains how these devices are connected when they are cascaded. If we wish to increase the word length that we can compare from 4 bits to 8 bits, two 74LS85s will be cascaded. The outputs from the least significant comparator (the first comparator) will be connected to the cascade inputs of the most significant comparator. The outputs of the second comparator will be used to indicate the final result of the 8-bit compare operation.

What should we do with the cascade inputs of the least significant magnitude comparator? The least significant comparator has no way of knowing that there are no other comparators upstream. The answer is simple: we tie the A = B input high and the A > B and A < B inputs low. We are telling the comparator to assume that any upstream comparisons have produced the result that A = B.

The Function Table Assume that we are comparing the relative magnitudes of the decimal numbers 945 and 768. What is the most efficient way to carry out the comparison? We should start our examination with the most significant digit. Obviously, 9 is greater than 7; our comparison is complete. There is no need to continue and examine the next digit. The only time we would do that is when the most significant digits are equal. With this approach in mind, examine the function table for the 74LS85 in Figure 7.1.

It is important to remember that an "×" in the function table denotes a don't-care condition. In our comparison of the numbers 945 and 768, the numbers in the 10's and 1's places are don't-care conditions. The only time the 10's place would be significant is when the numbers in the 100's place were equal. In the same vein, the only time the 1's place would not be a don't-care condition is when both the numbers and the 100's and 10's places were equal. That is the way the first eight lines in the function table are derived.

In the last three lines of the function table the cascade inputs become significant. If the 4 bits of word A and word B are equal, the comparator must look upstream to the next less significant comparator for information. Inspect the function table line by line in great detail. Assure yourself that you thoroughly understand the operation of this device.

7.1.2 The 4063B 4-Bit Magnitude Comparator

This is the CMOS version of the 74LS85. It has the same inputs and outputs as the 74LS85. Refer to Figure 7.2. This particular data sheet refers to the pinout as the terminal assignment. It also uses the term "truth table" rather than "function table." Manufacturers all have their own particular ways of making data sheets. You must learn to understand the information, regardless of what name it is called.

This data sheet also includes a functional diagram. This block diagram is an excellent overview of the device. The inputs are shown on the left-hand side and the outputs are shown on the right-hand side. Notice that the pins are all grouped by func-

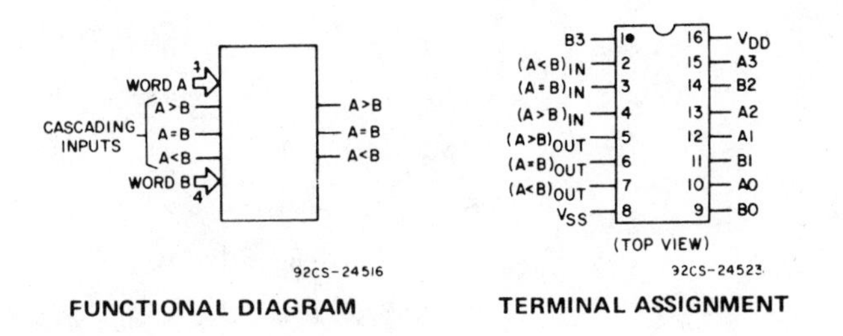

TRUTH TABLE

INPUTS							OUTPUTS		
COMPARING				CASCADING					
A3, B3	A2, B2	A1, B1	A0, B0	A < B	A = B	A > B	A < B	A = B	A > B
A3 > B3	X	X	X	X	X	X	0	0	1
A3 = B3	A2 > B2	X	X	X	X	X	0	0	1
A3 = B3	A2 = B2	A1 > B1	X	X	X	X	0	0	1
A3 = B3	A2 = B2	A1 = B1	A0 > B0	X	X	X	0	0	1
A3 = B3	A2 = B2	A1 = B1	A0 = B0	0	0	1	0	0	1
A3 = B3	A2 = B2	A1 = B1	A0 = B0	0	1	0	0	1	0
A3 = B3	A2 = B2	A1 = B1	A0 = B0	1	0	0	1	0	0
A3 = B3	A2 = B2	A1 = B1	A0 < B0	X	X	X	1	0	0
A3 = B3	A2 = B2	A1 < B1	X	X	X	X	1	0	0
A3 = B3	A2 < B2	X	X	X	X	X	1	0	0
A3 < B3	X	X	X	X	X	X	1	0	0

X = Don't Care   Logic 1 ≡ High Level   Logic 0 ≡ Low Level

Figure 7.2 4063B 4-bit magnitude comparator.

tion. Functional diagrams convey a great deal of information; study them closely whenever they are available.

Although they follow much different input patterns, the truth table in Figure 7.2 contains the exact information as the function table in Figure 7.1. Take a moment to compare and contrast these two tables. Assure yourself that they do indeed describe identical functions.

7.1.3 Applications of Magnitude Comparators

Magnitude comparators are widely employed as address decoding devices on memory boards. Chapter 13 will contain an application using a 74LS85 in a memory decoding circuit. Magnitude comparators are also useful devices in servomotor control circuits. By comparing a known reference quantity to a binary number that indicates the actual position of a servomotor, the outputs of the comparator can be used to drive the servomotor forward, backward, or to a halt.

A much less esoteric application employs the 4063B as the major component of an electronic lock. Consider a battery powered, electronic lock. Instead of having rows of numbers to set, as in a conventional mechanical lock, the input devices will be eight SPST switches. When the proper combination is entered

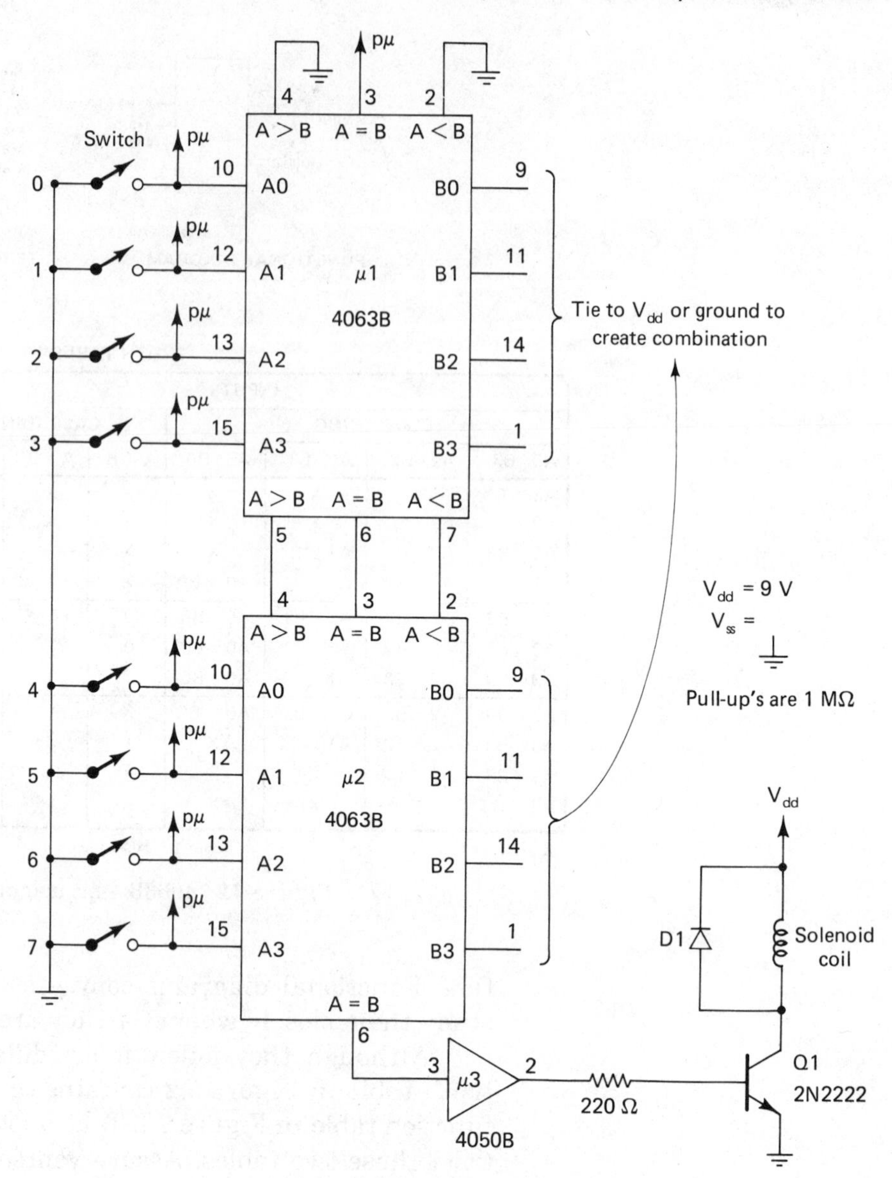

Figure 7.3 Combinational electronic lock.

into the switches, a solenoid bolt will retract and the lock will open. Consider Figure 7.3. This circuit employs two 4063Bs in cascade. The 8-bit combination will be entered via switches 0 through 7. Notice the new symbol that is used to identify a pull-up resistor, an arrow with the abbreviation "pu." This symbol will be used throughout this text. If every pull-up resistor was indicated with the normal resistor symbol, the schematic would get so congested that its readability would be impeded. The inputs A0 through A3 on U1 will be the least significant 4 bits, and the inputs A0 through A3 on U2 will be the most significant 4 bits.

The combination that opens this lock is internally set by pulling up or grounding the B inputs on U1 and U2. When all eight A inputs match all eight internally set B inputs, the A = B output on U2 will go to a logic 1. U3 is a 4050B noninverting buffer; its function is to provide enough current gain to drive the base of Q1. When the output of U3 goes high, Q1 will conduct and be driven into saturation. This will enable current to flow through the solenoid coil, retracting the bolt on the lock. D1 protects the solenoid coil and transistor from the inductive kickback voltage created when Q1 turns off. This is a simple application of magnitude comparators that illustrates the use of the cascading inputs. We could expand the input combination to any number of digits by adding additional comparators.

7.2 THE BCD-TO-DECIMAL DECODER

A decoder converts an encoded input into an intelligible output. We introduced the BCD code in Chapter 1. The BCD-to-decimal decoder inputs a BCD number and then activates one of its 10 output lines that relates to the decimal equivalent of the BCD input. The BCD decoder is also known as a 4-line-to-10-line decoder.

7.2.1 The 74LS42

Examine the pinout in Figure 7.4. This device has four inputs to accommodate a BCD number. Input A is the least significant bit

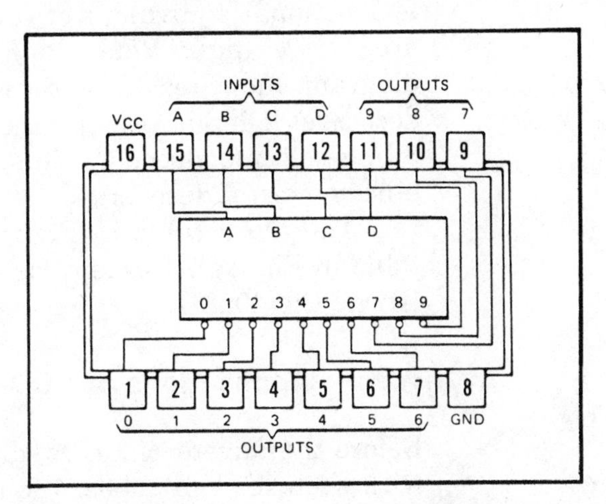

NO.	'42A, 'L42, 'LS42 BCD INPUT				ALL TYPES DECIMAL OUTPUT									
	D	C	B	A	0	1	2	3	4	5	6	7	8	9
0	L	L	L	L	L	H	H	H	H	H	H	H	H	H
1	L	L	L	H	H	L	H	H	H	H	H	H	H	H
2	L	L	H	L	H	H	L	H	H	H	H	H	H	H
3	L	L	H	H	H	H	H	L	H	H	H	H	H	H
4	L	H	L	L	H	H	H	H	L	H	H	H	H	H
5	L	H	L	H	H	H	H	H	H	L	H	H	H	H
6	L	H	H	L	H	H	H	H	H	H	L	H	H	H
7	L	H	H	H	H	H	H	H	H	H	H	L	H	H
8	H	L	L	L	H	H	H	H	H	H	H	H	L	H
9	H	L	L	H	H	H	H	H	H	H	H	H	H	L
INVALID	H	L	H	L	H	H	H	H	H	H	H	H	H	H
	H	L	H	H	H	H	H	H	H	H	H	H	H	H
	H	H	L	L	H	H	H	H	H	H	H	H	H	H
	H	H	L	H	H	H	H	H	H	H	H	H	H	H
	H	H	H	L	H	H	H	H	H	H	H	H	H	H
	H	H	H	H	H	H	H	H	H	H	H	H	H	H

Figure 7.4 74LS42 BCD-to-decimal decoder.

and input D is the most significant bit. There are ten outputs, one for each possible decimal number 0 through 9. Notice that the output lines are bubbled. You know that a bubble indicates an inversion. When an output line is bubbled, that output is said to be an "active-low" output. We normally think of an active output as indicating a logic 1 level. Many TTL devices have active-low outputs. The reason for this is simple: TTL is much better at sinking current than it is at sourcing current. If the active output level is a logic 0, the TTL circuitry can provide its best current capability.

Refer to the function table. Notice that the output which is the decimal equivalent of the BCD input will go to its active-low level, while the other outputs will be at their inactive-high level. We know that the BCD codes from 1010 to 1111 are invalid. For these six invalid combinations, all the outputs will retain their inactive-high levels.

7.2.2 The 4028B

CMOS sinks and sources current equally well; there is no advantage for CMOS devices to be active-low outputs. The 4028B is functional equivalent of the 74LS42 except that its outputs are active high. Refer to Figure 7.5. The functional block diagram indicates an interesting fact about 4-line-to-10-line decoders. If input D is grounded, this device will function as a 3-line-to-8-line decoder. This type of decoder is known as a binary-to-octal decoder.

The truth table in Figure 7.5 is identical to the function table in Figure 7.4 except for the active-high output characteristic of the 4028B.

7.2.3 Applications of BCD-to-Decimal Decoders

Before the advent of LEDs, digital displays were implemented with neon readout tubes. Ten neon tubes would be contained in one enclosure. Each tube was in the shape of a particular decimal digit. The 10 outputs of a BCD-to-decimal decoder were used to drive the neon tubes. Only one digit per package would ever be illuminated simultaneously. In this way digital displays were created. These tubes required high voltages and moderate amounts of power. These fluorescent displays have been replaced with LED and LCD indicators.

Like magnitude comparators, BCD-to-decimal decoders are also used in address decoding. Consider the situation where we wish to control ten separate dc motors. One and only one motor will ever be energized simultaneously. All we have to do is place the BCD code of the motor that we wish to energize on the inputs of a BCD-to-decimal decoder. By changing the BCD code, the motor that is energized will change. If we desire to turn off all the motors, we must place an illegal BCD code on the inputs of the decoder.

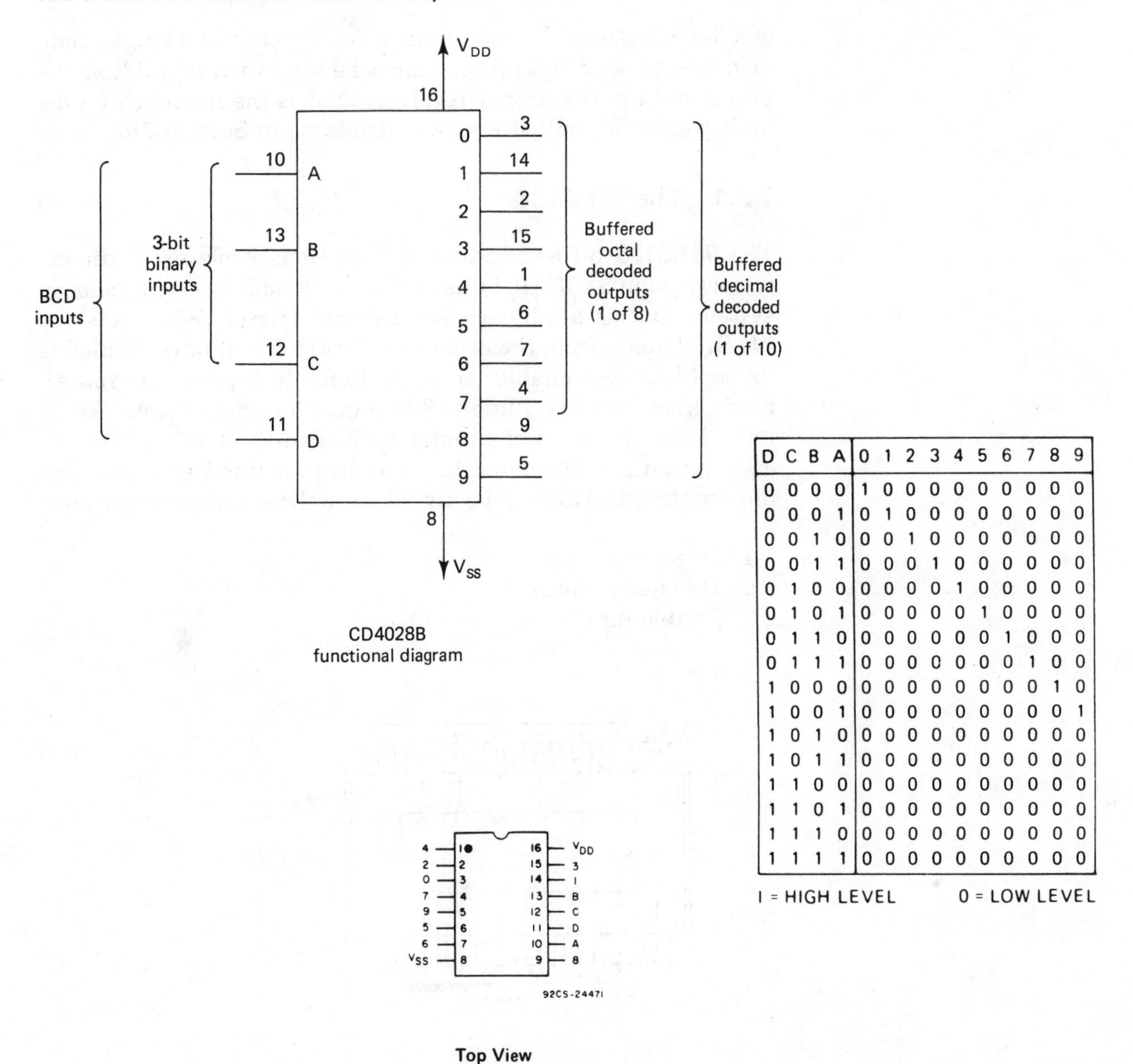

D	C	B	A	0	1	2	3	4	5	6	7	8	9
0	0	0	0	1	0	0	0	0	0	0	0	0	0
0	0	0	1	0	1	0	0	0	0	0	0	0	0
0	0	1	0	0	0	1	0	0	0	0	0	0	0
0	0	1	1	0	0	0	1	0	0	0	0	0	0
0	1	0	0	0	0	0	0	1	0	0	0	0	0
0	1	0	1	0	0	0	0	0	1	0	0	0	0
0	1	1	0	0	0	0	0	0	0	1	0	0	0
0	1	1	1	0	0	0	0	0	0	0	1	0	0
1	0	0	0	0	0	0	0	0	0	0	0	1	0
1	0	0	1	0	0	0	0	0	0	0	0	0	1
1	0	1	0	0	0	0	0	0	0	0	0	0	0
1	0	1	1	0	0	0	0	0	0	0	0	0	0
1	1	0	0	0	0	0	0	0	0	0	0	0	0
1	1	0	1	0	0	0	0	0	0	0	0	0	0
1	1	1	0	0	0	0	0	0	0	0	0	0	0
1	1	1	1	0	0	0	0	0	0	0	0	0	0

1 = HIGH LEVEL 0 = LOW LEVEL

Figure 7.5 4028B BCD-to-decimal decoder.

7.3 THE 3-LINE-TO-8-LINE DECODER/DEMULTIPLEXER

We have established that a BCD-to-decimal decoder can be used as a 3-line-to-8-line decoder by tying the most significant select bit to ground. With the BCD-to-decimal decoder we could drive all the outputs into their inactive state by applying an illegal BCD code to the select inputs. This would effectively "disable" the decoder. A 3-line-to-8-line decoder does not have any illegal select codes. Instead of using an illegal code to disable the outputs, 3-line-to-8-line decoders employ an enable input. When this enable input is active, the decoder will perform its normal function. When this enable input is inactive, the decoder will be disabled and all the outputs will go to their inactive level, regardless of the select code.

This enable input on decoders can also be used as a data input. In that case the decoder would be performing the function of a demultiplexer. If many data transmitters use a single, common line to send data, there must be a method of guiding the proper data to the proper receiver. That is the function of a demultiplexer. We will study demultiplexers in Section 7.9.

7.3.1 The 74LS138

The 74LS138 3-line-to-8-line decoder/demultiplexer is an extremely popular IC. It is used to decode addresses in memory systems and as a 1-line-to-8-line demultiplexer. What sets the 74LS138 apart from the other decoders that you have studied is its sophisticated enable circuitry. Refer to Figure 7.6. You already know what a 3-line-to-8-line decoder should look like. It must have three select inputs: A, B, and C. It will also have eight outputs, 0 through 7. Refer to the pinout in Figure 6.6. The pins on the 74LS138 can be divided into three functional groups:

1. Select inputs
2. Decoded outputs
3. Enable inputs

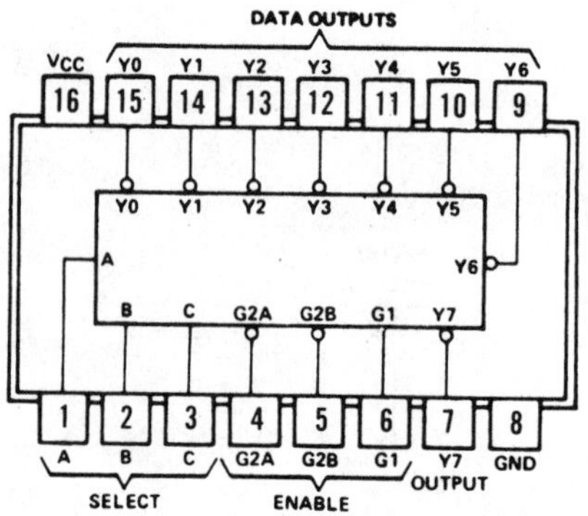

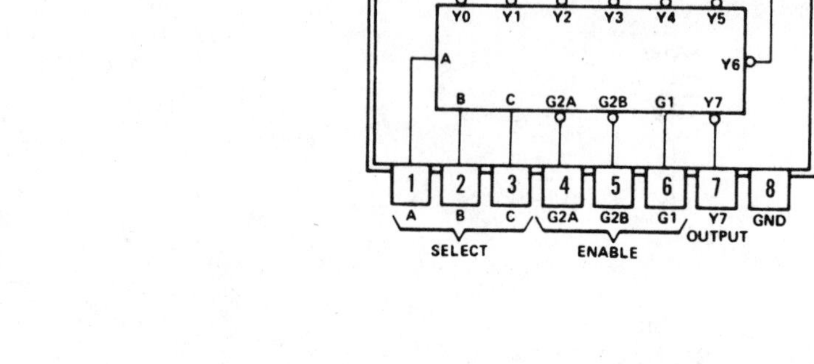

'LS138, 'S138
FUNCTION TABLE

INPUTS					OUTPUTS							
ENABLE		SELECT										
G1	G2*	C	B	A	Y0	Y1	Y2	Y3	Y4	Y5	Y6	Y7
X	H	X	X	X	H	H	H	H	H	H	H	H
L	X	X	X	X	H	H	H	H	H	H	H	H
H	L	L	L	L	L	H	H	H	H	H	H	H
H	L	L	L	H	H	L	H	H	H	H	H	H
H	L	L	H	L	H	H	L	H	H	H	H	H
H	L	L	H	H	H	H	H	L	H	H	H	H
H	L	H	L	L	H	H	H	H	L	H	H	H
H	L	H	L	H	H	H	H	H	H	L	H	H
H	L	H	H	L	H	H	H	H	H	H	L	H
H	L	H	H	H	H	H	H	H	H	H	H	L

*G2 = G2A + G2B
H = high level, L = low level, X = irrelevant

Figure 7.6 74LS138 decoder/demultiplexer.

The first two groups, the select inputs and decoded outputs, should appear exactly as you expected. Notice that the decoded outputs are bubbled; this indicates that they are active low. (You should expect TTL devices to have active-low outputs.) Notice that the decoded outputs are designated by the letters Y0 through Y7. Also the decoded outputs are called *data outputs* in the pinout diagram. When this device is being employed as a decoder, the outputs will be called *decoded outputs*. When this device is employed as a demultiplexer, the outputs will be called *data outputs*. When we refer to a device, it will always be analyzed in terms of its present application. Here we are examining this device as an isolated IC, separate from any application. Because the 74LS138 is most commonly used as a decoder, it is natural to refer to it as a decoder when it is not involved in a specific application.

The only unfamiliar pins on the 74LS138 should be the enables: G1, G2A, and G2B. The logic diagram indicates that G1 is active high, whereas G2A and G2B are active low. These enable inputs are effectively ANDed together to form an internal enable signal. For the 74LS138 to be enabled all the enable inputs must be in their active states:

If G1 is high AND G2A is low AND G2B is low, the decoder will be enabled.

If any of the enable inputs are not at their active levels, all the outputs, Y0 through Y7, will go to their inactive-high levels, regardless of the value of the select inputs. The logic symbol in the pinout is most often the circuit symbol that will be used to represent the 74LS138 in schematics. It is easy to forget how the enable inputs operate. Because of this, the symbol shown in Figure 7.7 is gaining wide usage as the preferred schematic representation of the 74LS138. The AND gate in Figure 7.7 is symbolic; it is not a physical device. It helps remind the technician reading the schematic that the enable inputs are ANDed together to derive an internal enable. For the same reasons that we sometimes use the negative-logic representations of simple gates, we will also try to use the most functional schematic symbols for MSI devices.

Refer to the function table in Figure 7.6. The first two lines indicate situations where an enable input is not at its active level. When this happens the select inputs have no effect, so they are designated don't cares and the outputs will all be driven to the inactive-high level. Notice that the second column is labeled G2. To save space in the function table, G2 is the ORed value of G2A and G2B. This says: If G2A is high OR G2B is high, the decoder will be disabled. The last eight lines in the function table are the normal decoded outputs for an enabled, active-low-output 3-line-to-8-line decoder.

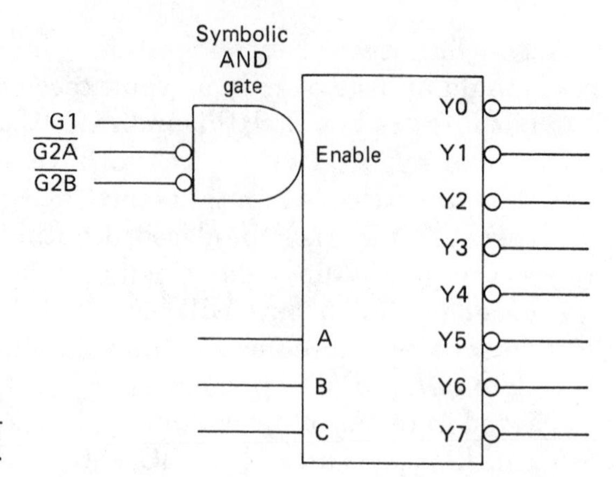

Figure 7.7 Functional schematic symbol for the 74LS138.

7.4 THE LED SEVEN-SEGMENT DISPLAY

7.4.1 The Seven-Segment Display

Seven-segment displays are the simplest method of displaying the decimal digits 0 through 9. Each of the seven segments is constructed from a bar-shaped LED. When a particular digit is to be displayed, the proper segments on the seven-segment display must be illuminated. Consider Figure 7.8. The LEDs are labeled a through g. By illuminating the proper combination of LEDs we can display any decimal digit. There are two different types of seven-segment displays: common anode and common cathode. A *common-anode* display has all the anodes of the LEDs tied together and connected to V_{cc}. To illuminate any particular segment, a logic 0 must be applied to its cathode. Common-cathode seven-segment displays have all the cathodes tied together and connected to ground. To illuminate any particular segment, a logic 1 must be applied to the anode.

7.4.2 The 74LS47 BCD-to-Seven-Segment Decoder/Driver

The seven-segment display must be driven by a device that has the ability to receive a digital character in BCD form and illuminate the proper segments to create the numeral. That is the function of a BCD-to-seven-segment decoder/driver. Refer to Figure 7.9. You already know that the pinout should indicate the provision for the BCD input on pins A through D. Furthermore, there must be seven outputs to drive the display, a through g. Notice that these outputs are active low. This indicates that the 74LS47 is designed to drive common-anode seven-segment displays. The outputs are open collector. Each segment output line will be pulled up to V_{cc} via a current-limiting resistor and the associated LED segment.

There are three pins on the 74LS47 that you are not yet familiar with: lamp test (LT), blanking in/ripple blanking out (BI/RBO), and ripple blanking in (RBI). They are all active-low signals. The best way to understand the function of these pins

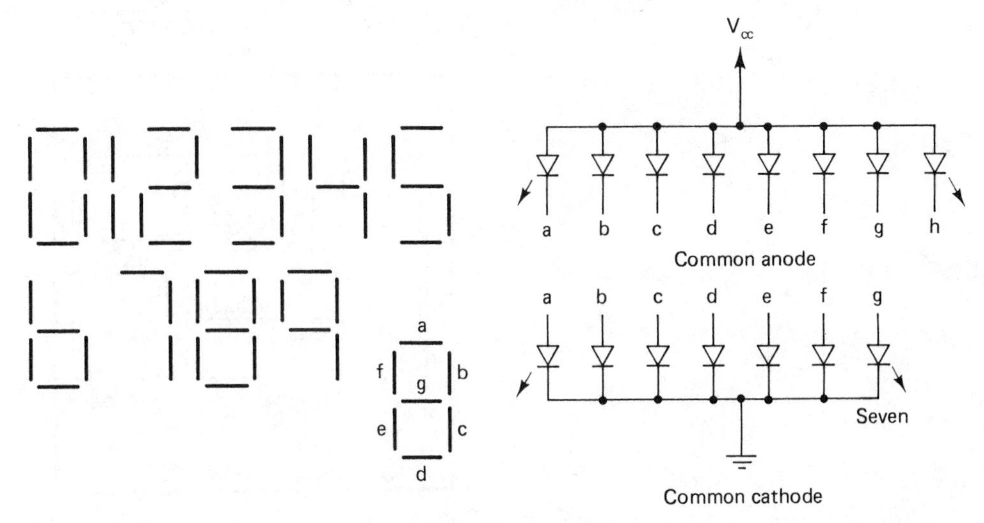

Figure 7.8 Seven-segment display.

is by examining the function table. Refer to the first line in the function table in Figure 7.9. The column labeled "Decimal or Function" indicates the decimal equivalent of the BCD input or a special function the 74LS47 is capable of performing. These special functions relate to those three undefined pins. The first 16 lines in the function table will concern normal operation for the BCD codes of 0000 through 1111. You may be surprised that the 74LS47 will react to the illegal BCD codes of 1010 through 1111. Figure 7.9 depicts the resultant display for all the BCD inputs. Notice that the codes 1010 through 1110 create nonstandard characters, while the input of 1111 will cause the seven-segment display to go blank. Most applications will not use these nonstandard characters, but they are available as an option.

Under inputs we see the values of LT, RBI, and the BCD inputs. Remember that LT and RBI are active-low inputs. BI/RBO is an open-collector pin that serves as both an input and an output. We will discuss the function of this pin when it goes active, in the seventeenth line of the truth table.

The output columns indicate, in active-low terms, which segments will be illuminated. The four notes in the last column will be explained as they are needed. For the first 16 lines of the truth table, the LT, RBI, and BI/RBO pins will all be high, in their nonactive state.

Lines 1–16 Line 1 indicates that the BCD code for the digit 0 is input to the 74LS47. Every segment is illuminated, except for g. This creates the numeral 0 on the display. Carefully examine the next 15 lines in the truth table, and associate each BCD code with the resultant display.

Line 17 In this line the BI/RBO pin goes to its active-low level. Whenever this happens all other inputs will be in don't-care states and all the segments will be extinguished! To blank a display simply means to turn it off; that is, make it go blank.

'46A, '47A, 'L46, 'L47, 'LS47
(TOP VIEW)

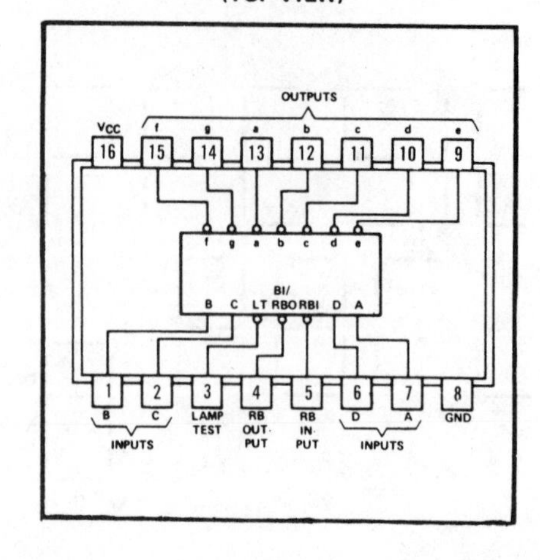

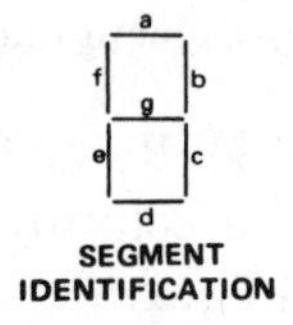

SEGMENT IDENTIFICATION

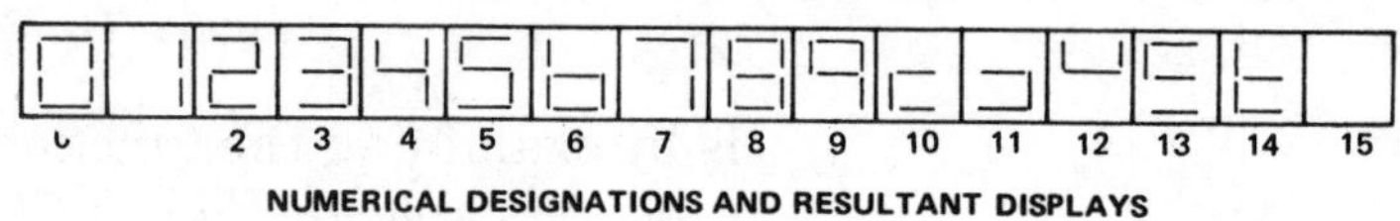

2 3 4 5 6 7 8 9 10 11 12 13 14 15

NUMERICAL DESIGNATIONS AND RESULTANT DISPLAYS

'46A, '47A, 'L46, 'L47, 'LS47 FUNCTION TABLE

DECIMAL OR FUNCTION	INPUTS						BI/RBO†	OUTPUTS							NOTE
	LT	RBI	D	C	B	A		a	b	c	d	e	f	g	
0	H	H	L	L	L	L	H	ON	ON	ON	ON	ON	ON	OFF	1
1	H	X	L	L	L	H	H	OFF	ON	ON	OFF	OFF	OFF	OFF	
2	H	X	L	L	H	L	H	ON	ON	OFF	ON	ON	OFF	ON	
3	H	X	L	L	H	H	H	ON	ON	ON	ON	OFF	OFF	ON	
4	H	X	L	H	L	L	H	OFF	ON	ON	OFF	OFF	ON	ON	
5	H	X	L	H	L	H	H	ON	OFF	ON	ON	OFF	ON	ON	
6	H	X	L	H	H	L	H	OFF	OFF	ON	ON	ON	ON	ON	
7	H	X	L	H	H	H	H	ON	ON	ON	OFF	OFF	OFF	OFF	
8	H	X	H	L	L	L	H	ON	ON	ON	ON	ON	ON	ON	
9	H	X	H	L	L	H	H	ON	ON	ON	OFF	OFF	ON	ON	
10	H	X	H	L	H	L	H	OFF	OFF	OFF	ON	ON	OFF	ON	
11	H	X	H	L	H	H	H	OFF	OFF	ON	ON	OFF	OFF	ON	
12	H	X	H	H	L	L	H	OFF	ON	OFF	OFF	OFF	ON	ON	
13	H	X	H	H	L	H	H	ON	OFF	OFF	ON	OFF	ON	ON	
14	H	X	H	H	H	L	H	OFF	OFF	OFF	ON	ON	ON	ON	
15	H	X	H	H	H	H	H	OFF	OFF	OFF	OFF	OFF	OFF	OFF	
BI	X	X	X	X	X	X	L	OFF	OFF	OFF	OFF	OFF	OFF	OFF	2
RBI	H	L	L	L	L	L	L	OFF	OFF	OFF	OFF	OFF	OFF	OFF	3
LT	L	X	X	X	X	X	H	ON	ON	ON	ON	ON	ON	ON	4

H = high level, L = low level, X = irrelevant

NOTES:
1. The blanking input (BI) must be open or held at a high logic level when output functions 0 through 15 are desired. The ripple-blanking input (RBI) must be open or high if blanking of a decimal zero is not desired.
2. When a low logic level is applied directly to the blanking input (BI), all segment outputs are off regardless of the level of any other input.
3. When ripple-blanking input (RBI) and inputs A, B, C, and D are at a low level with the lamp test input high, all segment outputs go off and the ripple-blanking output (RBO) goes to a low level (response condition).
4. When the blanking input/ripple blanking output (BI/RBO) is open or held high and a low is applied to the lamp-test input, all segment outputs are on.

†BI/RBO is wire-AND logic serving as blanking input (BI) and/or ripple-blanking output (RBO).

Figure 7.9 74LS47 BCD-to-seven-segment decoder/driver.

There are three principal times when we will wish to blank a display:

1. We can control the intensity of the display by switching it on and off quickly enough that flicker is not evident, but the display appears dimmer to the human eye. This will also

decrease the power that the display consumes. A variable-frequency oscillator can drive the BI/RBO pin. By adjusting a variable resistor the intensity of the display can be controlled. Many digital alarm clocks have this feature. When this feature is being used, the BI/RBO pin is an input.

2. We may desire to blank leading zeros to make the display more readable. For example, assume that a calculator with seven digits is to display the number 3002; if all seven displays are enabled, the calculator will display 0003002. The display would be much more readable and power would be saved if the most significant three zeros were blanked. After the first decimal digit greater than zero is displayed, all other zeros will then be significant and must be displayed (e.g., as in 3002). The BI/RBO and the RBI pins will be used to provide this leading zero blanking. When leading zero blanking is used, the BI/RBO pin will be used as an output to the RBI pin on the next, less significant display.
3. A scheme called *multiplexing displays* enables one BCD to seven-segment decoder/driver to control many displays simultaneously. When a multiplexing scheme is being used, the BI/RBO pin is an input pin. This will be covered in Chapter 9.

Line 18 In this line the lamp test input is in its inactive state. The ripple blanking input is in its active-low state. This enables the display to blank the decimal digit 0. When the BCD inputs equal 0000, the seven segments will all turn off to blank the display. The command to blank a 0 is rippled from the most significant display to the least significant display.

Consider a bank of four seven-segment displays. The ripple blanking input of the most significant decoder will be tied to ground. If this decoder receives a BCD input of 0000, it will always blank the display and force its BI/RBO line to go low. This output will be tied to the RBI of the next, less significant decoder. In this manner all zeros will be blanked until a decoder receives a nonzero input. That decoder will display the proper digit and pull its BI/RBO high to disable the blanking of any further displays.

Line 19 The lamp test (LT) input is at an active low. This will force the BI/RBO pin to go inactive (high) and all the segments will illuminate, regardless of the condition of any other input pin. It is important to be able easily to test the integrity of all the LED segments in the system. If the LT pin is pulled low and one or more segments do not illuminate, that indicates a problem in those unlit segments, the driver outputs, or the PCB connections between the driver outputs and the displays.

Take a moment to read Notes 1 to 4 provided with the function table. These notes are the manufacturer's description of the special functions. You should assure yourself that they agree with the author's description of the same events.

7.5 THE LCD SEVEN-SEGMENT DISPLAY

7.5.1 The Liquid Crystal Display

Seven-segment displays constructed from LEDs suffer from two major problems: they require a great deal of current and they are difficult to read outside or in any brightly lit area. LCDs (liquid crystal displays) were developed to overcome these limitations. LCDs consume minute amounts of power, and they are easily visible in brightly lit conditions.

LCDs do not emit light; instead, they either control reflected or transmitted light. Unlike LEDs, which can be driven with a dc voltage, LCDs must be driven with a square wave. The most popular type of LCD is the field-effect LCD. Field-effect LCDs are available in a reflective version, for high-ambient-light conditions, and a transmissive type. The transmissive type is used in low-level-light conditions and employs a backlight.

LCDs do have several disadvantages:

1. They must be used in well-lit areas or use a backlight that defeats the advantage of low-power operation.
2. They have a limited operating temperature.
3. They are more expensive than comparable LED displays.
4. They respond much more slowly than LEDs.
5. They cannot be multiplexed as easily as LEDs.

LCDs are used mainly in portable equipment, such as digital watches, calculators, and DMMs where power consumption is a major concern.

7.5.2 The 4055B BCD-to-Seven-Segment LCD Decoder/Driver

Because LCDs are used for their low-power characteristics, it is appropriate that CMOS devices are used to drive them. The 4055B serves the same function for seven-segment LCDs that the 74LS47 provided for seven-segment LEDs. Refer to Figure 7.10. Once again you should expect to see the four BCD inputs and the seven-segment outputs. Notice that the pinout indicates the significance of the BCD input pins by powers instead of letters. This is the best nonambiguous way to label BCD inputs. The segment outputs, a through g, are active high. Instead of seeing the special functions of lamp test, blanking in/ripple blanking out, and ripple blanking input, there is display frequency in, display frequency out, and V_{ee}.

These special pins will be discussed before the truth table is examined:

Display frequency in LCDs are usually driven at a frequency between 30 and 200 Hz. This input will cause the segments selected by the BCD input to become visible when it is low and blanked when it is high.

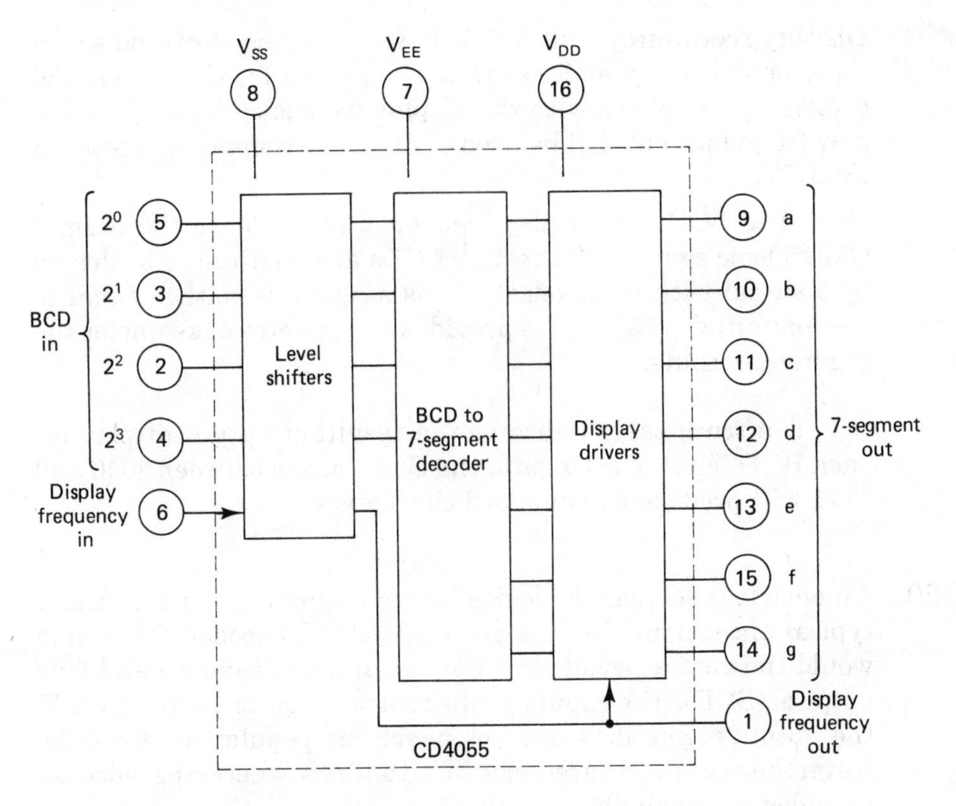

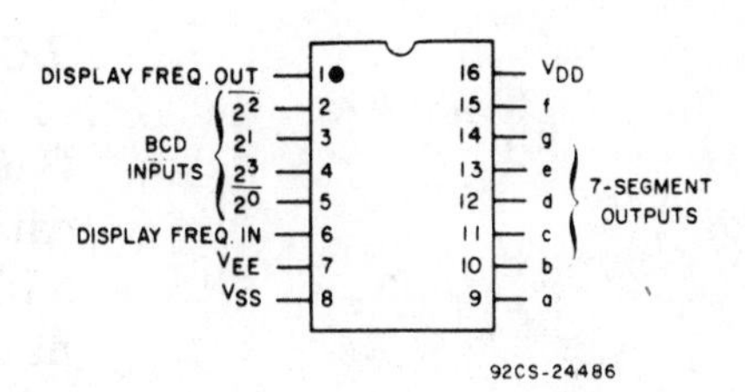

INPUT CODE				OUTPUT STATE							DISPLAY CHARACTER
2^3	2^2	2^1	2^0	a	b	c	d	e	f	g	
0	0	0	0	1	1	1	1	1	1	0	0
0	0	0	1	0	1	1	0	0	0	0	1
0	0	1	0	1	1	0	1	1	0	1	2
0	0	1	1	1	1	1	1	0	0	1	3
0	1	0	0	0	1	1	0	0	1	1	4
0	1	0	1	1	0	1	1	0	1	1	5
0	1	1	0	1	0	1	1	1	1	1	6
0	1	1	1	1	1	1	0	0	0	0	7
1	0	0	0	1	1	1	1	1	1	1	8
1	0	0	1	1	1	1	1	0	1	1	9
1	0	1	0	0	0	0	1	1	1	0	L
1	0	1	1	0	1	1	0	1	1	1	H
1	1	0	0	1	1	0	0	1	1	1	P
1	1	0	1	1	1	1	0	1	1	1	A
1	1	1	0	0	0	0	0	0	0	1	—
1	1	1	1	0	0	0	0	0	0	0	BLANK

Figure 7.10 4055 LCD decoder/driver.

Display frequency out LCDs have a common electrode which must be driven by a signal that is equal in frequency but 180 degrees out of phase with the display frequency input. The display frequency out will be used to drive the common electrode of the LCD.

Vee The BCD and display frequency inputs have the normal CMOS logic swing of V_{ss} to V_{dd}. LCDs are required to be driven by a signal with no dc offset. V_{ee} is a negative voltage equal to the magnitude of V_{dd}. This provides a symmetrical swing for the segment outputs.

The truth table indicates the resultant output display for each BCD input. Once again, the BCD inputs between 1010 and 1111 will produce nonstandard characters.

7.6 THE ENCODER

An encoder provides the logical inverse function of a decoder. A typical encoder might be a decimal-to-BCD encoder. The name would indicate a device that has ten inputs (decimal) and four outputs (BCD). The inputs could be active low or high, as could the outputs. Encoders are not nearly as popular as decoders. Nevertheless, there are certain situations where the encoder provides an invaluable function.

7.6.1 The 74LS148 8-Line-to-3-Line Priority Encoder

Consider the name of this device: "8-line-to-3-line." This indicates that it will have eight inputs which can be encoded into a 3-bit binary code ($2^3 = 8$). The word "priority" is also in the name. This indicates that the inputs are arranged in some scheme, from most important (highest priority) to least important (lowest priority). If more than one input goes active, that input with the highest priority will be encoded and output on A0 through A3, while the other lower-priority pins will be ignored. Refer to Figure 7.11. The pinout should hold few surprises. There are eight active-low inputs, 0 through 7, and the three encoded active-low outputs, A0 through A2. There is one unfamiliar input, EI, and two unfamiliar outputs: EO and GS. Let's tackle the explanation of these unknown pins before proceeding to the function table.

EI This is the enable input. If a high level is applied to this pin, the 74LS148 will be disabled: all the outputs will go to logic 1. A low level will enable the device to perform a normal encoding function.

EO The enable output pin will go low if the EI pin is low and none of the eight inputs are at an active level. The EI and EO pins are used to cascade 74LS148s. The EO pin from the highest-priority encoder will be connected to the EI pin of the next-lower-priority encoder. EI of the highest-priority encoder will be con-

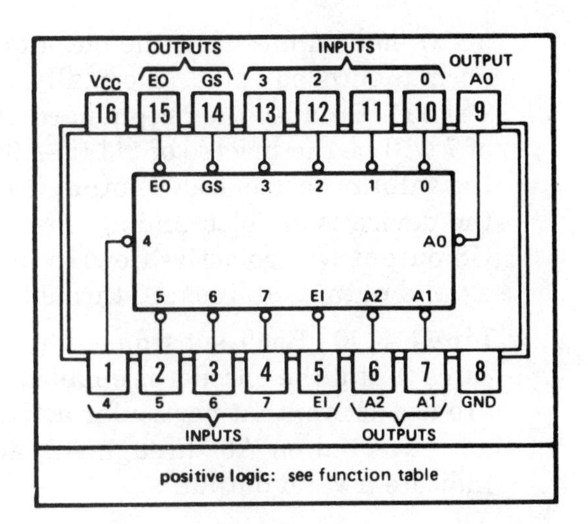

'148, 'LS148
FUNCTION TABLE

INPUTS									OUTPUTS				
EI	0	1	2	3	4	5	6	7	A2	A1	A0	GS	EO
H	X	X	X	X	X	X	X	X	H	H	H	H	H
L	H	H	H	H	H	H	H	H	H	H	H	H	L
L	X	X	X	X	X	X	X	L	L	L	L	L	H
L	X	X	X	X	X	X	L	H	L	L	H	L	H
L	X	X	X	X	X	L	H	H	L	H	L	L	H
L	X	X	X	X	L	H	H	H	L	H	H	L	H
L	X	X	X	L	H	H	H	H	H	L	L	L	H
L	X	X	L	H	H	H	H	H	H	L	H	L	H
L	X	L	H	H	H	H	H	H	H	H	L	L	H
L	L	H	H	H	H	H	H	H	H	H	H	L	H

Figure 7.11 74LS148 8-line-to-3-line priority encoder.

nected to ground or to the output of a gate that controls when the encoders are enabled.

GS The group select output goes low whenever the decoder is active. If the decoder is enabled (EI is low) and at least one input is at an active logic 0 level, the GS output will go low to indicate that the device has a valid encoded output on its A0 through A2 pins.

Refer to the function table in Figure 7.11. We will examine the function table on a line-by-line basis:

Line 1 The EI input is inactive at a high level. The inputs 0 through 7 become don't-care conditions and all the outputs will go high to their inactive state. When EI is high, the 74LS148 is disabled.

Line 2 The device is enabled with a low level at EI but none of the input lines (0 through 7) are active. All outputs will be inactive except the EO line, which will go low. If another encoder is connected in cascade, this low output level will be connected to its EI input and it will be enabled.

Line 3 The encoder is enabled and input 7 is active. The rest of the inputs assume don't-care conditions. This tells us that input 7 has the highest priority; if it is active, input 7 will override all other inputs (0 through 6). Notice that A0 through A2 all go to logic 0. Remember: The outputs are active low. Here active low implies that the outputs will be inverted from their normal encoded values. Normally, the encoded value of decimal 7 is 111. Because the outputs are active low, the encoded value of 7 will be the inverse of 111 (i.e., 000). This will be the case for the value of each encoded output. Also notice that in line 3, when the device is enabled and at least one input line is active, the GS output will go active (low) to indicate that the 74LS148 has a valid output on lines A0 through A2.

Lines 4–10 Each subsequent line in the function table indicates that if the encoder is enabled and the higher-priority lines are not active, that particular active input will appear encoded and inverted on A0 through A2, and the GS will go active to indicate a valid output.

7.6.2 The 4532B CMOS 8-Bit Priority Encoder

This is the CMOS equivalent of a 74LS148. All the inputs and outputs are active high. Refer to Figure 7.12. Notice the functional diagram. The input lines, D0 through D7, are NORed together; if any of these eight lines go to an active-high level, the output of the NOR gate will go low. This output is inverted and ANDed with the EI input to arrive at an active-high GS output.

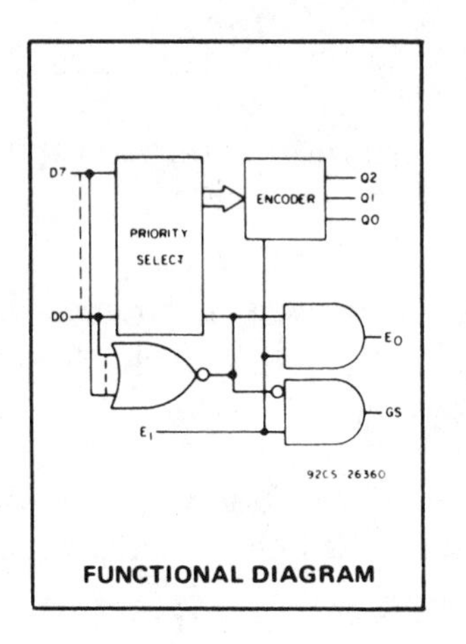

FUNCTIONAL DIAGRAM

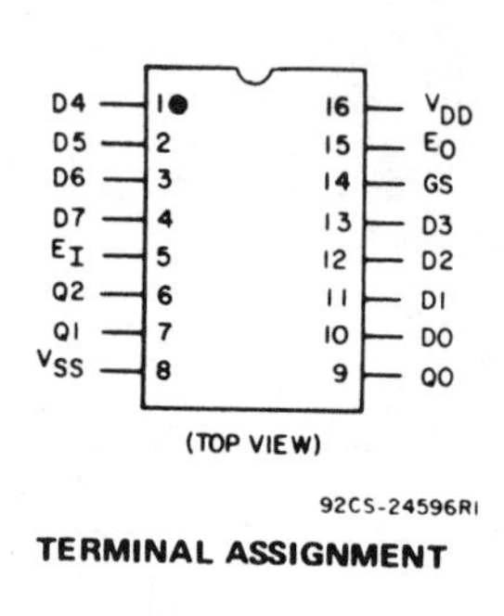

TERMINAL ASSIGNMENT

TRUTH TABLE

Input									Output				
E_I	D7	D6	D5	D4	D3	D2	D1	D0	GS	Q2	Q1	Q0	E_O
0	X	X	X	X	X	X	X	X	0	0	0	0	0
1	0	0	0	0	0	0	0	0	0	0	0	0	1
1	1	X	X	X	X	X	X	X	1	1	1	1	0
1	0	1	X	X	X	X	X	X	1	1	1	0	0
1	0	0	1	X	X	X	X	X	1	1	0	1	0
1	0	0	0	1	X	X	X	X	1	1	0	0	0
1	0	0	0	0	1	X	X	X	1	0	1	1	0
1	0	0	0	0	0	1	X	X	1	0	1	0	0
1	0	0	0	0	0	0	1	X	1	0	0	1	0
1	0	0	0	0	0	0	0	1	1	0	0	0	0

X = Don't Care   Logic 1 ≡ High   Logic 0 ≡ Low

Figure 7.12 4532B 8-bit priority encoder.

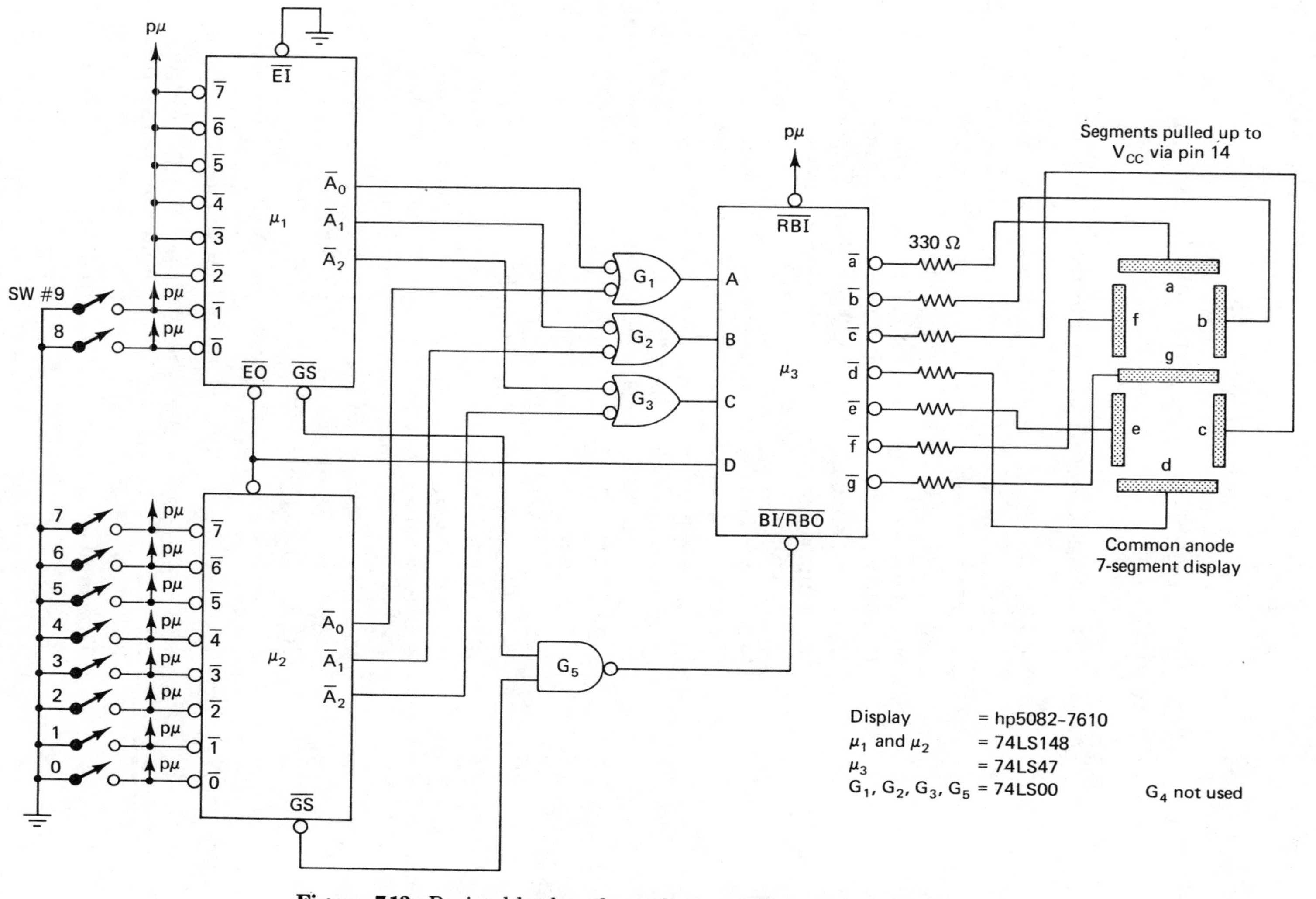

Figure 7.13 Decimal keyboard encoding/decoding and display circuit.

If the output of the NOR gate is high, which indicates that none of the eight input lines are active, and the EI is high, the EO line will go active to enable the next encoder in cascade. This functional diagram is only a symbolic indication of how the 4532B operates. Its real value is as an aid to help you understand the operation of the device.

The truth table in Figure 7.12 is identical to the function table in Figure 7.11 except that all levels are active high instead of active low.

7.6.3 Applications of the Priority Encoder

Priority encoders are often used in keyboard encoding/decoding schemes. Consider the simplified application shown in Figure 7.13 for employing 74LS148s to encode the depressed key in a 10-key keyboard. Switches 9 through 0 are the 10 keys that make up the decimal keyboard. U1 is the highest-priority encoder; its EI is hardwired to ground. If both switches 9 and 8 are open, U2 will be enabled via the EO pin of U1. Therefore, switch 9 is the highest-priority switch and switch 0 is the lowest priority.

It is important to realize that U1 and U2 will never have simultaneous valid outputs on their A0 through A2 pins. When U1 is active, outputs A0 through A2 on U2 will all be inactive (high). When U2 is active, outputs A0 through A2 on U1 will be inactive highs. The A input on the 74LS47 will be high if A0 of U1 or A0 of U2 goes active low. The same is true for inputs B and C of U3. If the EO from U1 goes high, this indicates that either switch 9 or 8 is closed. This high will disable U2 and cause input D of the 74LS47 to go high.

If GS of U1 is high and GS of U2 is high, this indicates that neither encoder is active and the seven-segment display will be blanked via the output of G5. Each segment output is current limited with a 330-Ω resistor. The segments are all pulled up to V_{cc} via pin 14 of the seven-segment display.

Redraw this circuit on a separate sheet of paper. Close each switch and derive the logic levels for each input and output of all circuit components.

7.7 THE DATA SELECTOR/MULTIPLEXER

The data selector is a digitally controlled switch. In Section 4.3 we designed a 2-bit data selector. It had two data inputs and a select line that controlled which input was steered to the output. MSI data selectors are available with 4, 8, and 16 data inputs. The number of select bits required on a data selector will be a function of the number of data bits:

$$2^{\text{(number of select bits)}} = \text{number of data inputs.}$$

Data selectors are also used to send many different channels of data on one common transmission line.

7.7.1 The 74LS151 One of Eight Data Selector/Multiplexer

From your previous knowledge of data selectors, you should be able to derive the basic pinout of the 74LS151. There will be eight data inputs, D0 through D7, and one data output. There must be three select bits ($2^3 = 8$) : A, B, and C. Refer to the pinout in Figure 7.14. Notice the two outputs, Y and W. Y is unbubbled and therefore the true output. W is bubbled and is the complement of Y. There is one pin that has not been discussed: the strobe input. In digital electronics the term *strobe* has several different meanings. Most often, it refers to a fast, low-duty cycle pulse that enables a device. On the 74LS151, the strobe is another name for an enable input. The strobe input is bubbled, which denotes that it is active low.

Refer to the function table in Figure 7.14. The first line indicates the situation where the strobe is at an inactive level. If the strobe input is high, the Y output will go low and the W output will go high, regardless of the values on the select or data inputs. The last eight lines of the function table illustrate each data input as it is selected and how it affects the Y and W out-

'151A, 'LS151, 'S151
FUNCTION TABLE

INPUTS				OUTPUTS	
SELECT			STROBE	Y	W
C	B	A	S		
X	X	X	H	L	H
L	L	L	L	D0	$\overline{D0}$
L	L	H	L	D1	$\overline{D1}$
L	H	L	L	D2	$\overline{D2}$
L	H	H	L	D3	$\overline{D3}$
H	L	L	L	D4	$\overline{D4}$
H	L	H	L	D5	$\overline{D5}$
H	H	L	L	D6	$\overline{D6}$
H	H	H	L	D7	$\overline{D7}$

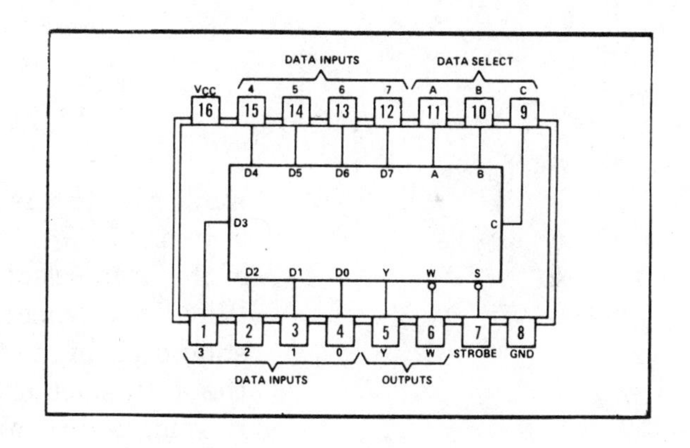

'151A, 'LS151, 'S151

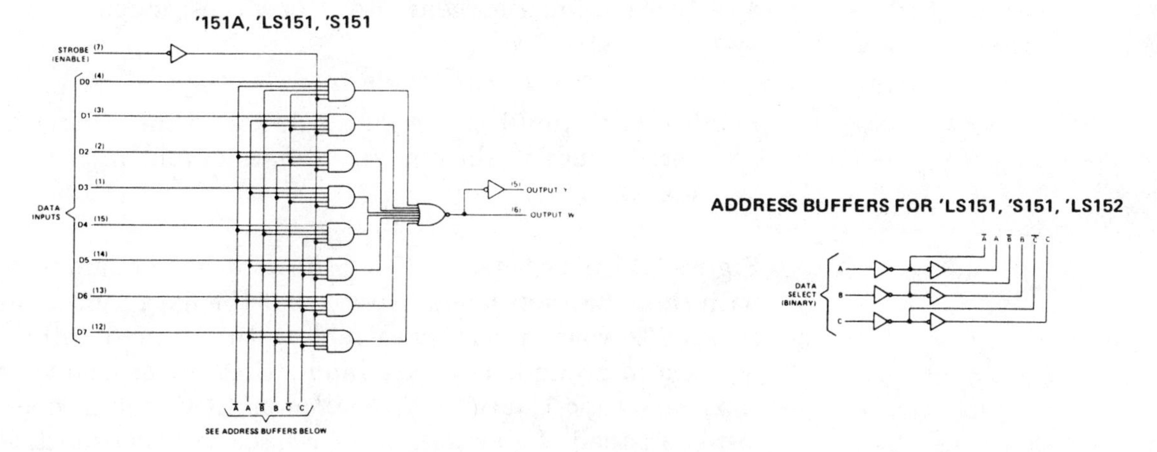

Figure 7.14 74LS151 1-of-8 eight data selector/multiplexer.

puts. A full line of data selectors is also available in the 4000B series.

7.7.2 Applications of Data Selectors

Data selectors are useful in many digital applications. Like many of the other devices in this chapter, data selectors find applications in advanced circuits, such as memory controller circuits. An interesting application of data selectors is in Boolean generators.

The 74LS151 can be used to implement any one- to three-variable Boolean equation. One 74LS151 can be used in place of a potentially large number of gates. Let's rebuild the room-with-three-doors circuit, from Section 4.2, using a 74LS151.

C	*B*	*A*	*Out*
0	0	0	0
0	0	1	1
0	1	0	1
0	1	1	0
1	0	0	1
1	0	1	0
1	1	0	0
1	1	1	1

$$\text{Light on} = A\overline{B}\,\overline{C} + \overline{A}B\overline{C} + \overline{A}\,\overline{B}C + ABC$$

Each row in the truth table corresponds to a particular select code on the 74LS151. When the select code is 000, the Y output of the data selector should go low, as in row 1 of the truth table. When the select code is 001, the output of the data selector should go high, as in row 2 of the truth table. In this manner the data selector will emulate the discrete gates that made up the original room-with-three-doors circuit. To use a data selector as a Boolean function generator follow these steps:

1. Construct the truth table.
2. Tie each input of the data selector to the logic level that corresponds to the equivalent data output in the truth table.

Figure 7.15 illustrates the 74LS151 as it implements the room-with-three-doors problem. Notice that the data inputs that correspond to rows in the truth table with a logic 0 output value are tied to ground. The data inputs that correspond with rows that have logic 1 outputs are pulled-up to V_{cc} via a common resistor. Instead of requiring an inverter to drive the LED, the complemented output, W, is used.

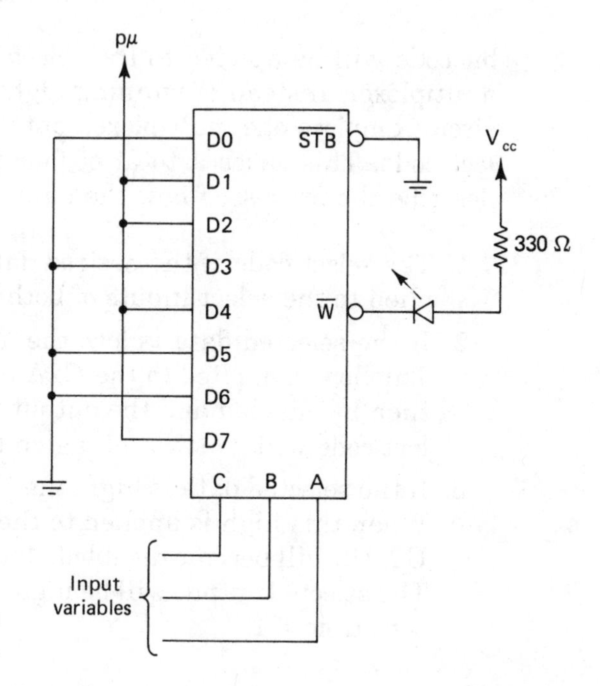

Figure 7.15 74LS151 as a Boolean function generator.

7.7.3 Using Multiplexers and Demultiplexers

We have stated that the data selector can also function as a multiplexer. The 74LS138 was described as a decoder/demultiplexer. We are going to employe a 74LS151 as a multiplexer and a 74LS138 as a demultiplexer in single-line digital communications circuit (see Figure 7.16). The 74LS151 functions as a one-of eight multiplexer. The 3-bit select code will command which of the data inputs will be steered to the Y output. The same 3-

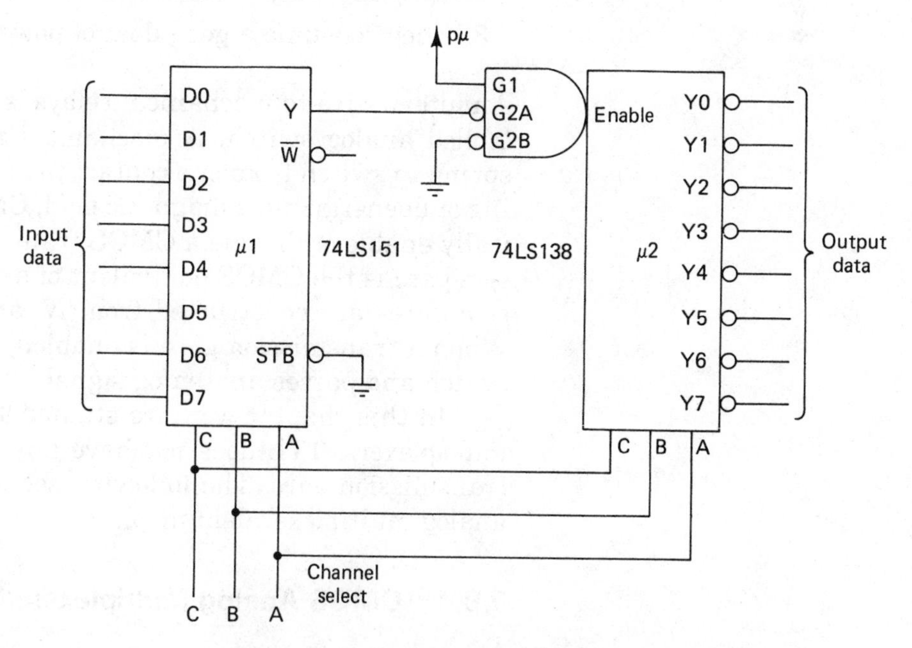

Figure 7.16 Multiplexer/demultiplexer circuit.

bit code will be applied to the select inputs of the 74LS138 demultiplexer. Instead of running eight separate data lines, this circuit employs one multiplexed data line and three lines of select code. This saves a total of four lines. The following steps describe the process of how the data are actually sent:

1. The select code of the desired data line (0 through 7) is applied to the select inputs of both U1 and U2.
2. If the selected data is low, the Y output of U1 will go low. This low is applied to the G2A enable input of U2. U2 will then be enabled and the output that corresponds to the select code will go low and match the original data from U1.
3. If the selected data is high, the Y output of U1 will go high. When this high is applied to the active-low enable input of U2, U2 will become disabled. All outputs of U2 will go high. The selected output will be high, matching the original data sent from U1.

Notice that the selected output will toggle high and low with the data sent from the multiplexer, while the nonselected outputs will stay inactive at a logic 1 level.

7.8 CMOS DIGITALLY CONTROLLED ANALOG SWITCHES

The digital control of analog quantities has traditionally been performed by magnetic relays. These mechanical relays have many limitations:

1. They take a millisecond or more to switch.
2. They are large and bulky.
3. They consume a good deal of power.

An alternative to mechanical relays is the CMOS digitally controlled analog switch. A mechanical relay employs a coil and spring to switch from one contact to another. Instead of energizing or deenergizing a magnetic field, CMOS analog switches digitally enable or disable a CMOS transmission gate. A transmission gate is the CMOS equivalent of an SPST switch. Transmission gates are constructed from *N*- and *P*-channel MOSFETs. When a transmission gate is enabled, it saturates like a closed switch and passes an analog signal.

In this chapter we have studied TTL multiplexers and demultiplexers. TTL does not have any equivalent to the CMOS transmission gate. The following section introduces the CMOS analog multiplexer/demultiplexer.

7.8.1 CMOS Analog Multiplexers/Demultiplexers

The 4051B, 4052B, and 4053B are CMOS devices that employ transmission gates to switch analog signals (see Figure 7.17). The 4051B is the CMOS equivalent of an SP8T switch. The three

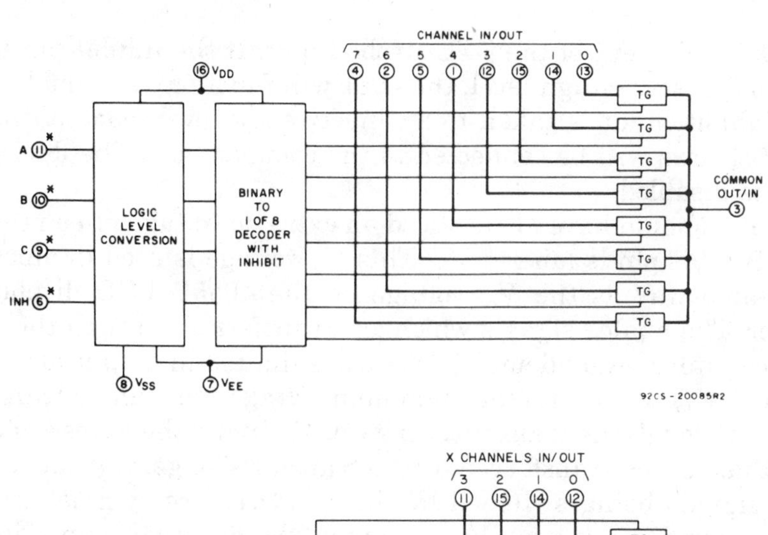

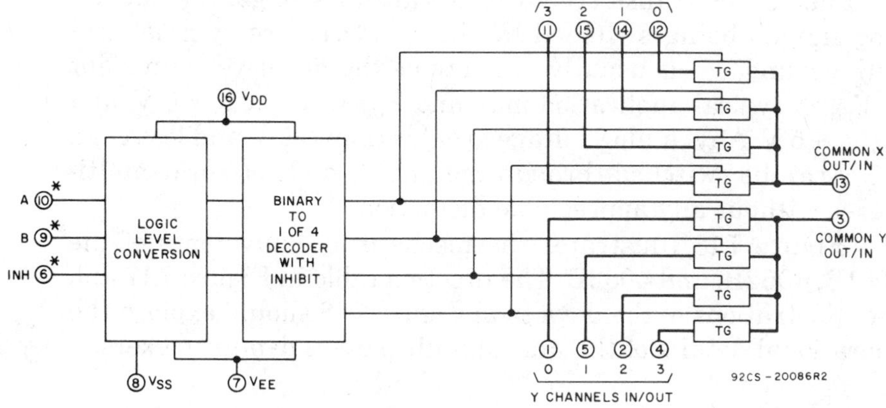

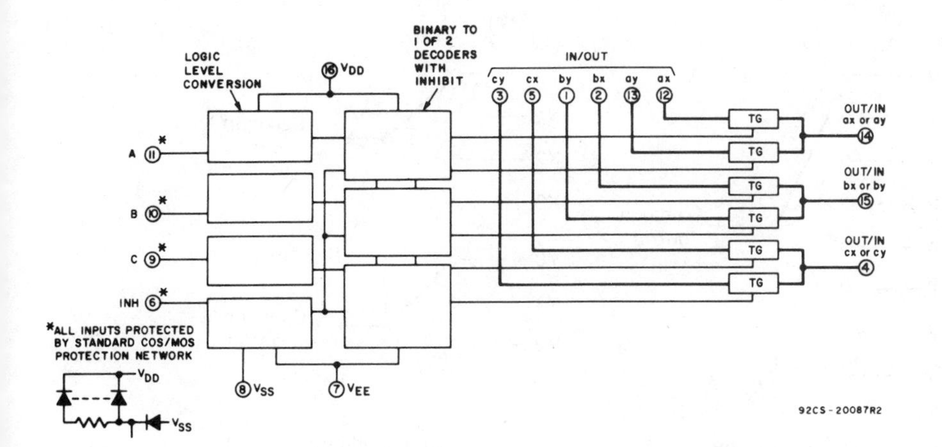

INPUT STATES				"ON" CHANNEL(S)
INHIBIT	C	B	A	
CD4051B				
0	0	0	0	0
0	0	0	1	1
0	0	1	0	2
0	0	1	1	3
0	1	0	0	4
0	1	0	1	5
0	1	1	0	6
0	1	1	1	7
1	X	X	X	NONE
CD4052B				
INHIBIT		B	A	
0		0	0	0x, 0y
0		0	1	1x, 1y
0		1	0	2x, 2y
0		1	1	3x, 3y
1		X	X	NONE
CD4053B				
INHIBIT	A or B or C			
0	0			ax or bx or cx
0	1			ay or by or cy
1	X			NONE

X = Don't care

Figure 7.17 4051B, 4052B, and 4053B analog multiplexers/demultiplexers.

select inputs, ABC, control which of the eight throws is connected to the common pole. If eight separate signals are placed on the throws, 0 through 7, and the output is taken from the common pole, the 4051B is functioning as an 8-line-to-1-line analog multiplexer. If an input is placed on the common pole and the output is taken from one of the eight throws, the 4051 is functioning as a 1-line-to-8-line analog demultiplexer.

There are two pins on the 4051B that require further discussion. Notice the pin labeled INH. This inhibit input is the

CMOS equivalent of the TTL strobe input. If the inhibit pin is held at an active-high level, the 4051 will function as normal. If the inhibit input is taken to an inactive-low level, none of the eight throws will be connected to the common pole. The device will be disabled.

You should have also noticed an extra dc voltage pin on the 4051B. This pin is labeled V_{ee}. This V_{ee} voltage is used in much the same way as the V_{ee} voltage on the 4055B LCD display driver. The analog signals which are transferred through these analog multiplexers/demultiplexers are limited in amplitude by V_{ee} and V_{dd}. V_{dd} limits the maximum voltage that can be transferred through the transmission gate; V_{ee} limits the lowest voltage that can be transferred by the transmission gate. If the analog signals being switched by these devices are symmetrical with ground, V_{ee} is usually taken to be the negative equivalent of V_{dd}. A typical application may find $V_{dd} = +5$, $V_{ss} = 0$ V, and $V_{ee} = -5$ V. Any analog voltage greater than -5 V and less than $+5$ V can be switched through the analog multiplexer/demultiplexer without any appreciable distortion.

Figure 7.18 illustrates the mechanical equivalents of the 4051B, 4052B, and 4053B. The function table in Figure 7.17 and the illustrations of the devices in Figure 7.18 should explain the operational details of the analog multiplexers/demultiplexers.

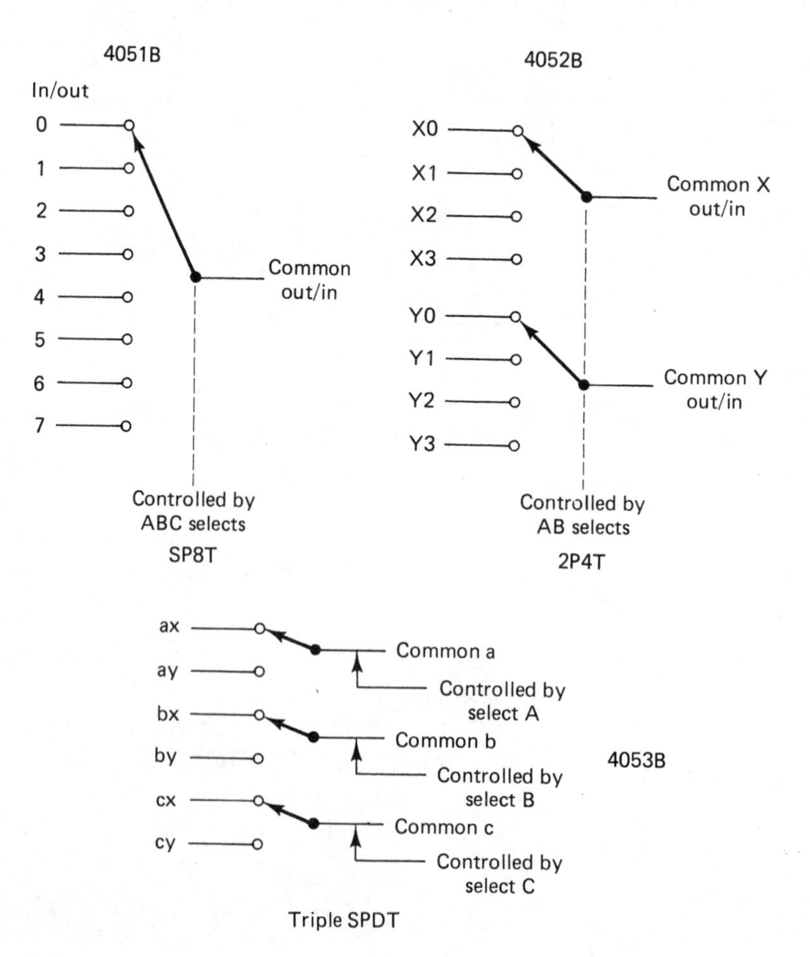

Figure 7.18 Mechanical equivalents of the 4051B, 4052B, and 4053B.

The "ON" resistance of a CMOS transmission gate is typically 125 Ω. This is far from the ideal ON resistance of 0 Ω. Because of this resistance, the use of CMOS transmission gates is limited to low-power signals. Where high-power switching is required, magnetic relays or solid-state relays are usually employed.

7.9 THE 4-BIT FULL ADDER

In a computer-based system all the math functions are performed by special-purpose LSI devices. However, there will be occasions when two binary numbers must be added. The 4-bit full adder can perform the addition of two 4-bit quantities, including a carry input and a carry output (see Figure 7.19).

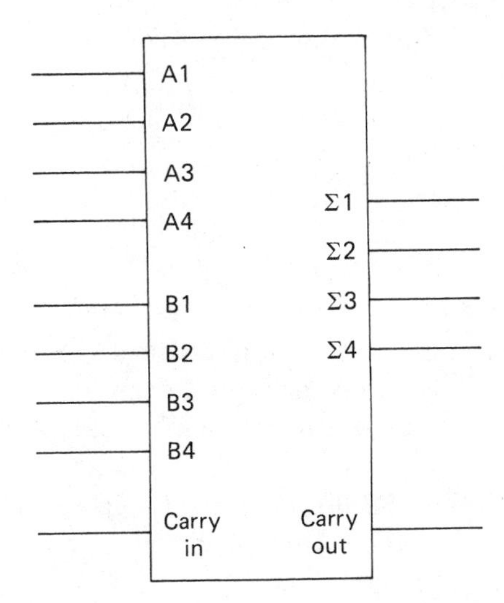

Figure 7.19 4-bit binary full adder.

Popular forms of 4-bit adders are the 74LS83/LS283 and 4008B. The 74LS83 and 74LS283 are functionally equivalent. The 74LS83 has nonstandard power connections; the pins on the 74LS283 have been rearranged to provide the standard power terminals: ground at pin 8 and V_{cc} at pin 16. The A and B inputs are arranged like the inputs on the 4-bit magnitude comparator. Instead of comparing the input values, the 4-bit adder will sum them. The symbol preceding the four outputs is the Greek capital letter sigma; sigma denotes summation. There is also a carry input. A carry produced from a less significant adder will drive this input. If the carry input is not used, it must be tied to ground. The two 4-bit inputs and carry input are added according to the normal rules of binary addition.

The 4-bit adder is often used in circuits where a 4-bit offset must be added to a particular binary quantity.

QUESTIONS AND PROBLEMS

Refer to Figure 7.3 for Questions 7.1 to 7.6.

7.1. How many different combinations are possible with this electronic lock?

7.2. The circuit in Figure 7.3 appears to be malfunctioning. What are the steps that you would take to isolate the problem. *Hint:* Try to check out the circuit in small blocks. Establish U1 is functioning correctly, then U2, and so on.

7.3. The trace between U1 pin 6 and U2 pin 3 is open. How does this affect the operation of the circuit? Why?

7.4. Pins 3 and 4 of U1 are shorted together with a solder bridge. How does this affect circuit operation? Why?

7.5. Switch 3 is open. How does this affect circuit operation? Why?

7.6. The 220-Ω resistor is accidentally stuffed with a 220-kΩ resistor. How does this affect circuit operation? Why?

7.7. A BCD-to-decimal decoder functions correctly for input values from 0000 to 0111. When the BCD code of 1000 is applied to the decoder, output 0 goes active instead of output 8. Furthermore, when 1001 is applied, output 1 goes active instead of output 9. What is the cause of the problem?

7.8. How can you blank an LED seven-segment display being driven with a 74LS47 without the use of the BI/RBO pin?

7.9. How can the BI/RBO pin on the 74LS47 be used as both an input and an output? In what situations is it an input? In what situations is it an output?

Refer to Figure 7.13 for Questions 7.10 to 7.17.

7.10. When all the keyboard switches (SW 9 through SW 0) are open, the seven-segment display is illuminating the decimal digit—0. What is the problem? What are some possible causes of this problem?

7.11. The EI pin on U1 is floating instead of being tied to ground. How does this affect circuit operation? Why?

7.12. The output of G1 is stuck low. How does this affect circuit operation? Why?

7.13. A technician closes SW 0 and a 0 is displayed on the seven-segment display. What will happen if the technician next closes SW 1 while leaving SW 0 closed? What would happen if this was done in the opposite order, closing SW 1, then SW 0? Why?

7.14. The 330-Ω resistors are accidentally stuffed with 33-Ω resistors. In relative terms, how would the display appear? What possible problems could occur?

7.15. The decimal digit "8" is displayed all the time, regardless of the position of SW 9 throught SW 0. What is the problem?

7.16. When SW 0 is closed and all other switches are open, the display goes blank. What is the possible problem?

7.17. SW 9 through SW 5 appear to function correctly. When all the switches are open, the digit "4" is displayed. Toggling SW 4 through SW 0 does not seem to affect the display. What is the problem?

7.18. Use a 74LS151 to implement the three-judges problem from Chapter 4.

Refer to Figure 7.16 for Questions 7.19 to 7.21.

7.19. How would the circuit be affected if G2B were tied to G2A instead of ground? Which method is better? Why?

7.20. How would the output data be affected if G2A and G2B were tied to ground and G1 were connected to the Y output of the multiplexer?

7.21. How would the output data be affected if G2A and G2B were tied to ground and the W output of the multiplexer were tied to the G1 input of the demultiplexer?

7.22. Modify the schematic in Figure 7.16 by using a seven-segment display to indicate the number of the channel that is presently active.

7.23. If a 16-line-to-1-line multiplexer and a 1-line-to-16-line demultiplexer were used in a scheme similar to Figure 7.16:

(a) How many select lines would be required?

(b) How many lines would be saved compared to a nonmultiplexed 16-line communications link?

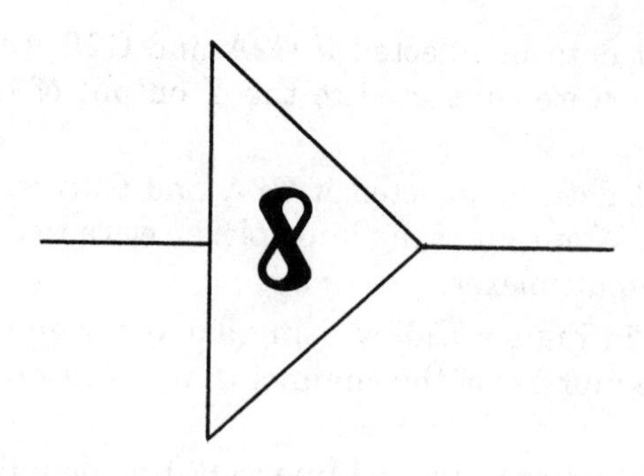

INTRODUCTION TO SEQUENTIAL DEVICES

The outputs of logic gates and MSI combinational devices depend only on the present state of the inputs. If we desire to know the output state of a combinational logic circuit, all we need do is substitute the logic values of the inputs into the device's truth table. This type of combinational logic operates much like the lock illustrated in Figure 8.1a. To open this lock, the proper digits must be aligned. This lock is the mechanical equivalent to the electronic combination lock that we constructed using magnitude comparators.

Contrast this combinational type of lock with the sequential lock illustrated in Figure 8.1b. To open this type of lock requires a sequence of events. By sequence, we mean that the correct number, in the correct order, must appear on the lock's dial. A typical sequential lock might require the following steps:

1. Turn the dial clockwise and stop at the number 24.
2. Turn the dial counterclockwise and stop at the number 16.
3. Finally, turn the dial clockwise and stop at the number 33.

If all three events are accomplished in the correct order, the lock should open. If an incorrect number is dialed in, one must start over at step 1.

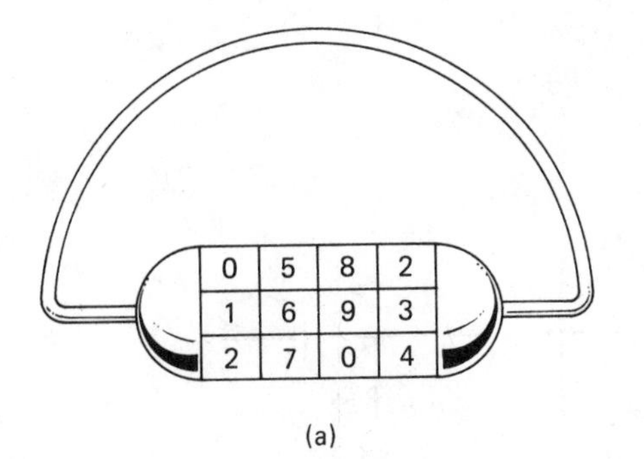

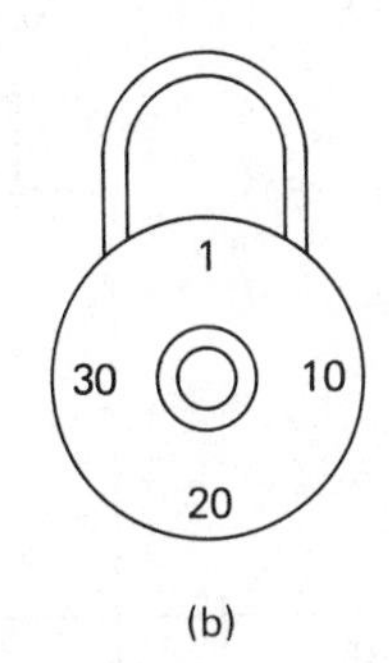

Figure 8.1 (a) Combinational lock; (b) sequential lock.

The most important difference between combinational and sequential devices is that sequential devices display the attribute of "memory." Carefully consider this concept of memory. The sequential lock effectively remembers whether the first two numbers in the sequence have been entered correctly. The combinational lock remembers nothing. All the numbers appear at the same time on the combinational lock's inputs.

The capacity of memory is a useful characteristic for a digital device to possess. As we proceed through this chapter, keep this concept of memory, as it relates to the sequential lock, in mind.

8.1 FEEDBACK AND DIGITAL CIRCUITS

You may be asking yourself: "How do we design digital devices to have this capacity for memory?" The answer to that question is feedback. You should already be familiar with the concept of feedback. A simple definition of feedback is:

the process of sending all or part of the output signal back to the input

In analog electronics there are two types of feedback: positive and negative. Positive feedback is used in oscillators and comparators with hysteresis (Figure 8.2). Op amps use negative feedback, to effectively reduce open-loop gain (Figure 8.3).

8.1.1 Cross-Coupled NAND Gates

Refer to Figure 8.4. This is our first example of a digital circuit employing feedback. Notice that output 1 is fed back to an input of gate 2 and output 2 is fed back to an input of gate 1. The only tool we need to analyze this circuit is the fact that:

The dynamic input level of a NAND gate is logic 0.

The negative-logic NAND symbol (Figure 8.5) was chosen to represent this circuit to indicate that the NAND gate should be considered as

an active-low-input, active-high-output OR gate.

To analyze this circuit we will complete the truth table in Fig-

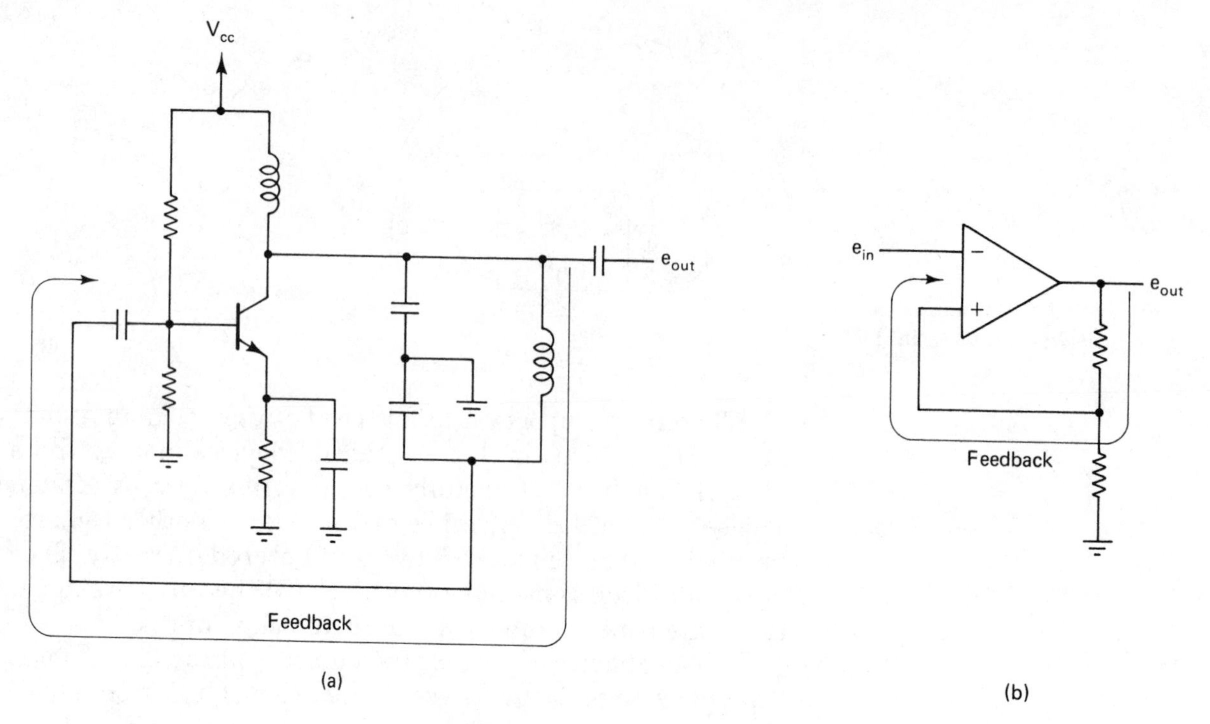

Figure 8.2 (a) Colpits oscillator; (b) comparator with hysteresis.

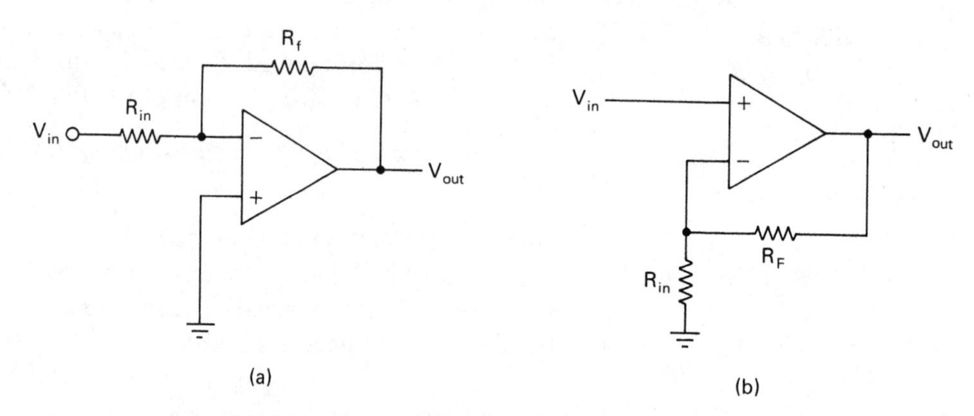

Figure 8.3 (a) Inverting op amp; (b) noninverting op amp.

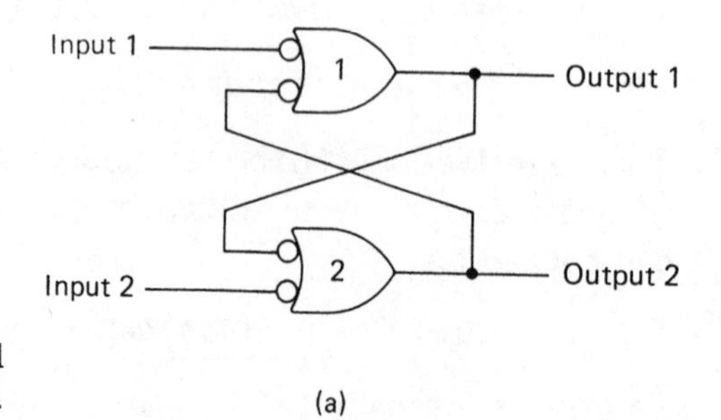

Input		Output	
2	1	1	2
1	0	?	?
0	1	?	?
1	1	?	?
0	0	?	?

(b)

Figure 8.4 Cross-coupled NAND gates.

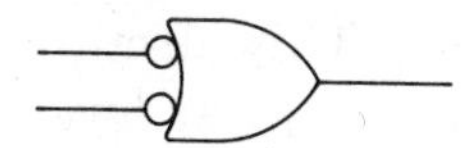

Figure 8.5 A low OR a low will output a high.

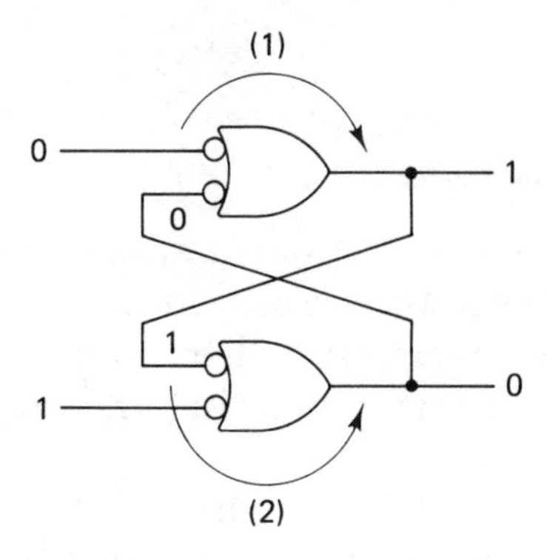

Figure 8.6 Input 1 = 0; input 2 = 1.

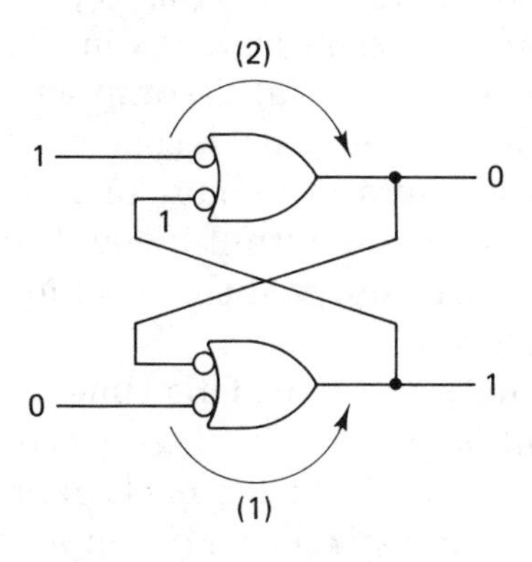

Figure 8.7 Input 1 = 1; input 2 = 0.

ure 8.4. Normally, the case of input 1 = 0 and input 2 = 0 would appear in the first line of the truth table. To simplify the circuit analysis, we will save this input combination for the last step in our analysis.

Line 1: input 1 = 0, input 2 = 1. The logic 0 on input 1 will force a logic 1 on the output of gate 1. This logic 1 is fed back to an input of gate 2. Because gate 2 does not have any logic 0's on its inputs, its output will be a logic 0. This logic 0 is fed back to an input of gate 1. Gate 1 is already outputting a logic 1, so the circuit appears stable. This circuit action is summarized in Figure 8.6.

Line 2: input 1 = 1, input 2 = 0. Refer to Figure 8.7. Here we find a dynamic input level on gate 2. This logic 0 will force the output of gate 2 high. This logic 1 will be fed back to an input of gate 1. Gate 1 does not have any logic 0's on its inputs, so the output of gate 1 will go to logic 0. This logic 0 will be fed back to the input of gate 2. The output of gate 2 is already high, so the circuit appears to be stable. Because the circuit is symmetrical, the two lines of the truth table that we have examined are mirror images of each other.

Line 3: input 1 = 1, input 2 = 1. This is an interesting case. Neither input 1 nor input 2 is at a dynamic level. We seem to be stuck. What should we do? We know that the outputs of the NAND gates must be at some logic level. We are forced to make some assumptions about the level of the outputs and analyze the circuit from that point.

Refer to Figure 8.8. Consider the first assumption—that both outputs 1 and 2 are equal to logic 0. Carefully examine Figure 8.9. We obviously have a problem. Both outputs are dynamic input logic 0's. As they are fed back to the inputs of gates 1 and 2, these logic 0's will force the outputs of both gates to logic 1. Now both gates will have two logic 1's on their inputs,

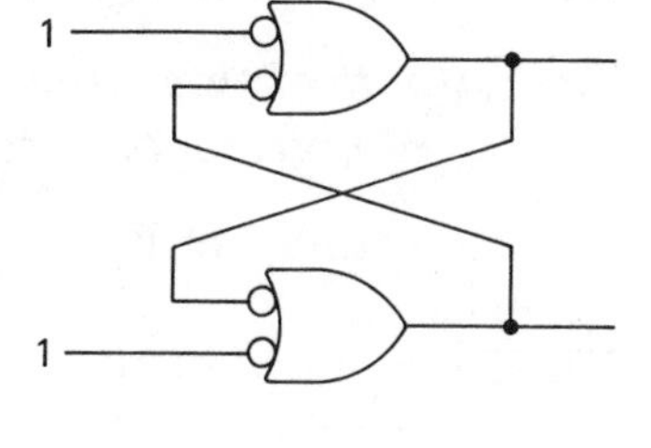

Outputs	
1	2
0	0
0	1
1	0
1	1

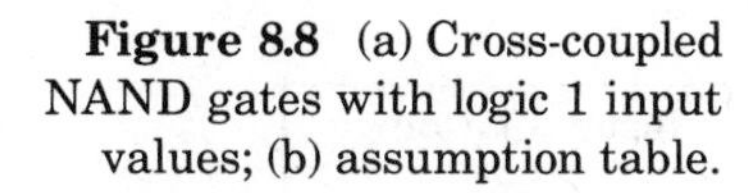

Figure 8.8 (a) Cross-coupled NAND gates with logic 1 input values; (b) assumption table.

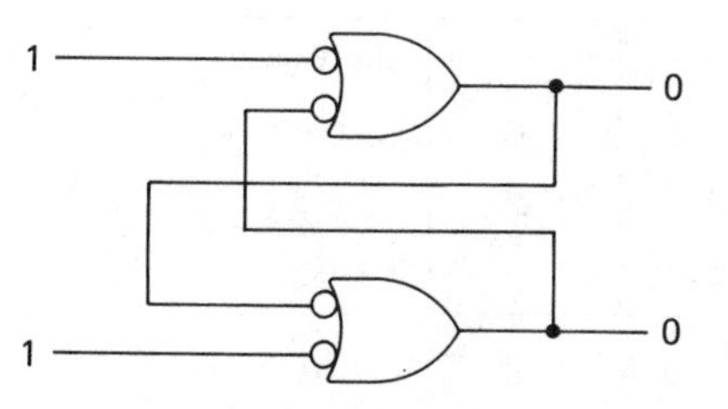

Figure 8.9 Assumption of both outputs equal to logic 0's.

which will force the outputs to logic 0's. This circuit appears to be unstable. It is impossible to predict at what level outputs 1 and 2 will finally settle.

If we had assumed that both outputs were at logic 1, the circuit action would be exactly what was just described: unstable and unpredictable. This part of the analysis has yielded an important fact:

The outputs of this device must always be complements of each other.

We have now established that lines 1 and 4 of our assumption table (Figure 8.8) are unstable conditions. We are still concerned with our initial problem of analyzing the circuit when both inputs are equal to logic 1's. Let's continue examining the assumption table.

Assume that output 1 = 0 and output 2 = 1. This assumption is easy to follow through the circuit. Output 1 is a logic 0 and is fed to an input of gate 2. This logic 0 forces the output of gate 2 high, which corresponds to our assumption. This high is fed back to an input of gate 1. Gate 1 has no logic 0's on its inputs, so its output must be a logic 0. Our initial assumptions concerning the state of the outputs have been proven correct.

Does this mean that when both inputs are equal to logic 1's, then output 1 will be equal to logic 0 and output 2 equal to logic 1? Absolutely not. We still have not analyzed the assumption where output 1 = 1 and output 2 = 0.

Assume that output 1 = 1 and output 2 = 0. This time we will follow the logic 0 from the output of gate 2. This logic 0 will force the output of gate 1 high. This high will be fed back to an input of gate 2. Gate 2 does not have any logic 0's on its inputs, so its output will be a logic 0. This analysis shows that our assumption of output 1 = 1 and output 2 = 0 is also true. Figure 8.10a summarizes our analysis of the circuit when inputs 1 and 2 are both nondynamic logic 1's. How are we to interpret the results in Figure 8.10a? Because unstable operation is something that we wish to avoid, we have already stated:

The outputs of this circuit must always be complements of each other.

What about lines 2 and 3 of Figure 8.10a? These lines demon-

Assumptions		
Output 1	Output 2	Result
0	0	Unstable
0	1	True
1	0	True
1	1	Unstable

(a)

Inputs		Outputs	
2	1	1	2
0	0	Undefined	
0	1	0	1
1	0	1	0
1	1	Memory	

(b)

Figure 8.10 (a) Summary of outputs when both inputs are high; (b) truth table for cross-coupled NAND gates.

strate this device's capacity of memory. Lines 2 and 3 tell us that if both inputs go to logic 1's (nondynamic levels), the logic values on the outputs of the device will remain unchanged: The device will enter a memory state. The Q outputs will retain the values induced in them by the last dynamic input level. In a few pages we will examine a practical circuit that employs this memory capacity.

Refer to Figure 8.4b. We must examine the last line in the truth table. This line states that both inputs 1 and 2 will be at logic 0. Can you see a problem with this? To avoid unstable conditions in this circuit, outputs 1 and 2 must never be the same value. This combination of inputs will force both outputs to logic 1, which will eventually lead to an unstable operating condition. We never want both inputs to be logic 0's at the same time; this will be considered as an *undefined state.* Undefined means that we cannot be sure how the device is going to react to this particular combination of inputs. Figure 8.10b illustrates the completed truth table for this device.

The device that we have been examining is called an S-R latch. The inputs are called S-not and R-not. The S and R letters stand for the operations of set and reset. The outputs are called Q and Q-not ($\overline{Q}$). (Recall that the outputs must always be complements for proper operation.) The word "latch" is the description of this circuit's behavior. It latches or grabs inputs and remembers them on the output. The following definitions are in order.

SET a latch Provide the proper input combination to force the Q output high and the $\overline{Q}$ output low. With the NAND type S-R latch, a set operation is accomplished by placing a logic 0 on the S input and a logic 1 on the R input.

RESET a latch Provide the proper combination of inputs to force the Q output low and the $\overline{Q}$ output high. On a NAND type S-R latch, this means that the S input should be at logic 1, and the R input at logic 0.

As their names imply, setting and resetting are inverse operations. Figure 8.11 summarizes the S-R latch.

8.1.2 Cross-Coupled NOR Gates

You know that any function accomplished by NAND gates can also be implemented using NOR gates. We are going to examine the operation of an NOR-type S-R latch (Figure 8.12). Unlike the NAND gate, the dynamic input level of an NOR gate is a logic 1.

A logic 1 on the input of an NOR gate will force a logic 0 on the output.

We need not start our analysis from ground zero. The NAND-

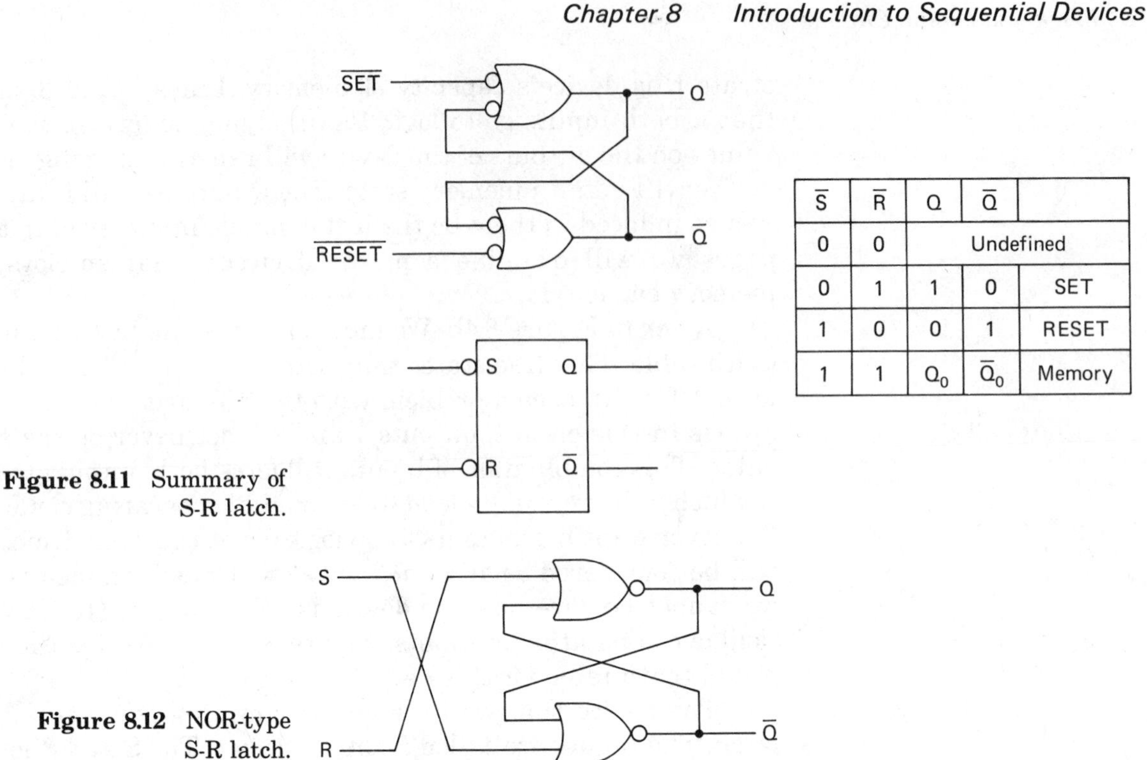

$\overline{S}$	$\overline{R}$	Q	$\overline{Q}$	
0	0	Undefined		
0	1	1	0	SET
1	0	0	1	RESET
1	1	Q_0	$\overline{Q}_0$	Memory

Figure 8.11 Summary of S-R latch.

Figure 8.12 NOR-type S-R latch.

type S-R latch principles can easily be applied to the operation of the NOR-type S-R latch. Most important, we learned that the outputs, Q and $\overline{Q}$, must always be complements. Because a NAND gate has a logic 0 dynamic input level, the input combination where both S and R were equal to logic 0 was undefined. We can take this knowledge and apply it to the NOR-type latch. It should seem reasonable that the input combination where both S and R are equal to logic 1's would be undefined for the NOR-type latch. Figure 8.13 illustrates this situation. Both Q and $\overline{Q}$ will be forced to a logic 0 level. This will create an unstable situation in the circuit:

An undefined state in an NOR-type latch is when both S and R inputs are high.

The memory state of an NOR-type latch (Figure 8.14) will be when both S and R inputs are at their nondynamic levels of logic 0. This corresponds to the memory condition of nondynamic logic 0's on the NAND-type latch. With the NAND-type latch, an active low on the Set input would set the latch. Because the NOR has a dynamic input level of logic 1, we set the NOR latch by

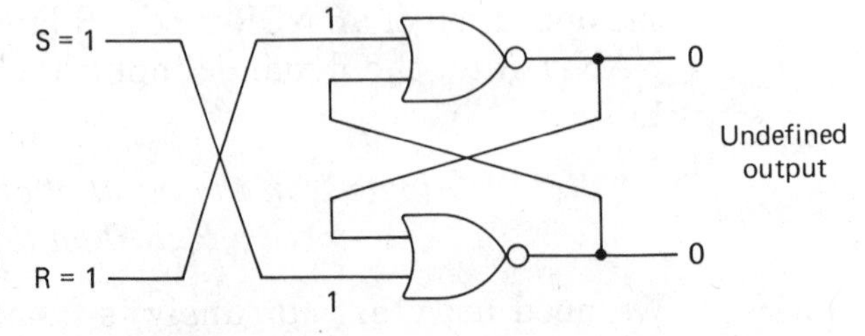

Figure 8.13 Inputs S and R are both equal to logic 1; output is undefined.

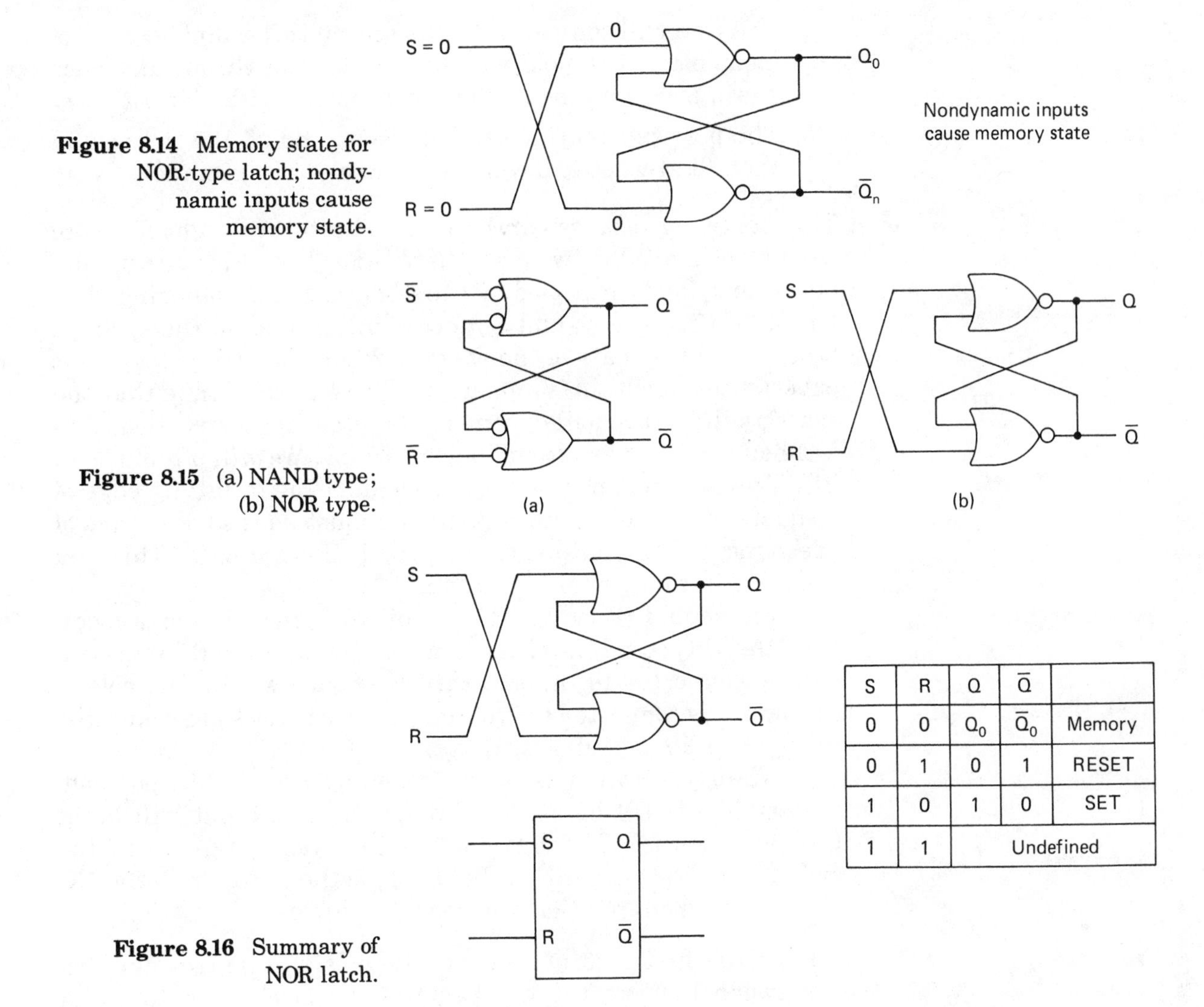

Figure 8.14 Memory state for NOR-type latch; nondynamic inputs cause memory state.

Figure 8.15 (a) NAND type; (b) NOR type.

S	R	Q	$\overline{Q}$	
0	0	Q_0	$\overline{Q}_0$	Memory
0	1	0	1	RESET
1	0	1	0	SET
1	1	Undefined		

Figure 8.16 Summary of NOR latch.

placing a logic 1 on the set input. As you have surely noticed, the set and reset inputs are applied to different gates in the NAND and NOR latches (Figure 8.15). The set input of the NOR latch is directed to the $\overline{Q}$ gate. A logic 1 will force $\overline{Q}$ low, which will in turn cause the Q output to go high. Figure 8.16 summarizes the NOR-type S-R latch.

8.1.3 Application of Latches as Mechanical Switch Debouncers

The most popular application for S-R latches is debouncing mechanical switches. Consider an SPDT (single-pole, double-throw) switch (Figure 8.17). The events that will happen when the switch is toggled are:

Pole — Throw 1 — Throw 2

Figure 8.17 Schematic symbol for SPDT switch.

1. When the switch is first toggled, the throw will leave the pole that it was connected to and start its travel toward the other throw. There will be a short period of time when the pole is not connected to either throw.

2. The pole will connect with the second throw and because of momentum the pole will bounce off the throw, and once again it will not be in electrical contact with either throw.
3. The pole will continue to bounce on and off the throw, and then finally settle down into solid electrical contact.

This bouncing process can be modeled after a weight hanging from a spring. When we drop the weight, it will bounce up and down for a certain period of time before finally stopping. The length of time the weight bounces will depend on the spring's tension and the mass of the weight. When a switch is toggled it experiences much the same action. The length of time that the switch will bounce will depend on its internal construction. Typical bounce times are from 50 μs to 25 ms. We will often use an SPDT switch to generate a good, clean rising or falling edge of a square wave. To accomplish this we must have some means of "debouncing" the output of the switch (Figure 8.18). This is a good application for the S-R latch.

You should recognize the circuit in Figure 8.19 as a modified NAND-type S-R latch. Instead of having the S-R inputs, we will be connected to the two throws of the switch. The pole of the switch is connected to ground. The two throws are connected to V_{cc} via 1-kΩ pull-up resistors.

When the switch is in the normally closed (NC) position, the set input will be at +5 V and the reset input will be at ground potential. The S-R latch is in a reset state when the switch is in its NC position. Let's toggle the switch into the NO (Normally Open) position and see what happens.

1. At the first instant in time, the pole will be traveling between the two throws. Because the pole is not touching either throw, the inputs to the latch are both equal to logic 1, pulled up to V_{cc}, via R1 and R2. This is the memory state of a latch of a NAND-type S-R latch; therefore, the latch is still in the reset condition.
2. The instant in time that the pole touches the normally open throw, the set input will go to logic 0, the reset input is already at logic 1. This corresponds with the set operation of a NAND type S-R latch; Q will go high and $\overline{Q}$ will go low.
3. The first bounce will now occur. As the pole bounces off the normally open throw, the outputs do not change states because the latch has entered the memory state. Both set and reset inputs are equal to logic 1's.
4. As the pole bounces back and forth off the NO throw, the latch will alternate between set and memory operations.
5. This process of memory state and set will continue until the bouncing finally stops. The output waveform is completely clean.

When the switch is toggled back into the normally closed position, the opposite action will result.

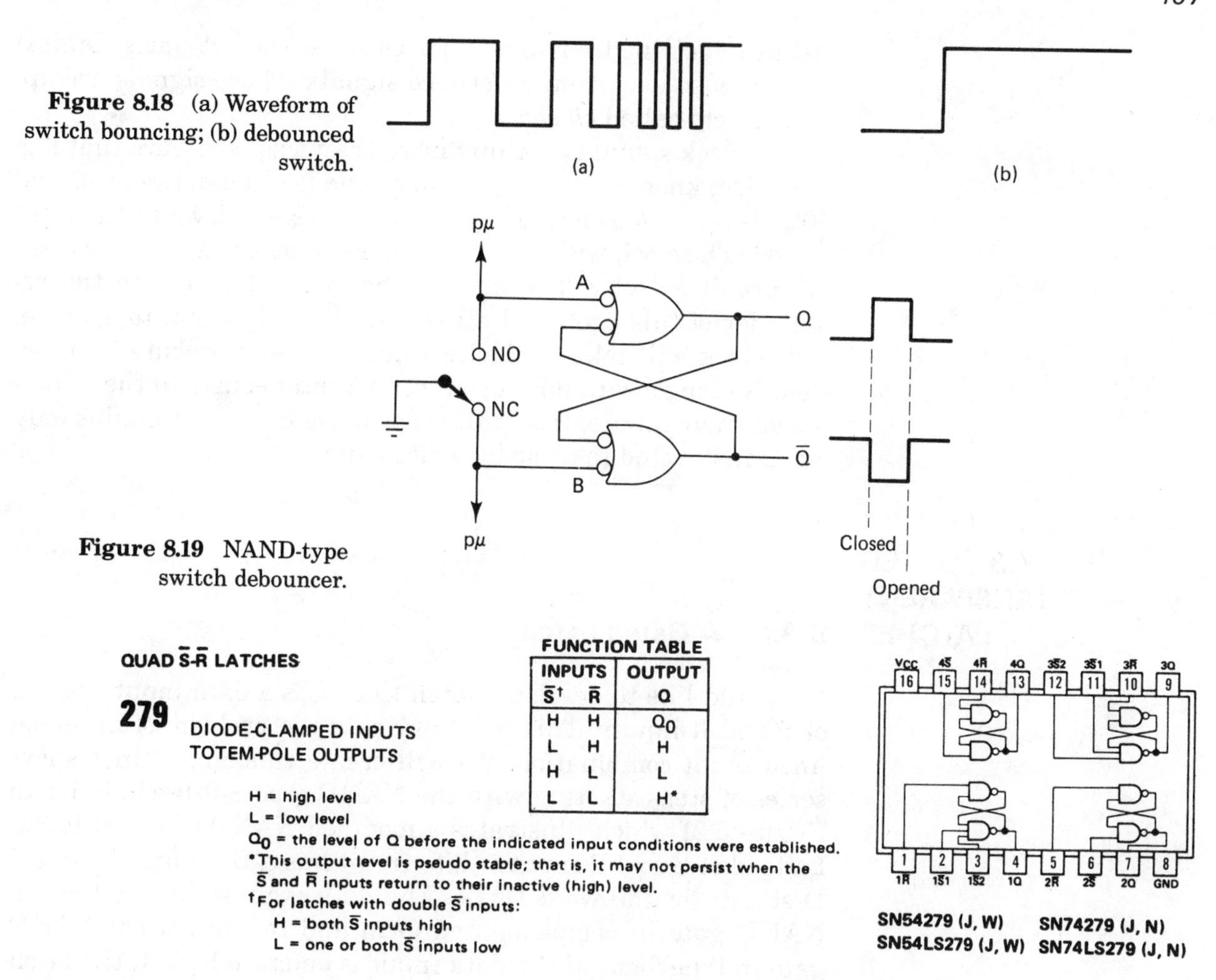

FUNCTION TABLE

INPUTS		OUTPUT
$\bar{S}$†	$\bar{R}$	Q
H	H	Q_0
L	H	H
H	L	L
L	L	H*

Figure 8.18 (a) Waveform of switch bouncing; (b) debounced switch.

Figure 8.19 NAND-type switch debouncer.

Figure 8.20 74LS279 quad S-R latches.

8.1.4 The 74LS279 Quad S-R Latch

When one or two debounced switches are needed, a quad NAND gate can be used. When many debounced switches are required the 74LS279, quad S-R Latches are employed. Refer to the function table in Figure 8.20. This is the standard truth table for a NAND-type S-R latch. The pinout indicates that two of the latches have double set inputs. The note under the function table explains that both set inputs must be high for the set input to go high; if either set input is low, the set input will be low. For most applications both set inputs will be tied together.

8.2 THE CLOCK AND DIGITAL ELECTRONICS

Until this point in our study of digital electronics, the outputs of the devices have reacted immediately to the stimulus of the inputs. Circuits such as these are called *asynchronous.* This term is a combination of Greek words that mean "without regard to time." In digital systems it is important that outputs change at precise points in time. Circuits that operate in this manner are called *synchronous* circuits. We rely on clocks to assist us in getting from and to places at the correct times. Digital circuits also need time reference signals. These signals are appropriately called *clocks*.

A clock signal is nothing more than a square wave that has a precise, known period. The duty cycle is often between 10 and 50%. *Duty cycle* refers to the ratio of a logic 1 level to the total period. The clock will be the timing reference that synchronizes all circuit activity. The devices that we will study in the remainder of this book will all require a clock signal to operate. This clock will tell the device when it should execute its function. No longer will devices react instantaneously to the stimulus on their inputs; they will react to their input stimulus only when instructed to do so by a clock signal.

8.3 GATED TRANSPARENT LATCHES

8.3.1 A Gated Latch

We would like to develop a latch that uses a data input instead of S and R inputs. This will eliminate the problem of an undefined input combination. We will create this circuit in a short series of steps starting with the NAND-type S-R latch. Refer to Figure 8.21 which illustrates a modified NAND-type S-R latch. Instead of S and R inputs, this circuit has a D input. The letter D stands for the word DATA. The data is presented to the top NAND gate in complemented form and to the bottom NAND gate in true form. If the data input is equal to logic 1, the latch will SET; if the data is equal to logic 0, the latch will RESET. Because it is impossible to present logic 1's to both the set and reset inputs, this latch does not have a memory state. Figure 8.22 is a further modification of Figure 8.21. NAND gates, G1 and G2, have been added to this circuit. G1 and G2 function as electronic switches. Driven by the input called "GATE," G1 and G2 will either block the data input or act as inverters of the data input. Refer to the truth table in Figure 8.22.

Line 1 If the gate input is at logic 0, the data input is a don't-care state and the output of the latch will be in the memory state. Notice that the gate input goes to both G1 and G2. If this input is low, the outputs of both G1 and G2 will be forced high. This will make the internal S-R inputs both equal to logic 1's,

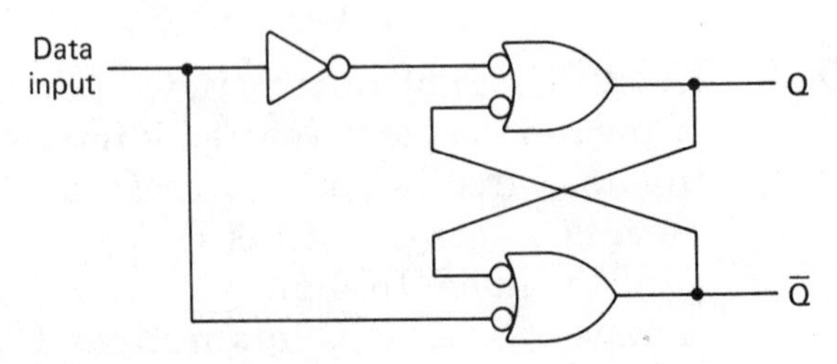

Figure 8.21 Modified S-R latch.

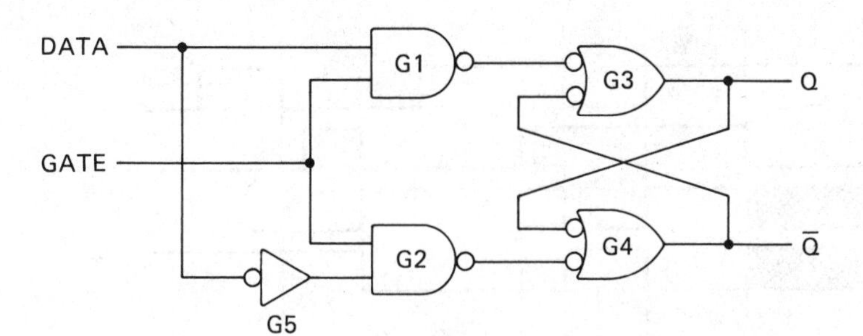

GATE	DATA	Q	$\overline{Q}$	
0	X	Q_0	$\overline{Q}_0$	Memory
1	0	0	1	RESET
1	1	1	0	SET

Figure 8.22 Logic diagram and truth table for discrete gated latch.

which is a memory state. In this situation, the data is in a don't-care state because G1 and G2 are disabled by the logic 0 input of the gate input.

Lines 2 and 3 The gate input level is equal to logic 1. Because logic 1 is the nondynamic input level of a NAND gate, G1 and G2 will now function as inverters. That is why the location of the inverter was switched in Figures 8.21 and 8.22.

If the gate input is at logic 1 and the data input is equal to logic 0, the latch will RESET. Remember that G1 and G2 are acting as inverters to the data input signal. If the gate input is at logic 1 and the data input is equal to logic 1, the latch will SET.

We have now accomplished the construction of a device with a data input that can SET, RESET, and REMEMBER. The key to this device is the gate input. The gate is often referred to as an enable. If it is high, the latch will be enabled to accept a data input; if it is low, the latch will be disabled and in a memory state.

8.3.2 The 74LS75 4-Bit Bistable Latch

The 74LS75 is an example of a gated latch. Another popular name for gated latches is *transparent latches.* Transparent is a good description of how the latch appears when it is enabled.

4-BIT BISTABLE LATCHES

75

FUNCTION TABLE
(Each Latch)

INPUTS		OUTPUTS	
D	G	Q	$\overline{Q}$
L	H	L	H
H	H	H	L
X	L	Q_0	$\overline{Q}_0$

H = high level, L = low level, X = irrelevant
Q_0 = the level of Q before the high-to-low transistion of G

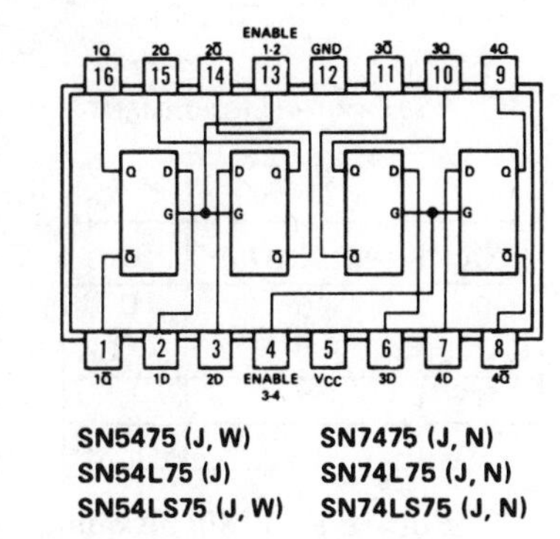

Figure 8.23 74LS75 4-bit bistable latches.

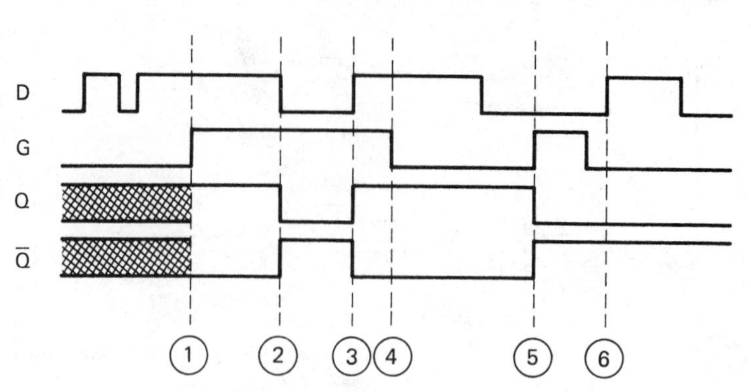

Figure 8.24 74LS75 timing diagram.

Refer to the function table in Figure 8.23. If the G input (gate), is high, the logic level that appears on the D input will also appear on the Q output. The device appears to be transparent to the data input (i.e., the data flow straight through the latch). When the G input goes low, the data most recently on the data input before the high-to-low transition of the gate signal will be saved on the Q output. Notice on the pinout that the gate signal is called the enable. Latches 1 and 2 share one enable and latches 3 and 4 another enable.

Refer to the timing diagram of the 74LS75 (Figure 8.24). Previous to the gate going high, at event 1, we do not know what level the Q and $\overline{Q}$ outputs are. (Crossed lines on timing diagrams are used to indicate an unknown state.) When the gate goes high, the latch becomes transparent. We see that the Q output follows the D input for events 1, 2, and 3. Event 4 indicates that the gate has gone to logic 0; the latch is now disabled to further data inputs and is in a memory state. The last level of data to go through the latch while the gate was still high was a logic 1. The Q output will stay at logic 1 until the gate once again goes high and some low data are presented at the D input. That is exactly what happens at event 5. Event 6 shows that even though high data are presented to the latch, the gate is disabled, so the Q output will not respond.

8.3.3 The 4042B Quad Clocked D Latch

Figure 8.25 illustrates the functional diagram and truth table for the 4042B. This latch is interesting because the polarity of the enable signal is programmable. With the 74LS75, a logic 1 on the gate passed data and a logic 0 latched data. Refer to pin 6 of the 4042B's functional diagram. This polarity pin controls the active level of the enable. In this 4042B the enable is called the clock. If the polarity pin is low, the latch will be transparent when the clock is low and in a memory state when the clock is high. If the polarity pin is held high, the opposite conditions are true. The truth table in Figure 8.25 summarizes the operation of the 4042B. The word "latch" is another way of referring to the memory state of a latch.

CD4042B
FUNCTIONAL DIAGRAM

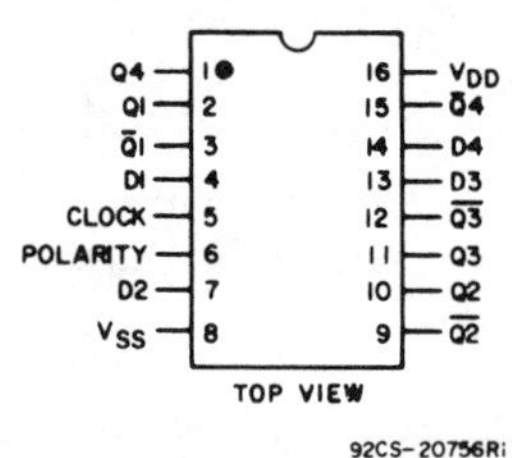

TERMINAL ASSIGNMENT

CLOCK	POLARITY	Q
0	0	D
↗ (rising edge)	0	LATCH
1	1	D
↘ (falling edge)	1	LATCH

Figure 8.25 4042B quad clocked D latch.

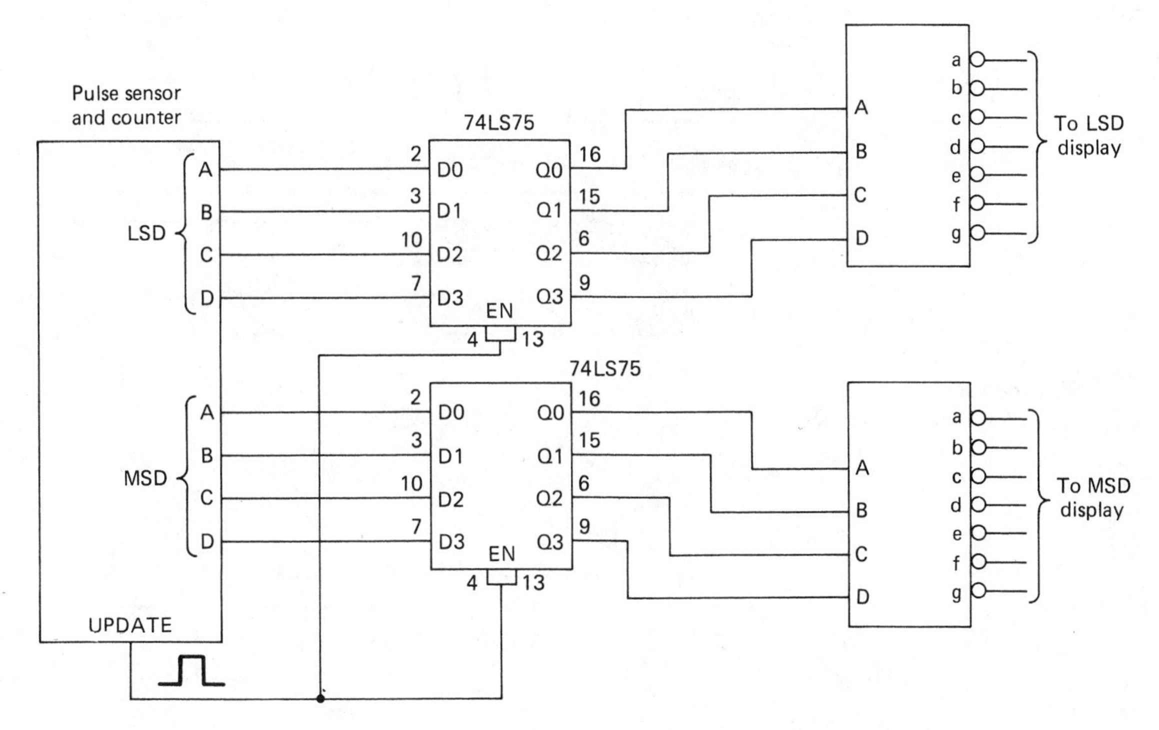

Figure 8.26 Application of 4-bit latches.

8.3.4 Applications of Clocked Latches

Consider the following problem: An electronic pulse meter senses and calculates pulse rates. Every 10 seconds the pulse meter outputs a narrow strobe indicating that the updated BCD code for the two decimal digits is on the output lines. We must have some method of saving the BCD codes when the strobe goes active. Refer to Figure 8.26. While the pulse meter is busy counting the next 10-second sample of pulses, the displays are unaffected. They are driven by the two 74LS75 4-bit latches. The latches will receive the new count every time the strobe goes to a logic 1 level. Both TTL and CMOS MSI display drivers are available with on-board latches. An example of a display driver incorporating an on-board latch is the CMOS 4511B.

8.4 CLOCKS AND TRANSITIONS: RISING AND FALLING EDGES

We have talked about the clock and how it relates to digital circuits as a timing reference. Consider the transparent latches that we have just examined. Data will flow through the latch whenever the enable is active. If we wanted to catch some data at an exact moment in time, we would have to make the enable active for a very short period. Refer to Figure 8.27. The enable pulses in Figure 8.27 are narrow, but there still is a short amount of time when the latch is transparent. Event 2 shows us that the wrong data have slipped through because of the width of the gate. Instead of letting data through on the active level of the enable, why not only let the data through on the rising or

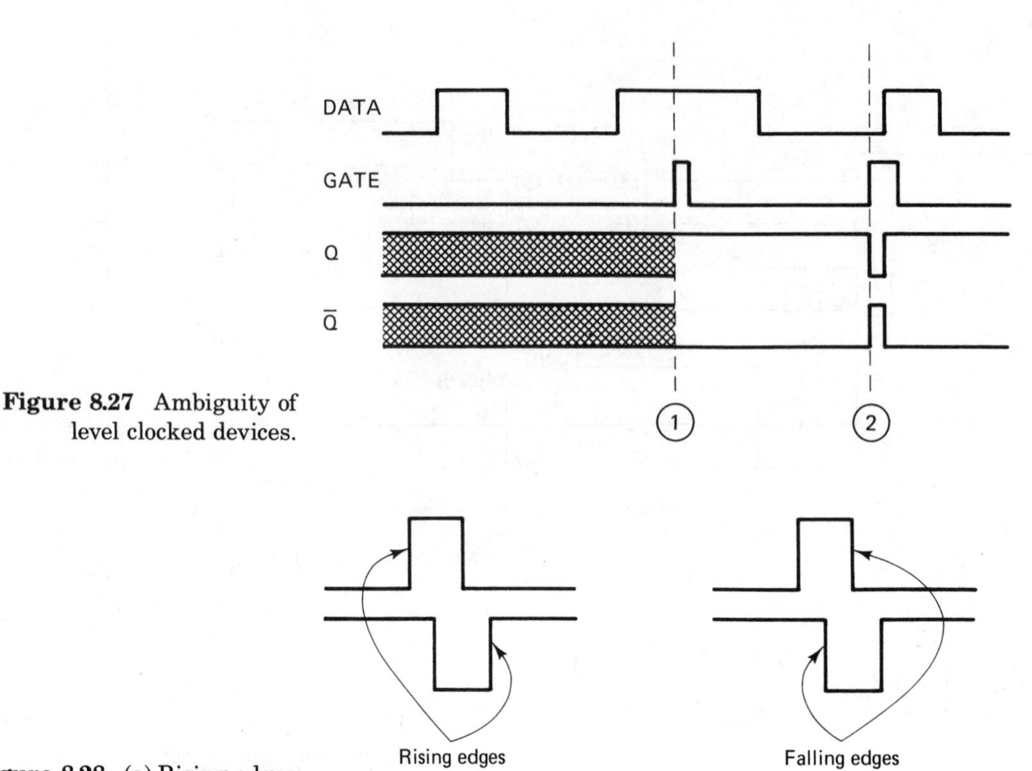

Figure 8.27 Ambiguity of level clocked devices.

Figure 8.28 (a) Rising edges; (b) falling edges.

falling edge of the enable? A square wave is either a logic 0 or a logic 1. This statement assumes that the rise and fall times of a square wave must be infinitely fast. We know, of course, that it does take a finite amount of time for a square wave to go from high to low or low to high. This time is so fast that it appears to be instantaneous compared to the speed of the other signals in the circuit. We are now going to study an important group of digital devices called flip-flops. Flip-flops have many similarities to gated latches. There will be one important difference. In gated latches data is transferred during the active level of the enable; in flip-flops, data is only transferred during the active edge of the enable. The enables on flip-flops are called *clocks.*

Rising edge A logic 0-to-logic 1 transition of a digital signal.

Falling edge A logic 1-to-logic 0 transition of a digital signal.

Refer to Figure 8.28 for examples of rising and falling edges.

8.5 THE D FLIP-FLOP

The "D" in "D flip-flop" stands for the word "data." The function of a D flip-flop is simple:

On the rising edge of the clock signal the logic level that is residing on the D input will be transferred to the Q output.

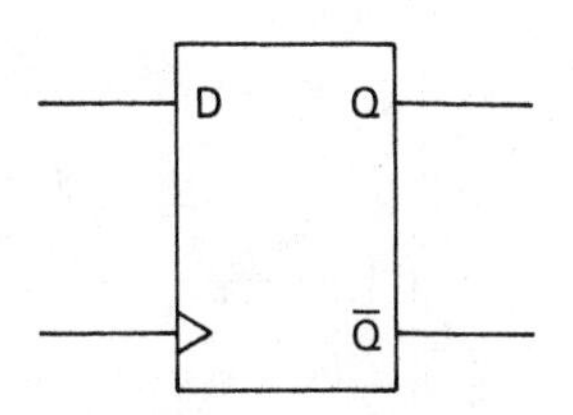

Figure 8.29 Logic symbol for data flip-flop.

Notice that this is an edge-, not a level-triggered device. Figure 8.29 illustrates the standard logic symbol for a D flip-flop. The clock input is indicated by a symbol that looks like half a diamond. This symbol means that the clock is edge active. Because this symbol is not preceded by a bubble, you should assume that it means a positive edge-triggered clock. Figure 8.30 illustrates a simple timing diagram for this device. The vertical dashed lines represent the times when the clock has a rising edge. It is only at these rising edges that the output will have the potential to change. Until the first rising edge, we do not know the values of Q and $\overline{Q}$. At rising edge 1, we see that the data input is equal to logic 0. This logic 0 will be transferred into the Q output, and its complement, a logic 1, will be transferred into the $\overline{Q}$ output. The outputs will stay in a memory state until the next rising edge of the clock. At this point the data input is logic 1; the Q output will go to logic 1. At the third rising edge the data input is logic 0 and the Q output will follow this logic 0. The fourth rising edge also occurs when the input data is logic 0; therefore, the Q output will stay at logic 0. The last rising edge in the timing diagram happens when the data is high, so the Q output will go high.

The D flip-flop is the most elementary form of digital memory. Whenever a rising edge of the clock occurs, the data will be stored at the Q output of the flip-flop.

8.5.1 The 74LS74 Dual D-Type Positive Edge-Triggered Flip-Flop

The 74LS74 is a popular D flip-flop. It also has two added functions: a direct preset and clear. Refer to Figure 8.31. You should recognize this device as a positive edge-triggered D flip-flop. Even though you do not know what the preset and clear functions are, you can tell from the pinout and function table that they are active-low inputs. The best way to learn a new device is to spend

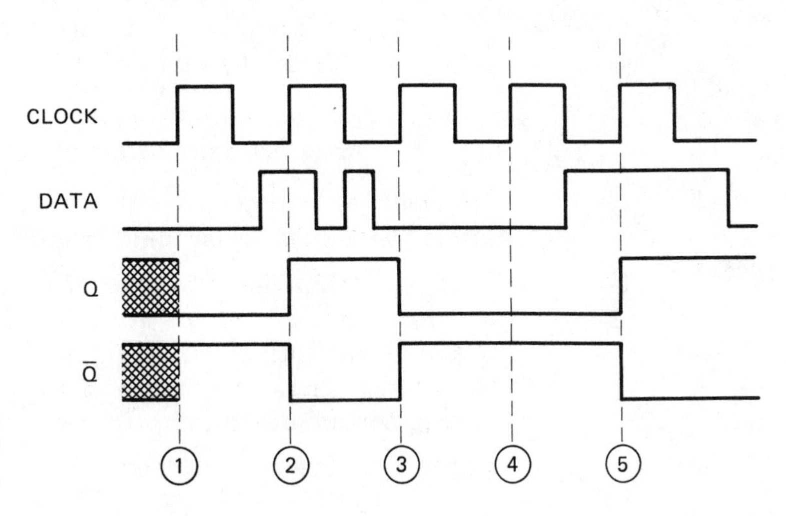

Figure 8.30 Timing diagram for positive edge-triggered D flip-flop.

DUAL D-TYPE POSITIVE-EDGE-TRIGGERED FLIP-FLOPS WITH PRESET AND CLEAR

74

FUNCTION TABLE

INPUTS				OUTPUTS	
PRESET	CLEAR	CLOCK	D	Q	$\overline{Q}$
L	H	X	X	H	L
H	L	X	X	L	H
L	L	X	X	H*	H*
H	H	↑	H	H	L
H	H	↑	L	L	H
H	H	L	X	Q_0	$\overline{Q}_0$

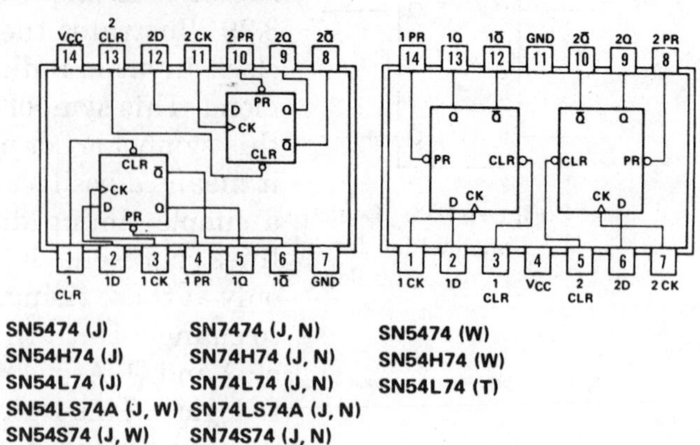

Figure 8.31 74LS74 dual D-type positive edge-triggered flip-flop.

time trying to understand its truth table. Let's examine the truth table line by line.

Line 1 Preset is low, clear is high, and the clock and D inputs are don't cares. This line of the truth table tells us many things. Because the preset and clear inputs are bubbled, they must be active low. Here the preset input is at an active-low level, and the clear input is inactive high. This line represents an operation called *presetting.* This presetting operation must have a higher priority than the normal data clocking operation. This is true because the D and clock inputs are don't cares. "Preset" sounds like the term "set" that we used in describing the S-R latch. You will discover that the terms "preset" and "set" are used interchangeably. According to line 1 of the truth table:

When the preset input is active, the Q output will go high and the $\overline{Q}$ output will go low, regardless of the logic levels on the D and clock inputs.

Line 2 Preset high, clear low, and the clock and D inputs are don't cares. We now see that the clear input is active. The truth table indicates for this situation that:

The Q output will go low and the $\overline{Q}$ output will go high, regardless of the state of the clock or D input.

To "clear" a flip-flop means the same thing as "resetting" an S-R latch. The terms "clear" and "reset" are used interchangeably.

Line 3 Both preset and clear are active; the clock and D input are don't cares. This is a situation that must be avoided. Both Q and $\overline{Q}$ will both be forced high—this output state is unstable. Circuit designers will ensure that this situation will never occur during normal operation.

Lines 4–6 Preset and clear will be inactive for these lines.

Line 4 This flip-flop should now behave exactly like the one that we have previously examined. With a rising edge at the clock and high data on the D input, the Q output will go high and the $\overline{Q}$ output will go low.

Line 5 The same as line 4, but with low data on the D input. Q output will go low and the $\overline{Q}$ output will go high.

Line 6 This line shows that the preset and clear are both high, and the clock does not have a rising edge: nothing is active. The symbols Q_0 and $\overline{Q}_0$ indicate that the output does not change; it is in a memory state.

8.5.2 The 4013B Dual CMOS D-Type Flip-Flop

Refer to Figure 8.32. There are two major differences between the 4031B and the 74LS74. The preset input is called set and it is active high, and the clear input is called reset and it is also active high. Aside from these differences, the truth tables indicate that 74LS74 and the 4013B operate identically.

Some D-type flip-flops do not have preset and clear inputs. If a device has these inputs but they are not being used, they must be tied to their inactive state.

8.5.3 Timing Diagrams of D Flip-Flops

When analyzing timing diagrams of D-type flip-flops, you must remember that preset and clear override the data and clock inputs. These presets and clears are often called "asynchro-

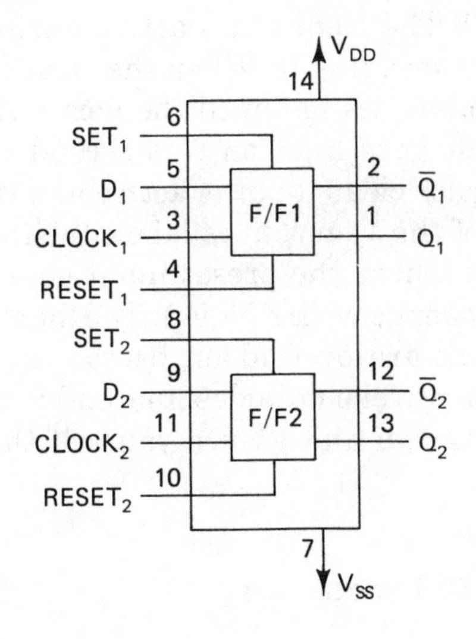

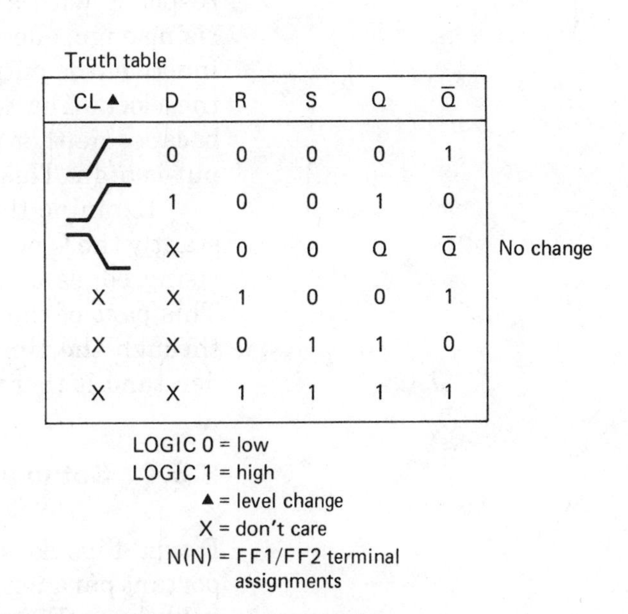
Truth table

CL ▲	D	R	S	Q	$\overline{Q}$	
_/‾	0	0	0	0	1	
_/‾	1	0	0	1	0	
‾_	X	0	0	Q	$\overline{Q}$	No change
X	X	1	0	0	1	
X	X	0	1	1	0	
X	X	1	1	1	1	

LOGIC 0 = low
LOGIC 1 = high
▲ = level change
X = don't care
N(N) = FF1/FF2 terminal assignments

Figure 8.32 4013B CMOS D-type positive edge-triggered flip-flops.

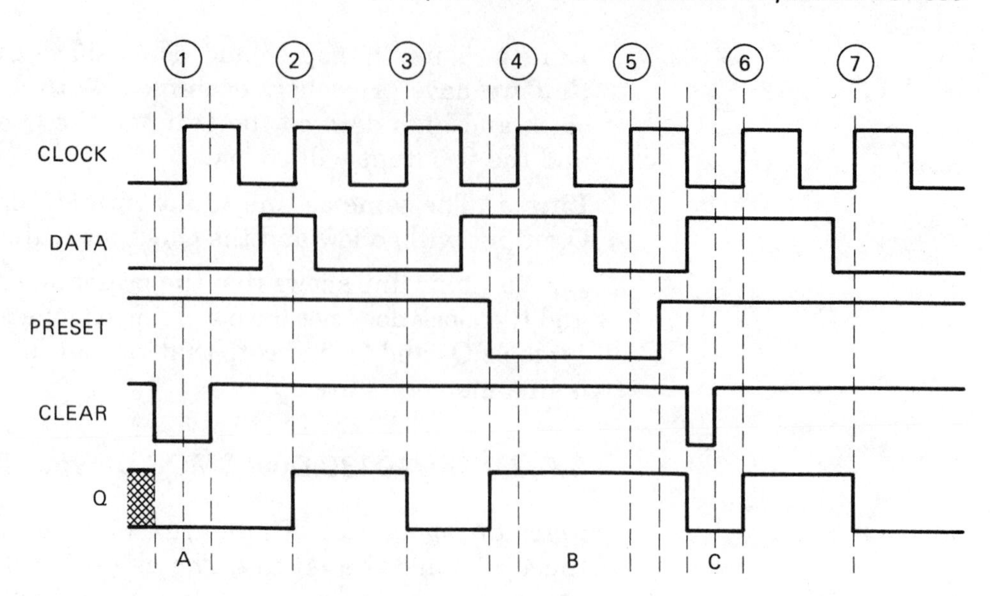

Figure 8.33 Timing diagram for D flip-flop with preset and clear.

nous" inputs because they occur independently of the clock. Refer to Figure 8.33. The best approach to analyzing this timing diagram is to start with the times when the preset or clear inputs are active. The timing diagram is for a 74LS74 with active-low preset and clear. Whenever a preset or clear goes active, the output will automatically respond and stay at that particular level at least until the preset or clear goes inactive and another rising edge of the clock occurs. Refer to the part of the timing diagram where the clear input first goes active. The Q output responds with a logic 0. The clear stays active during the first rising edge, effectively overriding it. When the clear finally goes inactive, the output cannot change until the next rising edge of the clock. The Q output goes high on this second rising edge because neither preset nor clear input is active and the data input is high. This part of the Q output is labeled with an A.

Examine the point where the preset input goes active. Instantly the Q output responds with a logic 1. The fourth and fifth rising edges of the clock are overridden by the active preset. This part of the output waveform is designated by a B. Follow through the timing diagram and assure yourself that you understand it thoroughly.

8.5.4 Setup and Hold Times

D-type flips-flops have many ac parameters. The two most important parameters for technicians to understand are setup and hold times (Figure 8.34).

Setup time The length of time that the data on the D input

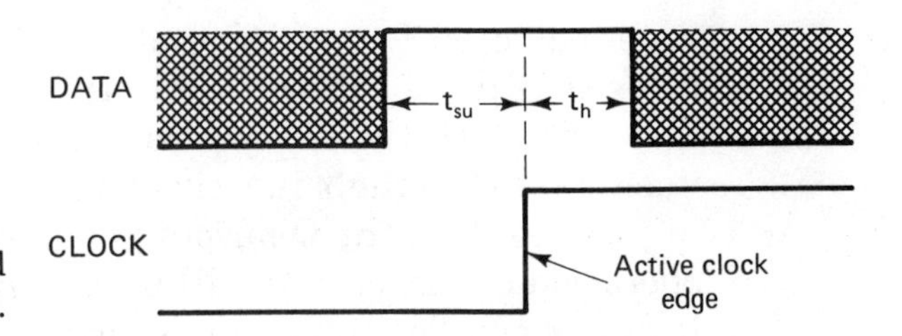

Figure 8.34 Setup and hold times.

must be present and stable before the active edge of the clock signal.

Hold time The length of time that data on the D input must remain stable after the active edge of the clock.

The setup time for a 74LS74 is 25 ns, and the hold time is 5 ns. The setup and hold times for CMOS flip-flops depend on the value of V_{dd}. With a V_{dd} of +5 V, the setup and hold times for a 4042B are 50 and 120 ns, respectively. As V_{dd} increases, the setup and hold times will decrease proportionally.

If a flip-flop circuit is displaying erratic behavior (working sometimes and other times malfunctioning), use a dual-trace oscilloscope to monitor the D-input and clock. Assure yourself that the setup and hold time specifications are being met.

8.5.5 Two Elementary D Flip-Flop Applications

D-type flip-flops are used in a wide variety of applications. In practical circuits, they provide a simple means of saving a bit of data at specified time. Refer to the circuit and timing diagram in Figure 8.35. The flip-flop in Figure 8.35 can be any positive

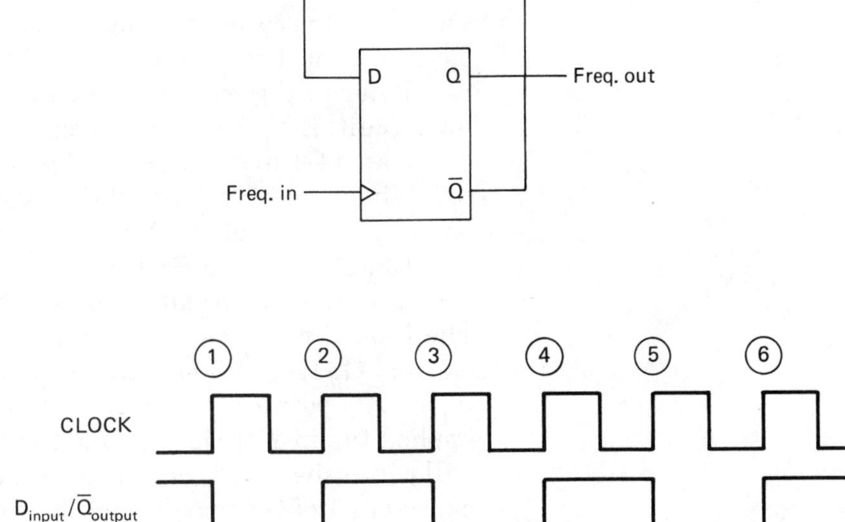

Figure 8.35 D-type flip-flop, configured to toggle.

edge-triggered D type. The absence of set and clear pins indicates that these functions are not being utilized in this particular circuit. (If the flip-flop does have set and clear inputs, they must be tied to their inactive level.)

Notice that the $\overline{Q}$ output is fed back to the data input. The operation of this circuit will be analyzed by constructing a timing diagram. The timing diagram will have three lines: the clock input, Q output, and the data input/$\overline{Q}$ output. All the signals in the circuit will be referenced to the clock. When we create a clock signal for the timing diagram, its absolute frequency and duty cycle are not important. Most clock signals on timing diagrams will be a 50% duty cycle square wave with just enough cycles to analyze the circuit operation fully. We will focus our attention on the rising edges of the clock. Initially assume that the Q output is low and the $\overline{Q}$ output is high. The $\overline{Q}$ output is fed back to the data input of the flip-flop.

Edge 1 With a high on the data input, this rising edge will cause the Q output to go high and the $\overline{Q}$ output to go low.

Edge 2 The data input is now at a logic 0 level. This rising edge will cause the Q output to go low and the $\overline{Q}$ output to go high.

Edge 3 The data input is once again at logic 1. This rising edge will cause the Q output to go high and the $\overline{Q}$ output to go low.

By now you should realize that each time there is a rising edge of the clock, the output of the flip-flop will toggle states. Compare the frequency of the clock signal to the frequency of the signals at Q and $\overline{Q}$.

The output frequency of this flip-flop is exactly one-half of the clock frequency.

This circuit is called a divide-by-2-circuit. It is the basic building block of digital counters.

Refer to Figure 8.36, which illustrates a pulse synchronizing circuit. Let's analyze this circuit in functional parts.

Gates G1 and G2 are configured as an S-R switch debouncer. When the SPDT switch is toggled from the NC to the NO position, the output of G1 will be a clean, debounced rising edge.

G3, G4, G5, and G6 function as an electronic inverting differentiator. This circuit takes advantage of propagation delay. The typical propagation delay of a 74LS04 is 10 ns. Gates G4, G5, and G6 form a 30-ns inverting delay. With the SPDT switch in the NC position, the output of G1 is a logic 0. This logic 0 is applied to pin 2 of G3 and pin 1 of G4. A logic 0 on pin 3 of G3 will cause the preset on U1 to be high. The three inverters will cause pin 1 of G3 to be normally high. When the S-R latch outputs a rising edge, pin 2 of G3 will instantly go high. This will cause the preset input of U1 to go low. Approximately 30 ns after the rising edge out of G1, pin 1 of G3 will finally go low. This

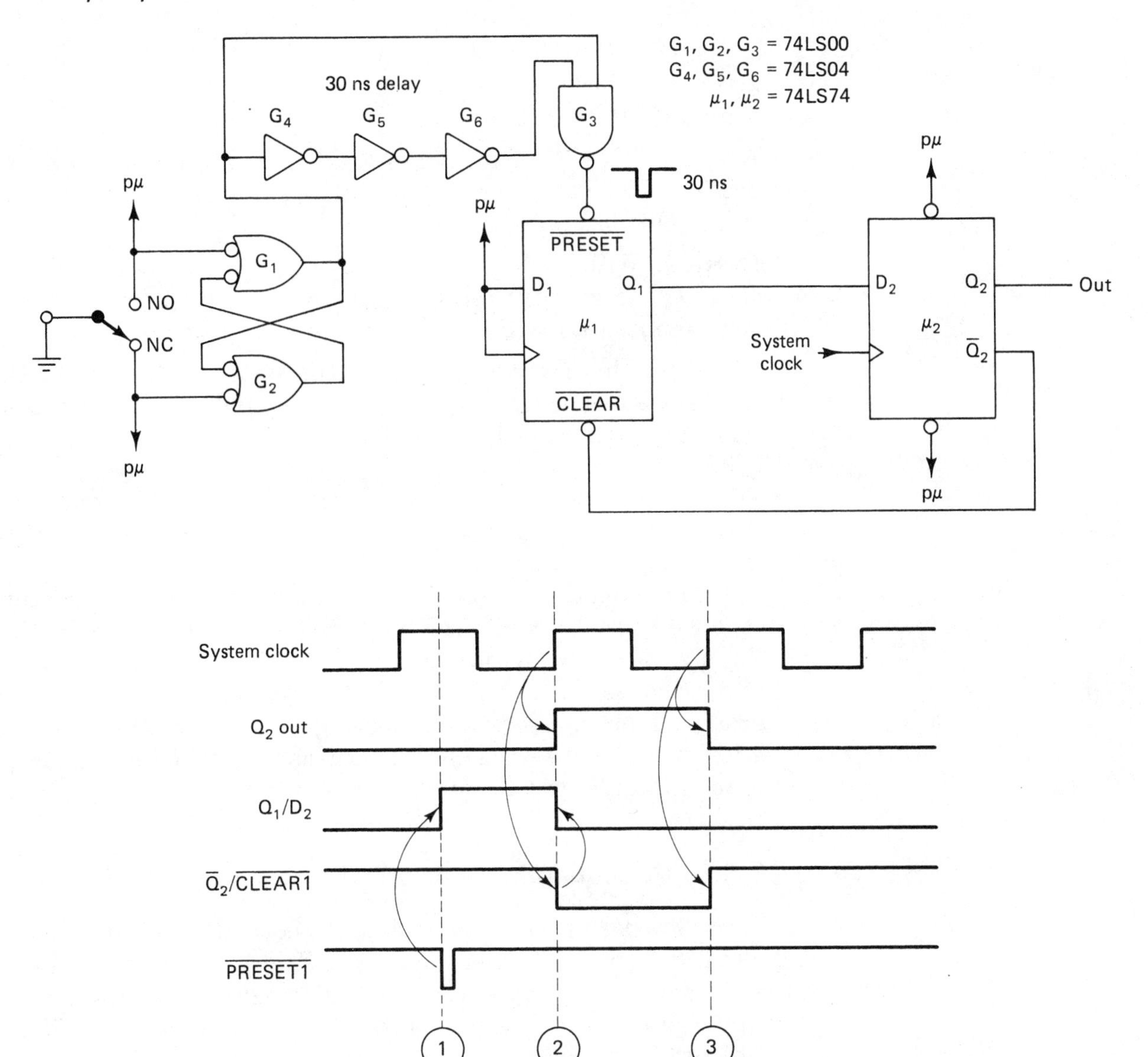

Figure 8.36 Pulse synchronizing circuit.

will cause the preset input of U1 to go high. Therefore, each time the SPDT switch is toggled from the NC to NO position, a 30-ns active-low pulse will appear on the preset input of U1.

U1 is configured as a set/reset flip-flop. Because the clock is tied to V_{cc}, the data input will always be a don't-care condition. The only time this flip-flop will be active is when a low level is applied to the preset or clear inputs. We have already seen that the preset input is driven by the 30-ns negative pulse from the electronic differentiator. The clear input is driven by $\overline{Q}$ from U2. Anytime Q1 is high and a rising edge from the system clock occurs, Q2 will go high and $\overline{Q}2$ will go low. This low from $\overline{Q}2$ will clear flip-flop U1.

Now that we have discussed the functional blocks in this system, it is time to examine the timing diagram. Assume that the SPDT switch is in the NC position and that both U1 and U2 are reset. There are three events of interest on the timing diagram in Figure 8.36.

Event 1 The SPDT switch is toggled from the NC to the NO position. This creates a 30-ns negative pulse on the preset input of U1. This low pulse will cause Q1 to go high.

Event 2 With D2 high, the rising edge of the system clock will cause Q2 to go high and $\overline{Q2}$ to go low. When $\overline{Q2}$ goes low, the clear input of U1 will go active, forcing Q1 low.

Event 3 The next rising edge of the system will transfer the low data onto the Q2 output. $\overline{Q2}$ will go high, releasing the active clear input on U1.

The circuit is now ready for another positive edge from G1. Notice that the Q2 output was one positive pulse synchronized with the system clock.

This simple example demonstrates the important concept of cause and effect in digital circuits. Carefully examine the timing diagram. The arrows at each event illustrate the stimulus and response. This is the manner in which digital circuits operate. When you are troubleshooting a circuit, you must understand all the cause-and-effect relationships. If you see a particular stimulus occur, you must know the circuit's proper response.

8.5.6 A Sequential Electronic Lock

The following circuit is an application of the D flip-flop. We have already examined an electronic lock. That lock used two 74LS85 4-bit magnitude comparators. It was an example of a lock using only combinational logic circuits. The electronic lock in Figure 8.37 uses both combinational and sequential logic circuits. It is an example of a sequential lock that demonstrates the capacity to remember the previous digits that were entered. It is the electronic equivalent of the mechanical lock illustrated in Figure 8.1b. Refer to the block diagram in Figure 8.37. The rising edge of the clock will transfer the four binary combination bits into the lock. When the LED illuminates the lock will be open.

To operate the lock, proceed as follows:

1. Set the first digit of the combination into switches SW 0 through SW 3.
2. Clock this digit into the lock by toggling the SPDT switch.
3. Repeat steps 1 and 2 for the next two digits and the lock will open.

If an incorrect number is entered, the operator must start over at step 1.

Now that you understand how to operate the lock, let's analyze how it actually functions:

The combination is set with the three banks of switches

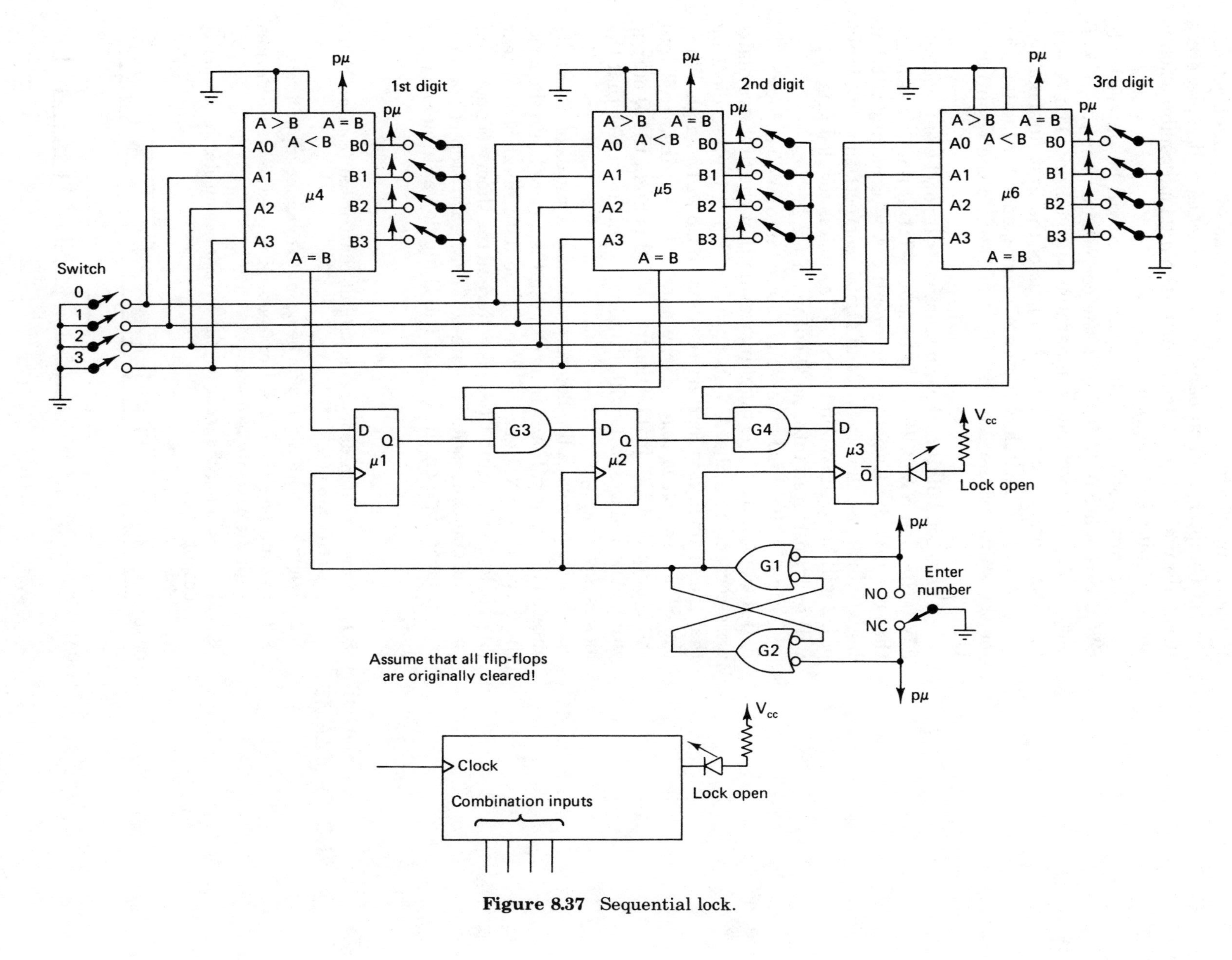

Figure 8.37 Sequential lock.

connected to the magnitude comparators (B0 through B3). This is the same way that the combination was set in the electronic lock of Chapter 5. Each bank of four switches will represent a decimal number from 0 to 15. After the internal combination is set, the unit will be sealed.

The first number of the combination will be entered into SW 0 through SW 3. If this is the correct number the A = B output of the magnitude comparator will go high. The A = B output is connected to the data input of the first flip-flop. Toggling the SPDT switch will create a rising edge that will transfer the status of the A = B output of U4 into the Q output of U1.

The second digit of the combination will now be entered into SW 0 through SW 3. If the Q output of U1 is high (indicating that the first digit was correct) and the A = B output of U5 is high (indicating that the second digit is correct), the data input of U2 will be high. When the SPDT switch is toggled this value will be transferred into the Q output of U2.

Finally, the third number is entered into SW 0 through SW 3. If the output of U2 is high (indicating that the first two digits were correct) and the A = B output of U6 is high (indicating that the third digit is correct), the data on U3 will be high. When the SPDT switch is toggled, the data will be transferred into the Q output of U3. If the Q output of U3 is high, the $\overline{Q}$ output will be low and the LED will illuminate. This indicates that the lock is now open.

Notice how each flip-flop remembers the status of the digits entered previously. If any incorrect digit is entered into the lock, the respective AND gate will force the data input of the next flip-flop low, indicating this error. Reread the theory of operation for this sequential lock. Be sure to follow each step of the process. Concentrate on the idea of the D flip-flop as a memory device.

QUESTIONS AND PROBLEMS

8.1. Draw a switch debouncing circuit constructed from an NOR-type S-R flip-flop.

8.2. Why can't both S-R inputs be equal to dynamic input levels?

8.3. Explain how the concept of feedback is used to create devices with the capacity of memory.

8.4. In the timing diagram for the 74LS75, complete the Q and $\overline{Q}$ outputs.

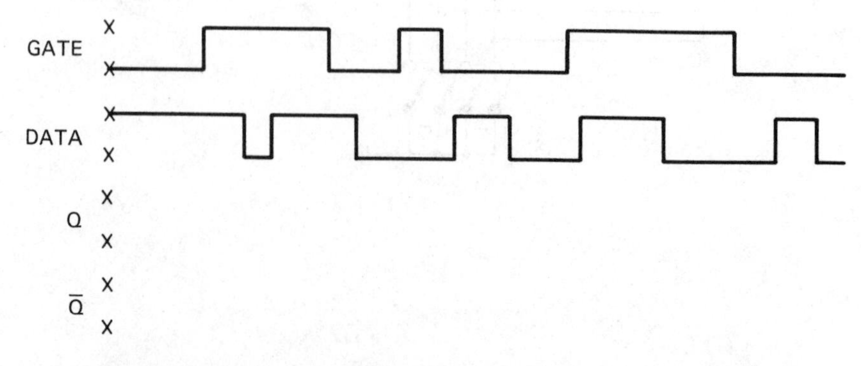

8.5. Construct a function table that describes the circuit shown.

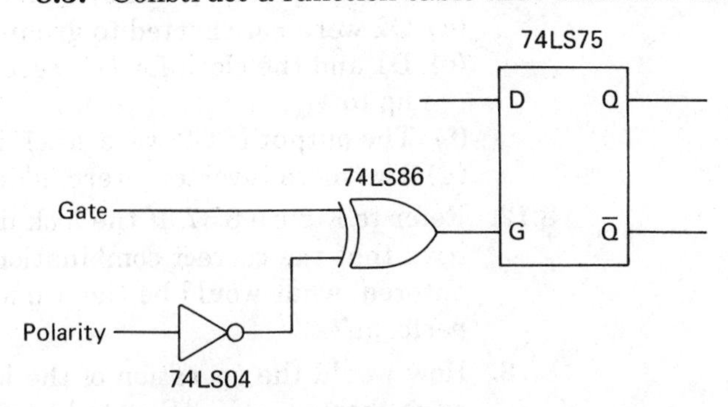

8.6. Refer to Figure 8.26. The D0 input of the top 74LS75 is floating because of a cold solder joint. How will this affect the LSD display? What would happen if the update pulse was shorted to ground?

8.7. In the timing diagram for a positive edge-triggered D-type flip-flop, complete the Q and $\overline{Q}$ outputs.

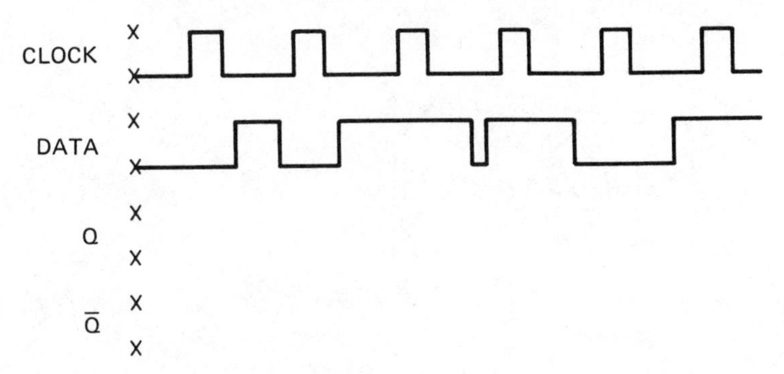

8.8. In the timing diagram for a 4013B positive edge-triggered flip-flop with active high preset and clear, complete the Q output.

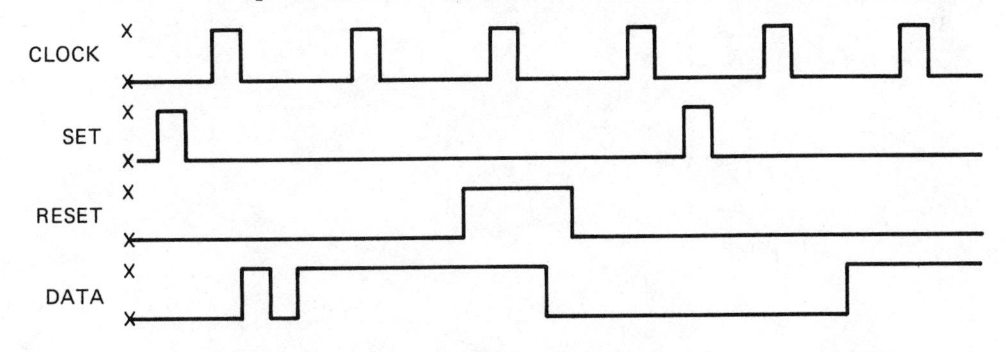

8.9. Refer to Figure 8.35. What would happen if the Q and $\overline{Q}$ outputs shorted together? How would the circuit operate if the Q output, instead of the $\overline{Q}$ output, was connected to the data input? Construct a circuit that divides an input frequency by a factor of 4, and another circuit that divides an input frequency by a factor of 8.

8.10. Construct a timing diagram that illustrates the operation of the circuit shown.

8.11. Refer to Figure 8.36. How would the circuit operation be affected if:

(a) The input to G6 were floating?
(b) D2 were shorted to ground?

(c) The frequency of the system clock were doubled?
(d) Q2 were not shorted to ground?
(e) D1 and the clock for U1 were left floating instead of pulled up to V_{cc}?
(f) The output of G2 were stuck internally low?
(g) Two more inverters were added to the delay circuit?

8.12. Refer to Figure 8.37. If the lock did not open after you are positive that the correct combination in the correct sequence was entered, what would be the sequence of checks that you would perform?

8.13. How would the operation of the lock be affected if Q out of U1 were shorted to V_{cc}? Shorted to ground?

8.14. The clock input to U3 is open via an open feedthrough. How does this affect circuit operation?

8.15. What checks would you perform if the lock appeared to always be open, regardless of the input combination?

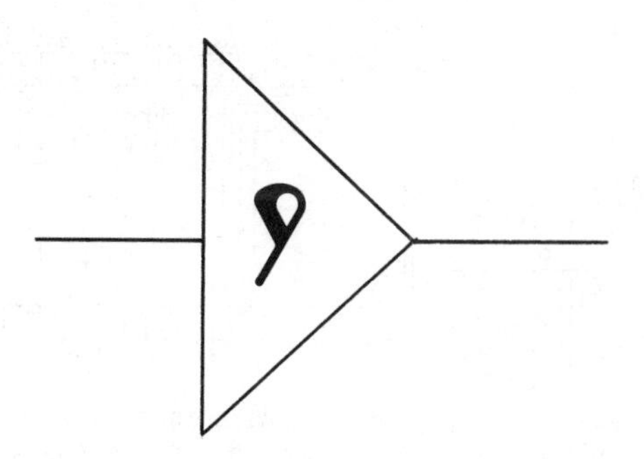

ADVANCED SEQUENTIAL DEVICES: THE J-K FLIP-FLOP AND ONE-SHOT

9.1 THE J-K FLIP-FLOP

In Chapter 8 we introduced sequential devices by illustrating the S-R latch. The S-R latch had one major problem—one combination of S and R inputs would result in an undefined output. With this limitation in mind, gated latches and edge-triggered D flip-flops were developed. Refer to Figure 9.1. The logic symbol indicates a negative edge-triggered clock. Instead of S and R inputs or a data input, we see J and K inputs. (Unlike the letters "S" and "R" on the S-R flip-flop, the letters "J" and "K" give no clue about the functions of the pins that they label. The letters "J" and "K" are merely identifiers.) Because of the J and K inputs and the edge-triggered clock, the J-K flip-flop appears to be a hybrid of the S-R latch and the D flip-flop.

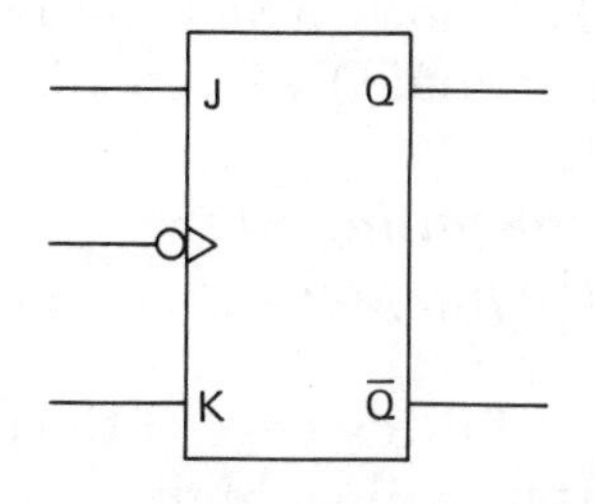

Figure 9.1 Basic logic symbol for the J-K flip-flop.

9.1.1 The 74LS76 Dual J-K Flip-Flop

Refer to Figure 9.2. Notice that the function table for the 7476 and 74H76 has a positive pulse symbolized in the clock column, whereas the function table for the 74LS76 has a falling edge under its clock column. The logic symbol in the pinout diagram is applicable to all three J-K flip-flops, and it indicates that the devices are negative edge-triggered. This discrepancy is easy to explain: J-K flip-flops are internally constructed from two dif-

DUAL J-K FLIP-FLOPS WITH PRESET AND CLEAR

76

'76, 'H76
FUNCTION TABLE

INPUTS					OUTPUTS	
PRESET	CLEAR	CLOCK	J	K	Q	$\bar{Q}$
L	H	X	X	X	H	L
H	L	X	X	X	L	H
L	L	X	X	X	H*	H*
H	H	⎍	L	L	Q_0	$\bar{Q}_0$
H	H	⎍	H	L	H	L
H	H	⎍	L	H	L	H
H	H	⎍	H	H	TOGGLE	

'LS76A
FUNCTION TABLE

INPUTS					OUTPUTS	
PRESET	CLEAR	CLOCK	J	K	Q	$\bar{Q}$
L	H	X	X	X	H	L
H	L	X	X	X	L	H
L	L	X	X	X	H*	H*
H	H	↓	L	L	Q_0	$\bar{Q}_0$
H	H	↓	H	L	H	L
H	H	↓	L	H	L	H
H	H	↓	H	H	TOGGLE	
H	H	H	X	X	Q_0	$\bar{Q}_0$

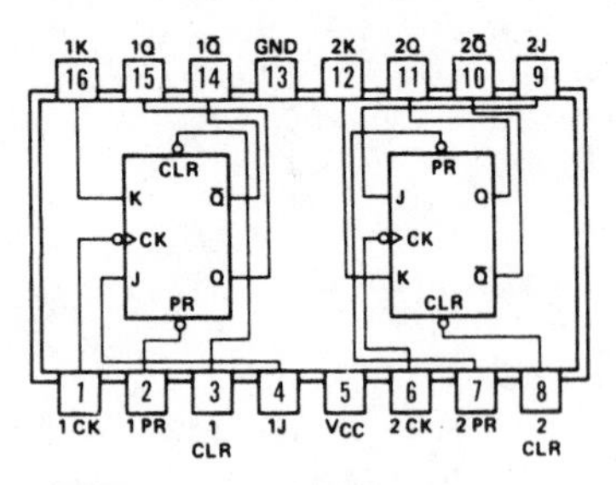

SN5476 (J, W) SN7476 (J, N)
SN54H76 (J, W) SN74H76 (J, N)
SN54LS76A (J, W) SN74LS76A (J, N)

Figure 9.2 74LS76 dual J-K flip-flop with preset and clear.

ferent flip-flops called the *master* and the *slave*. In older technology J-K flip-flops, such as the 7476 or 74H76, the information on the J-K inputs was transferred into the master portion of the flip-flop on the rising edge of the clock, and into the slave (output) portion of the flip-flop on the falling edge of the clock. Because the output would react on the falling edge of the clock, these devices appeared to be negative edge-triggered. The newer technology J-K flip-flops, such as the 74LS76, are true edge-triggered devices. The important thing to remember is that whether the clock symbolized in the function table is a positive pulse or a falling edge, the device will appear to be negative edge-triggered.

Let's follow the function table for the 74LS76 line by line:

Lines 1–3 The preset and clear functions work exactly the same way as they did on the D flip-flop. The logic diagram shows that they are active low. As always, both preset and clear should not be active simultaneously.

Line 4 Both preset and clear are inactive, and the clock is showing a falling edge. With the J and K inputs both at logic 0 the output will not change. Therefore:

*When J and K are both at logic 0,
the flip-flop is in the memory state.*

Line 5 In this line J is high and K is low. When the falling edge of the clock occurs, Q will go high and $\bar{Q}$ will go low.

*When J = 1 and K = 0, the falling edge
of the clock will set the flip-flop.*

Line 6 J is low and K is high. On the falling edge of the clock, the Q output goes low and the $\bar{Q}$ output will go high.

*When J = 0 and K = 1, the falling edge of the clock
will reset the flip-flop.*

Line 7 Both J and K inputs are high. This is the operation that makes the J-K flip-flop such a flexible device. Under the outputs column we see the word "toggle." Each output, Q and $\overline{Q}$, will change to the complement of their value before the falling edge of the clock. This is how the J-K flip-flop handles the previously undefined operation.

When J = 1 and K = 1, the falling edge of the clock will cause the Q and $\overline{Q}$ outputs to toggle.

Line 8 This line symbolizes an inactive clock. Unless the clock has a falling edge, or the preset or clear inputs go active, the Q and $\overline{Q}$ outputs will not change.

9.1.2 The 4027B Dual J-K Flip-Flop

This CMOS J-K flip-flop operates in much the same manner as the 74LS76. The set and reset inputs are active high and it is positive edge-triggered. Other than these slight differences the 4027B should be easy for you to understand. Let's follow the function table (Figure 9.3) line by line.

Line 1 J = 1 and K = don't care, and the present state of the Q output is low. On the rising edge of the clock, the Q output will go high and the $\overline{Q}$ output will go low. This line symbolizes both the set (if K = 0) and the toggle (if K = 1) operations. The initial assumption in the function table was that the Q output started low.

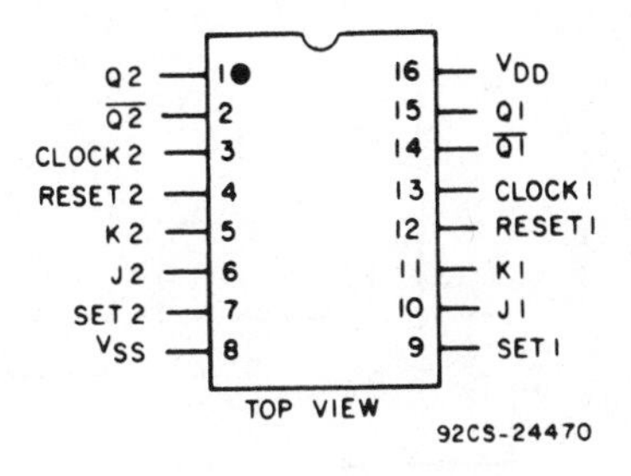

TERMINAL ASSIGNMENT

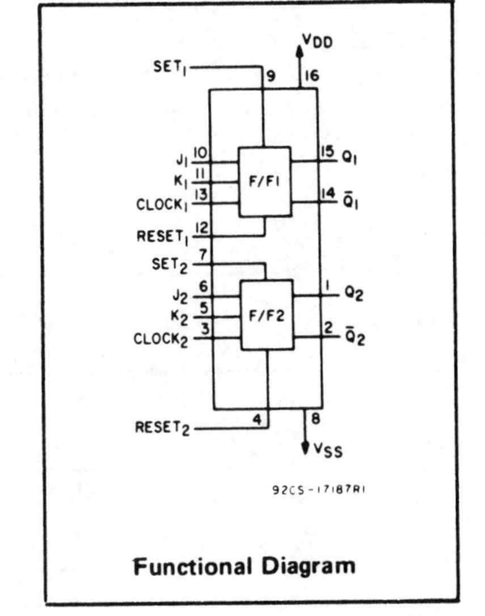

Functional Diagram

PRESENT STATE						NEXT STATE		
INPUTS				OUTPUT	CL▲	OUTPUTS		
J	K	S	R	Q		Q	$\overline{Q}$	
1	X	0	0	0	↑	1	0	
X	0	0	0	1	↑	1	0	
0	X	0	0	0	↑	0	1	
X	1	0	0	1	↑	0	1	
X	X	0	0	X	↓			← NO CHANGE
X	X	1	0	X	X	1	0	
X	X	0	1	X	X	0	1	
X	X	1	1	X	X	1	1	

LOGIC 1 = HIGH LEVEL
LOGIC 0 = LOW LEVEL
▲ LEVEL CHANGE
X DON'T CARE

92CM-27551R1

Figure 9.3 4027B dual J-K flip-flop.

Line 2 J = don't care, K = 0, and Q is high. After the rising edge of the clock, Q will still be high. This line symbolizes both the memory state (if J was low) and the set operation (if J was high). So far the TTL and CMOS J-K flip-flops are performing exactly alike. Do not let the differences in the format of the truth tables throw you. They are both saying the same thing!

Line 3 J = 0, K = don't care, and Q is low. After the rising edge, Q will stay low. This line symbolizes the memory state (K = 0) and the reset operation (K = 1). If J = 0 and the Q output starts low, this puts the K input into a don't-care state. Think about this concept, it is a subtle, but important point.

Line 4 J = don't care, K = 1, and Q starts high. On the rising edge of the clock, the Q output will go low. This line symbolizes the toggle operation (J = 1) and the reset operation (J = 0).

Line 5 Unless a rising edge occurs or the set or reset inputs are active, the Q and $\overline{Q}$ outputs will not change.

Lines 6–8 These lines describe the operation of active-high set and reset inputs. They behave exactly as the set and reset inputs of the 4013B.

9.1.3 J-K Timing Diagrams

Other than the active edge and the active level of the set and reset inputs, the TTL and CMOS J-K flip-flops operated exactly alike. The timing diagram in Figure 9.4 describes the 4027B. The events of interest will be whenever the direct set or reset are active high, or the clock has an active edge.

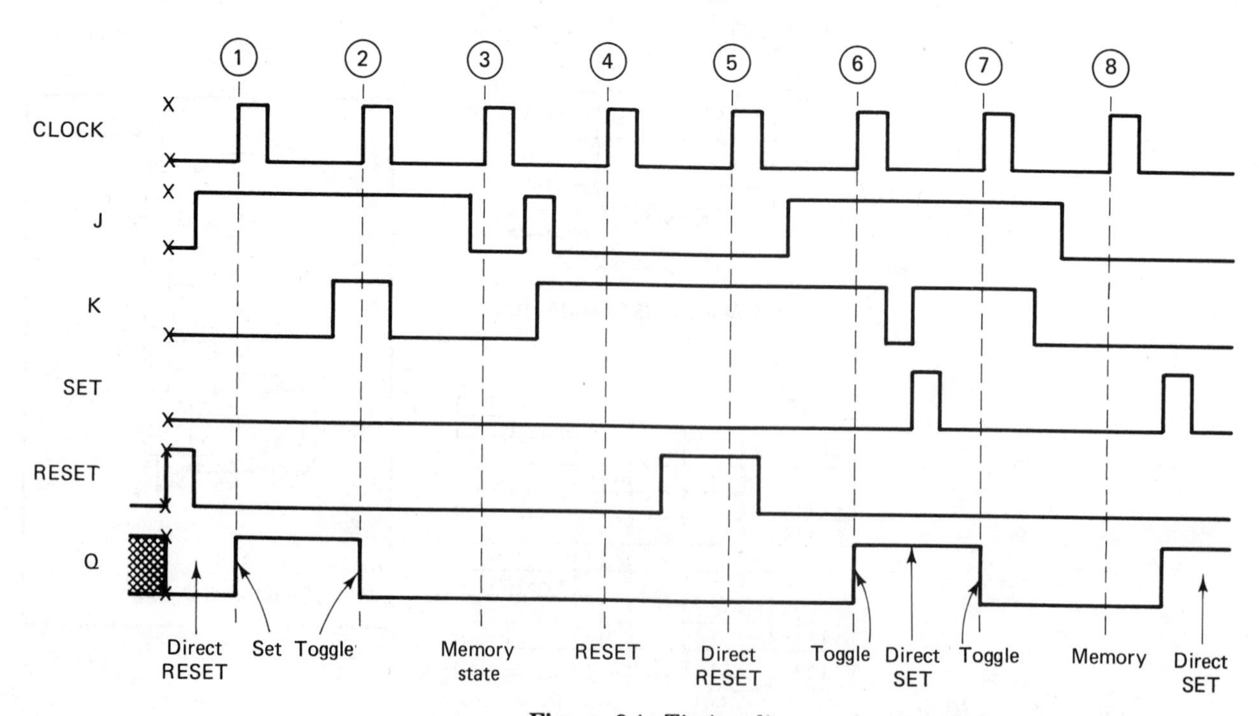

Figure 9.4 Timing diagram for the 4027B flip-flop.

We will follow the Q output as it responds to each input stimulus.

Event 1 The direct reset line goes active, forcing the Q output low. Until this active reset, it was impossible to determine the state of the Q output.

Event 2 J = 1, K = 0, the set and reset inputs are inactive, and the clock has a rising edge. This is the clocked set operation of the J-K flip-flop.

Event 3 Both J and K are high, set and reset are inactive, and the clock has a rising edge. This is the toggle operation of the J-K flip-flop.

Event 4 J and K are both low, set and reset are inactive, and the clock has a rising edge. This is the memory state operation of the J-K flip-flop.

Event 5 J = 0, K = 1, the set and reset are inactive, and the clock has a rising edge. This is the reset operation of the J-K flip-flop.

Event 6 The direct reset line is active, overriding the J, K, and clock inputs.

Event 7 On the rising edge of the clock, the Q output toggles.

Event 8 The direct set line goes active, but Q is already high.

Event 9 The Q output toggles on the rising edge of the clock.

Event 10 On the rising edge of the clock, the Q output stays low because of the memory operation.

Event 11 The direct set goes active, forcing the Q output high.

Review the timing diagram and assure yourself that you understand every event. As digital devices increase in complexity manufacturers use extensive timing diagrams to explain their operations. It is extremely important that you become thoroughly comfortable with these simple timing diagrams before you attempt to study more complex devices!

9.1.4 Applications of the J-K Flip-Flop

J-K flip-flops are the basic building block of many advanced circuits. In the early days of digital electronics, J-K flip-flops were used to construct digital counters and shift registers. Digital counters and shift registers are now available as MSI devices. Nonetheless, there are many applications that J-K flip-flops are still called on to perform (see Figure 9.5).

In Chapter 7 we constructed a data transmission system using an 8-line-to-1 line multiplexer (74LS151) and a 1-line-to-8 line demultiplexer (74LS138). The channel that was to be transmitted was selected by applying the appropriate inputs to the A, B, and C selects. We would like to use the circuit illustrated in Figure 9.5 to supply the select inputs to the multiplexer and

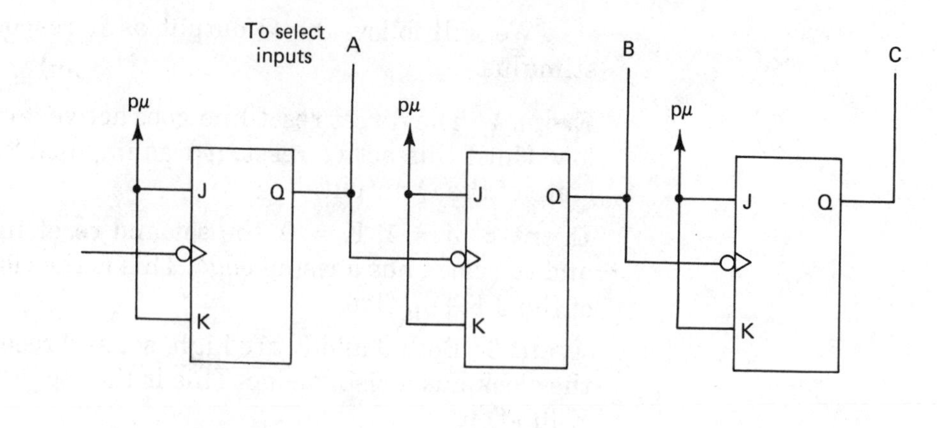

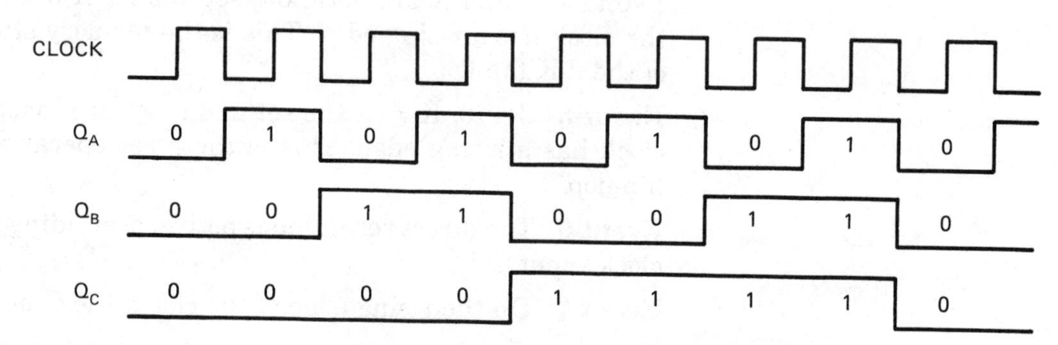

Figure 9.5 J-K flip-flops configured to supply select pulses to multiplexer and demultiplexer.

demultiplexer. In this manner all eight channels will appear to be sent simultaneously.

In Figure 9.5 all three J-K flip-flops are configured to toggle. A system clock will clock the first flip-flop. The Q output from the first flip-flop will clock the second flip-flop, and the Q output from the second flip-flop will clock the third flip-flop.

Unlike previous timing diagrams, we will complete this timing diagram one full line at a time. We will initially assume that the Q outputs of the three flip-flops are at a logic 0.

Q1 output Each time the clock has a negative edge, Q1 will toggle. As we discovered in Chapter 8 with the toggling D flip-flop, the frequency at Q1 will be one-half of the system clock frequency.

Q2 output We will now use the Q1 output in the timing diagram to clock the second flip-flop. Each time Q1 has a falling edge, Q2 will toggle. The frequency at Q2 will be one-fourth the system clock's frequency.

Q3 output Flip-flop 3 will be clocked by the Q output of flip-flop 2. Each time Q2 has a falling edge, Q3 will toggle. The frequency at Q3 will be one-eighth of the system clock's frequency.

If we choose Q1 to be the least significant bit and Q3 to be the most significant bit, these flip-flops will count from 0 (000 binary) to 7 (111 binary). This binary count will be applied to the select inputs of the multiplexer and demultiplexer.

Many advanced digital devices require a two-phase clock. Instead of just one clock input, these devices require two clock inputs, where the clocks are both the same frequency but are out of phase with each other. Consider the circuit in Figure 9.6 carefully. The clock input is common to both J-K 1 and J-K 2. The circuit is connected as follows:

1. Q2 is fed back to K1.
2. $\overline{Q2}$ is fed back to J1.
3. Q1 output is connected to J2.
4. $\overline{Q1}$ output is connected to K2.

Output 1 is $\overline{Q1}$, and output 2 is Q2.

Take a moment to consider the feedback. When Q2 is high, the next rising edge of the clock will force flip-flop 1 to perform a reset operation. When Q2 is low, the next rising edge will cause flip-flop 1 to perform a set operation. The feedback scheme is always the key to how the device operates. Remember this information as we go into the circuit analysis.

To analyze this circuit we must make an assumption regarding the starting values of outputs 1 and 2. Let's assume that both outputs are initially at logic 0. When we find another

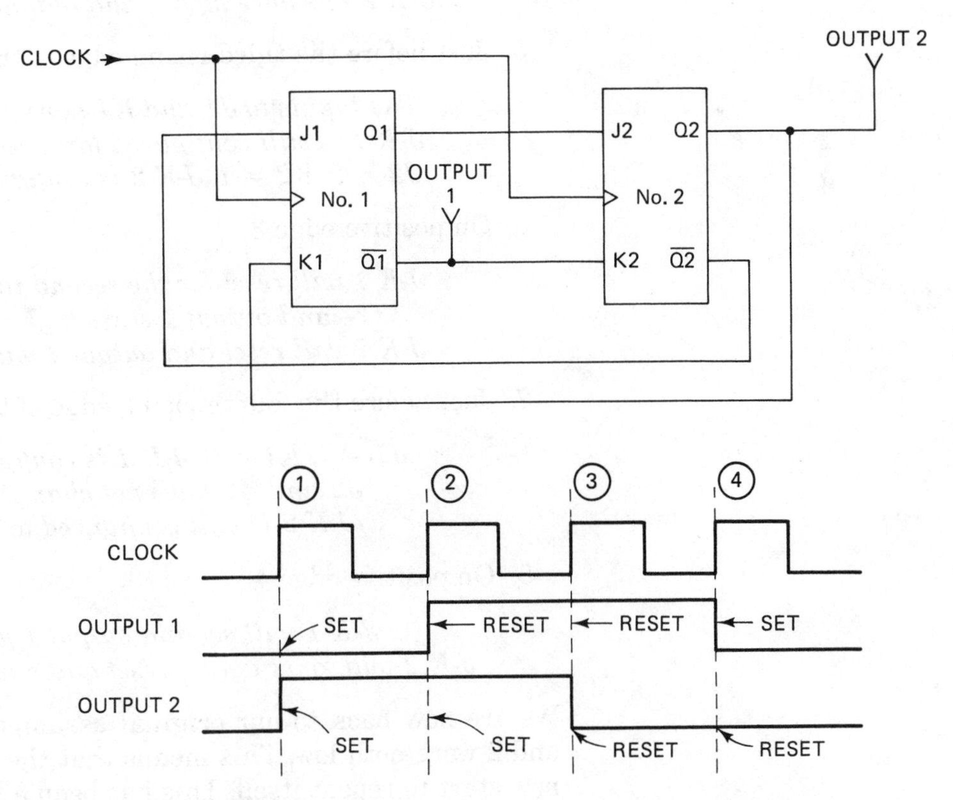

Figure 9.6 Using J-K flip-flops to create a two-phase clock.

spot in the timing diagram where both inputs are back at logic 0, we will then have finished the analysis. The method that we will use to examine the output waveforms will be to freeze time just before a rising edge of the clock. We will then establish what function the particular flip-flip will perform when the rising edge of the clock actually occurs.

1. Just before the first rising edge of the clock:

   J1 = 1, K1 = 0, J-K 1 is configured for a set operation.
   J2 = 1, K2 = 0, J-K 2 is also configured for a set operation.

2. On the positive edge 1:

   J-K 1 sets and output 1 remains low, and
   J-K 2 sets and output 2 goes high.

3. Just before the second rising edge of the clock:

   J1 = 0, K1 = 1, J-K 1 is configured for a reset operation.
   The inputs to J2 and K2 did not change after the first clock, so J-K 2 is still configured for a set operation.

4. On positive edge 2:

   J-K 1 resets and output 1 goes high.
   J-K 2 sets once again, and output 2 stays high.

5. Just before the third rising edge of the clock:

   The inputs to J1 and K1 have not changed;
   J-K 1 is still configured for a reset operation.
   J2 = 0, K2 = 1, J-K 2 is configured to reset.

6. On positive edge 3:

   J-K 1 will reset for the second time in a row, and output 1 stays high.
   J-K 2 will reset and output 2 will go low.

7. Just before the fourth rising edge of the clock:

   J1 = 1, K1 = 0, J-K 1 is configured to set.
   J2 and K2 have not changed and J-K 2 is still configured to reset.

8. On positive edge 4:

   J-K 1 will set and output 1 goes low.
   J-K 2 will, once again, reset and output 2 stays low.

We are now back to our original assumption where outputs 1 and 2 were both low. This means that the timing diagram will now start to repeat itself. This has been a tricky analysis. Read through the previous steps a few more times.

9.2 THE ONE-SHOT

The formal name for the one-shot is *monostable multivibrator.* The one-shot provides a simple function: it delivers one pulse of predetermined length for each trigger pulse it receives. In digital electronics, timing parameters are important. Sometimes a pulse may reach a particular input a little ahead of time. In such a case, the pulse must be delayed for a short period of time. The one-shot is good at this type of operation. Sometimes a pulse may be too short or too wide, and may need to be widened or shortened, respectively. This is another function that the one-shot provides. Often it is important that a pulse arrive at an input within a particular time frame; the one-shot can also be used to sense this situation.

9.2.1 Retriggerable and Nonretriggerable One-Shots

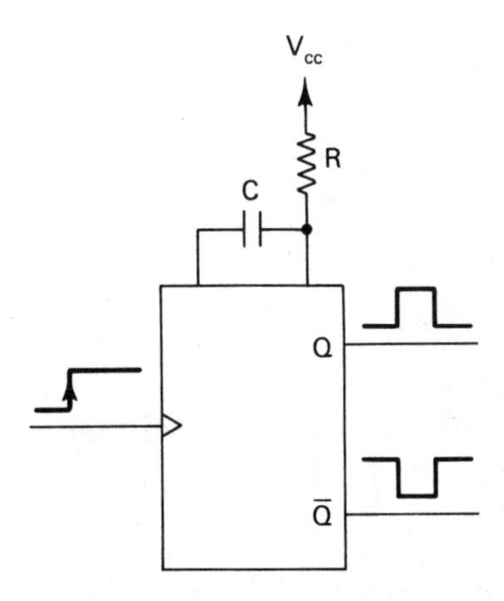

Figure 9.7 Logic symbol for the one-shot.

One-shots can be divided into two different types. The first type, the retriggerable one-shot, can continually sense input triggers. Figure 9.7 illustrates a simplified logic diagram of a one-shot. The input is obviously sensitive to rising edges. Instead of calling this input a clock, when discussing one-shots this input will be called a trigger input. When this device receives a rising-edge trigger, the Q output will respond with a positive pulse. The width of this positive pulse will depend on the *RC* time constant of R1 and C1; the equation that describes the output pulse width varies with each particular device. In a retriggerable one-shot, if the Q output is high and another trigger pulse occurs, the Q output will stay high for another total time period. For example, the output of a particular one-shot is set to be 10 ms. Five milliseconds after an initial trigger pulse is received, another trigger pulse occurs. Another 10 ms will be added to the pulse length, making it a total of 15 ms. If we trigger a one-shot at shorter periods than its output pulse width, the Q output will stay high. We will use this feature of a retriggerable one-shot for an important application.

Once a nonretriggerable one-shot is fired, it will ignore any further triggers until its Q output is once again at a logic 0 level. If the one-shot in the previous example was a nonretriggerable type, its output pulse width would be 10 ms regardless of how many extra triggers occur after it is initially fired.

9.2.2 The 74LS123 Dual Retriggerable One-Shot

Refer to Figure 9.8. The triggering schemes of one-shots are more sophisticated than a single rising edge. Examine the logic symbol for the 74LS123. Notice the three-input AND gate that is attached to the rectangle. Whenever a positive transition occurs at the output of this internal AND gate, the one-shot will receive a trigger. This symbolic AND gate is similar to the symbolic AND gate on the enable input of the 74LS138. Its function

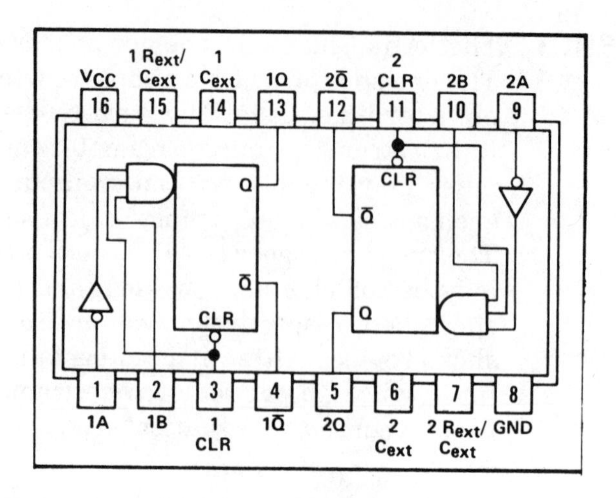

'123, 'L123, 'LS123
FUNCTION TABLE

INPUTS			OUTPUTS	
CLEAR	A	B	Q	Q̄
L	X	X	L	H
X	H	X	L	H
X	X	L	L	H
H	L	↑	⊓	⊔
H	↓	H	⊓	⊔
↑	L	H	⊓	⊔

Figure 9.8 74LS123 dual retriggerable one-shot.

is to help remind the technician of all the ways that a rising edge can trigger the one-shot. Examine the function table in Figure 9.8.

Line 1 It appears that this one-shot also has a direct clear of the type that we have seen in the D and J-K flip-flops. The logic diagram shows that this clear input is bubbled, indicating that it is active low. When this direct clear is active, the A and B inputs will be don't-care conditions, and the Q output will go low. This direct clear is used to effectively shorten an output pulse or to disable the one-shot.

Line 2 Notice that the A input is bubbled; A is an active-low trigger input. This line of the truth table tells us that if A is high, it is impossible to trigger this device. A high on the A input would be inverted to a logic 0 (via the internal inverter). This logic 0 would disable the internal AND gate from having any low-to-high transitions.

Line 3 Input B is not bubbled, it is an active-high input. If B is low, it has the same result as we examined in line 2.

Line 4 The clear input is inactive and the A input is active low. When the B input has a rising edge, the one-shot will be fired. The positive and negative pulses in the function table symbolize the output pulses. To fire this one-shot, a positive transition must occur at the output of the internal three-input AND gate. Figure 9.9 illustrates how the conditions in line 4 of the truth table cause this transition to occur.

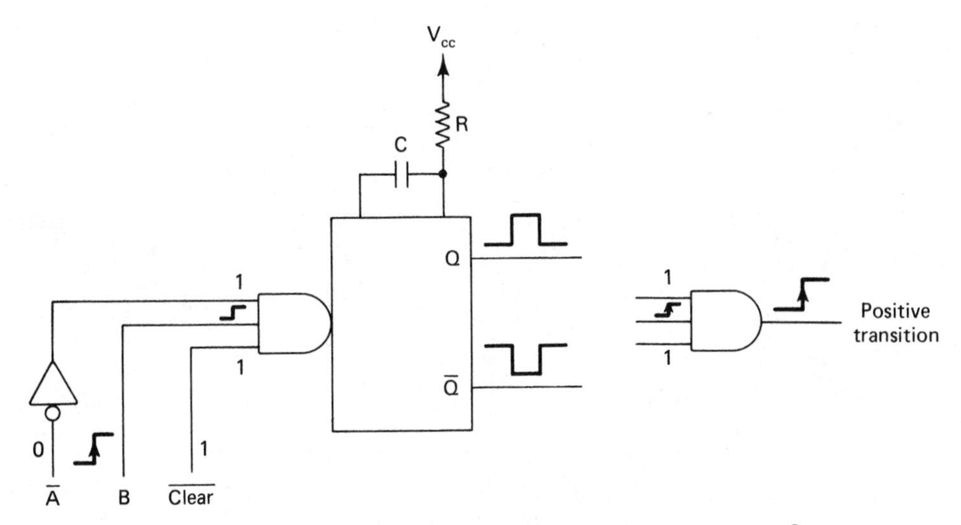

Figure 9.9 Using a rising edge on the B input to fire the one-shot.

Line 5 Because input A is active low, this line is the A input version of line 4. If clear is inactive and input B is active high, a negative transition on input A will cause a positive transition on the output of the internal AND gate; this will fire the one-shot. We now see that the 74LS123 can be fired with negative or positive edge triggers.

Line 6 The third way to get a positive transition on the output of the internal AND gate is illustrated in this line. If inputs A and B are at their active levels, a low-to-high transition on the clear input will fire the one-shot. This is the least commonly used method of firing the one-shot.

Circuit designers will always indicate on the logic symbol of the one-shot how wide the output pulse should be. The equations that describe the length of the output pulse for the 74LS123 depend on several different factors. If the capacitor used is less than 1000 pF, the specifications sheet will contain a graph that illustrates the output pulse width. If the capacitor used is greater than 1000 pF, the following equation describes the output pulse width:

$$\text{pulse width (PW)} = R \times C \times 0.33$$

where R is a value in kilohms, C is in picofarads, and the output pulse width is expressed in nanoseconds.

9.2.3 An Application of the Retriggerable One-Shot.

The TV signal in the United States operates at a vertical rate of 60 Hz. This means that every 16 ms the CRT beam must be moved from the bottom right-hand corner of the screen back to the top left-hand corner. This operation is called *vertical retrace.* We can use a retriggerable one-shot to monitor vertical retrace.

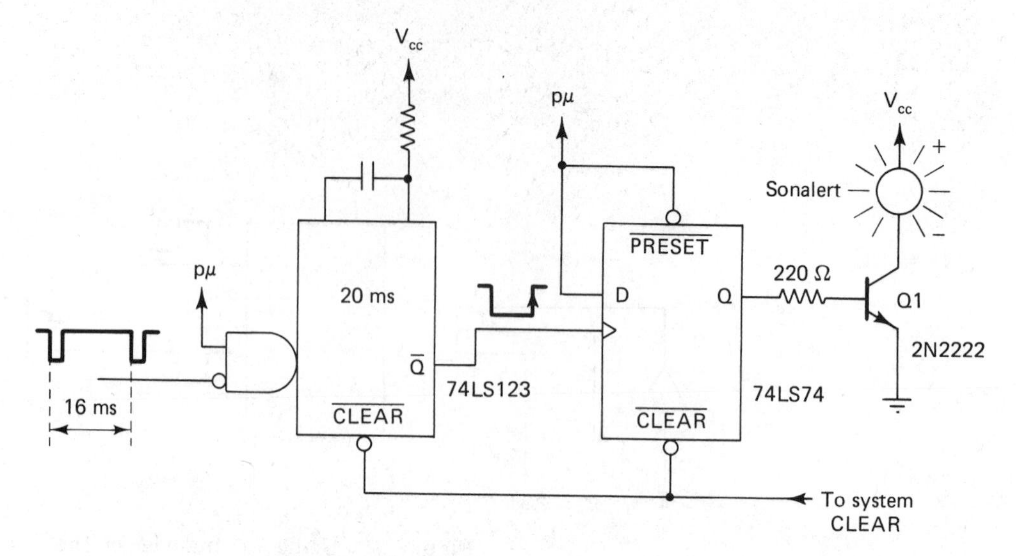

Figure 9.10 Circuit to detect loss of vertical retrace.

If this pulse ever fails to occur, an alarm should be set off (see Figure 9.10). Notice the schematic symbol used to represent the one-shot. It is a simplified version of the logic symbol appearing in Figure 9.8. Remember that the AND gate is contained within the 74LS123 IC.

The active-low trigger input is connected to the vertical sync signal. The Sonalert is a device that will output a loud tone when it receives a ground from Q1. The D flip-flop controls the state of Q1. The circuit is initialized by an active-low system reset pulse. This sets $\overline{Q}$ of the one-shot high, and Q of the D flip-flop low. This turns off the transistor and the Sonalert. The output pulse width of the one-shot is 20 ms. When the vertical sync arrives every 16 ms, the negative edge will retrigger the one-shot and keep the Q output high. If the vertical sync ever fails to arrive, $\overline{Q}$ of the one-shot will go high. The rising edge of $\overline{Q}$ will clock a logic 1 into the output of the D flip-flop. This logic 1 will saturate Q1, providing the Sonalert with a ground. Even if another vertical sync comes along to trigger the one-shot, the D flip-flop will stay high until a system reset signal occurs.

9.2.4 The 74LS221 Dual Nonretriggerable One-Shot.

The 74LS221 is an extremely popular one-shot. It can be used for many different types of applications. Refer to Figure 9.11. The logic diagram of the 74LS221 appears to be a bit more complicated than the logic diagram of the 74LS123. Notice the cross-coupled NAND gates that form an S-R latch. Examine the function table. It is almost identical with the function table of the 74LS123. Therefore, we can ignore the complicated logic diagram and treat this device like a nonretriggerable version of the 74LS123. The schematic symbol that represents the 74LS221 will be the same logic symbol used to represent the 74LS123. There will be no indication on the logic symbol to designate

FUNCTION TABLE
(EACH MONOSTABLE)

INPUTS			OUTPUTS	
CLEAR	A	B	Q	$\bar{Q}$
L	X	X	L	H
X	H	X	L	H
X	X	L	L	H
H	L	↑	⊓	⊔
H	↓	H	⊓	⊔
↑	L	H	⊓	⊔

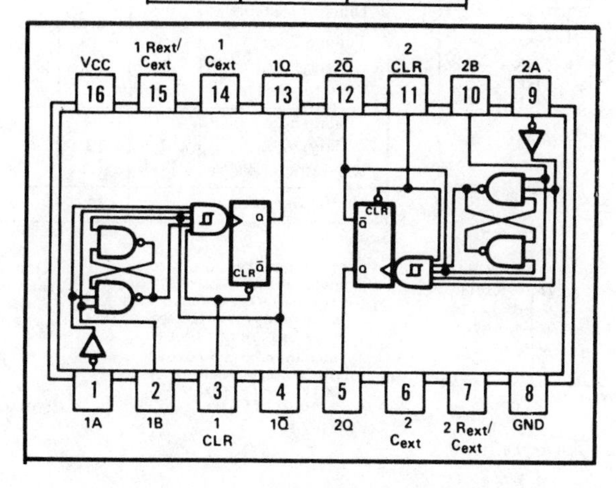

Figure 9.11 74LS221 dual nonretriggerable one-shot.

whether a particular one-shot is retriggerable or nonretriggerable. It will be up to the technician to refer to the data sheets of the device to discover its specific characteristics.

The equation that describes the output pulse width for the 74LS221 is

$$\text{PW} = 0.69 \times R \times C$$

The constant 0.69 is the natural log of 2. This constant is used in many circuits that deal with an *RC* network. If you have studied the 555 timer, this constant should be familiar.

9.2.5 The 4098B Retriggerable/Nonretriggerable One-Shot.

Refer to Figure 9.12. The triggering scheme for this device is the same as that for the 74LS123. The different ways in which this device can be used are:

1. Rising edge, retriggerable
2. Rising edge, nonretriggerable
3. Falling edge, retriggerable
4. Falling edge, nonretriggerable

Each line of the table illustrated in Figure 9.12 explains how to configure the 4098B for a particular option. Notes 1 and 2 under the function table in Figure 9.12 provide an excellent explanation of the difference between a retriggerable and nonretriggerable one-shot.

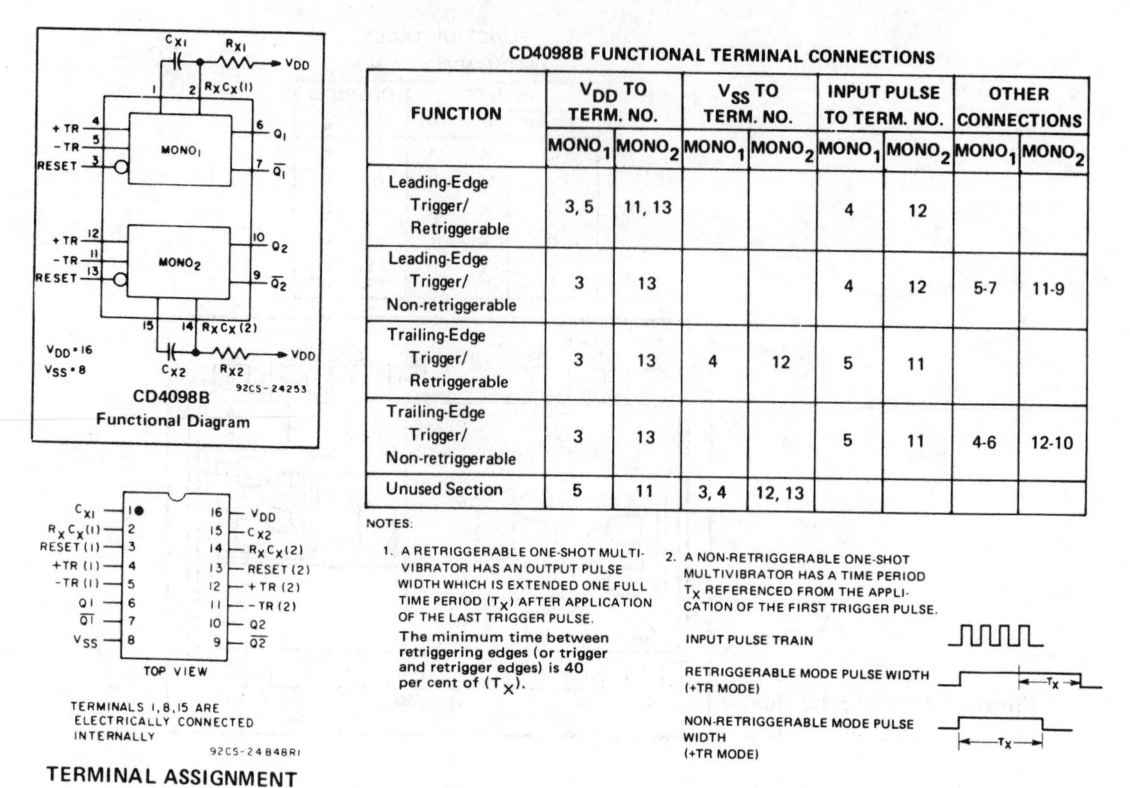

CD4098B FUNCTIONAL TERMINAL CONNECTIONS

FUNCTION	V_{DD} TO TERM. NO.		V_{SS} TO TERM. NO.		INPUT PULSE TO TERM. NO.		OTHER CONNECTIONS	
	MONO1	MONO2	MONO1	MONO2	MONO1	MONO2	MONO1	MONO2
Leading-Edge Trigger/ Retriggerable	3, 5	11, 13			4	12		
Leading-Edge Trigger/ Non-retriggerable	3	13			4	12	5-7	11-9
Trailing-Edge Trigger/ Retriggerable	3	13	4	12	5	11		
Trailing-Edge Trigger/ Non-retriggerable	3	13			5	11	4-6	12-10
Unused Section	5	11	3, 4	12, 13				

NOTES:

1. A RETRIGGERABLE ONE-SHOT MULTIVIBRATOR HAS AN OUTPUT PULSE WIDTH WHICH IS EXTENDED ONE FULL TIME PERIOD (T_X) AFTER APPLICATION OF THE LAST TRIGGER PULSE.
   The minimum time between retriggering edges (or trigger and retrigger edges) is 40 per cent of (T_X).
2. A NON-RETRIGGERABLE ONE-SHOT MULTIVIBRATOR HAS A TIME PERIOD T_X REFERENCED FROM THE APPLICATION OF THE FIRST TRIGGER PULSE.

Figure 9.12 4098B retriggerable/nonretriggerable one-shot.

9.2.6 Timing Diagrams of One-Shots

Let's examine a typical timing diagram for a 74LS221 nonretriggerable one-shot (Figure 9.13). The values of R and C were chosen to provide a 100-μs output pulse. Input B is pulled up to V_{cc}; input A will be used to fire the one-shot.

On the first falling edge of input A, Q goes high and stays high for 100 μs. The second series of triggers on the A input demonstrate that the 74LS221 is nonretriggerable. The third output pulse is cut short by the direct clear input.

9.2.7 Applications of One-Shots

The first application that we will examine is the one-shot as a pulse-delaying circuit (Figure 9.14). One-shot 1 is positive edge triggered and delays the input pulse by 50 μs. Notice that pulse width of the input pulse has no effect on the outputs of either one-shot; only the rising edge of the trigger input is important. Fired by the rising edge of the input signal, Q1 will go high for 50 μs. When Q1 has a falling edge it will trigger one-shot 2. This one-shot will provide the output pulse width of 10 μs.

Refer to the timing diagram in Figure 9.14. Follow the active edges of all the signals through the timing diagram. This circuit can delay input pulses and can also lengthen or shorten them, depending on the pulse width chosen for Q2.

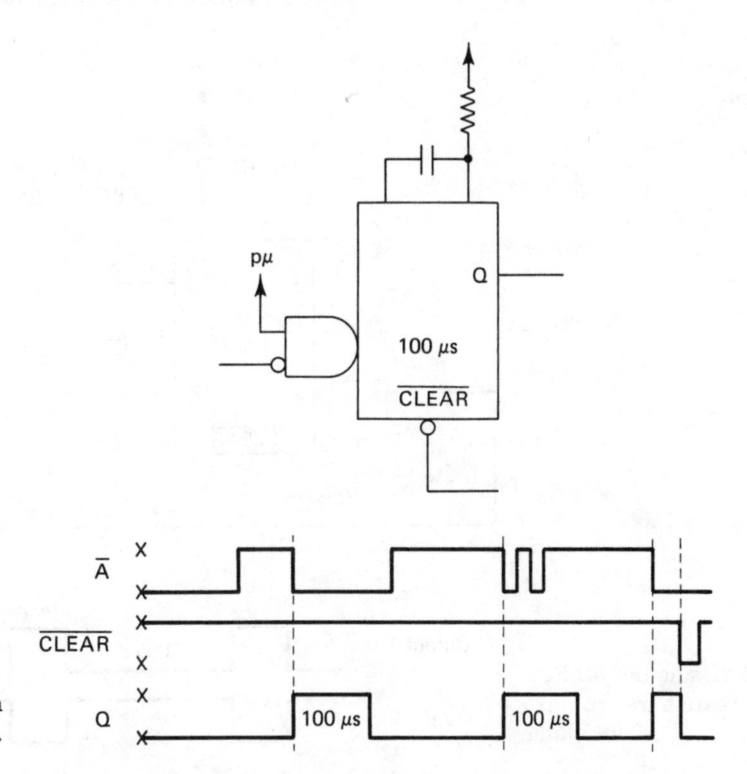

Figure 9.13 Timing diagram for 74LS221.

Figure 9.15 illustrates a free-running square-wave oscillator. The output square wave can be designed for a wide range of frequencies and duty cycles, all dependent on the two *RC* time constants. Both one-shots will be triggered on rising edges. One-shot 1 will be triggered when Q2 goes from high to low. This will cause $\overline{Q}2$ to go from low to high, which provides the rising edge. One-shot 2 is triggered in a similar manner by $\overline{Q}1$.

It looks as if we have the basis of a good oscillator; the only question is: How does this process of self-triggering begin? That is a good question and can be explained by the process of "powering up" the circuit. When V_{cc} is first applied to the one-shots,

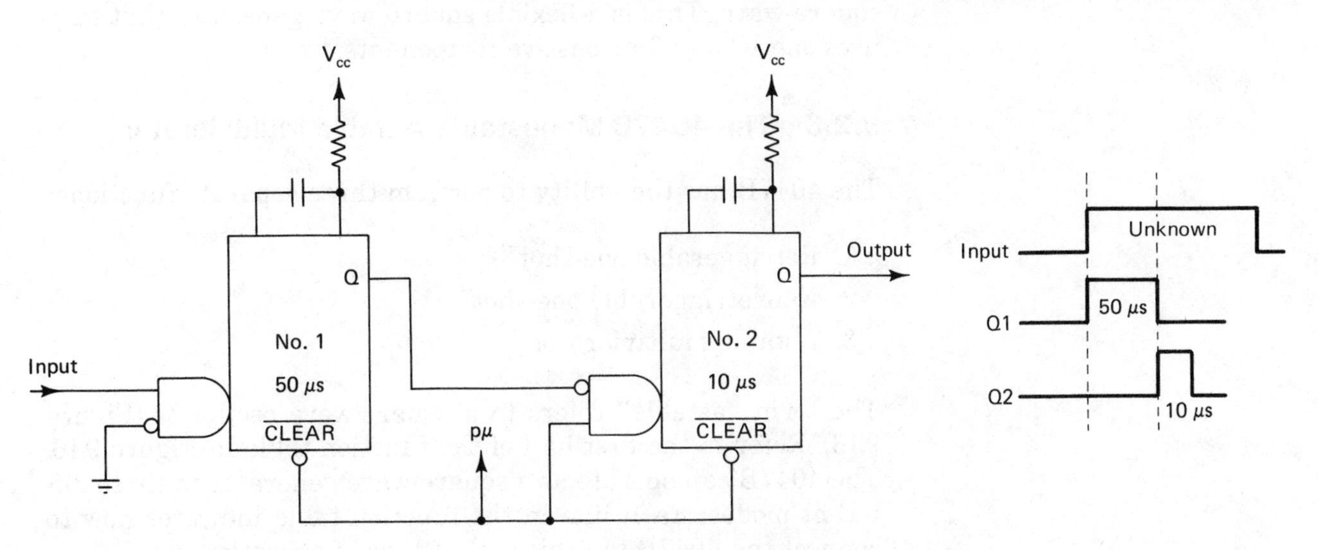

Figure 9.14 Pulse-delaying circuit.

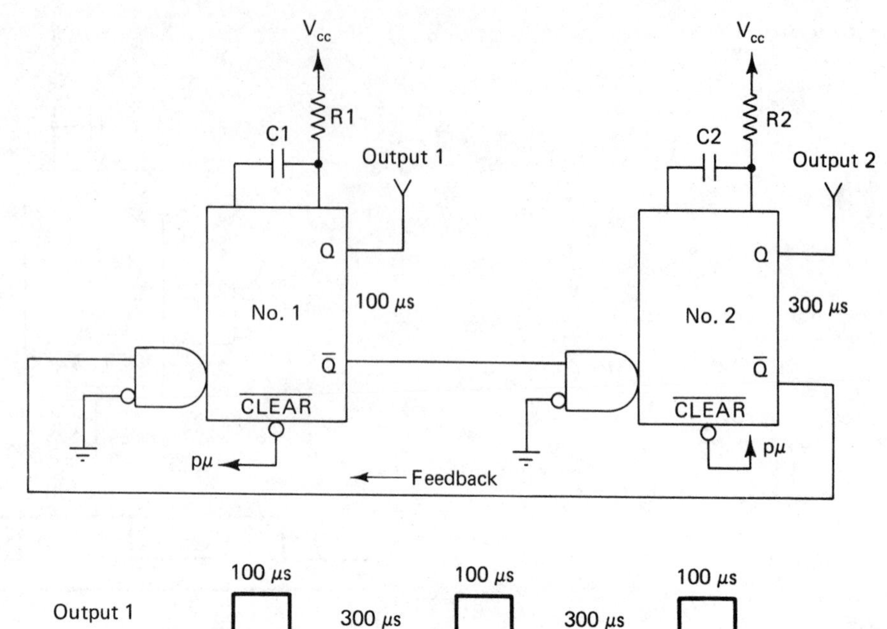

Figure 9.15 Using the 74LS221 to create a free-running oscillator.

the charge on the timing capacitors is 0 V. Because of this, the action of applying power actually starts the oscillations.

Let's follow a couple of cycles of the oscillations. Refer to the timing diagram in Figure 9.15. Let's assume that Q1 is high and Q2 is low. When Q1 finally goes low, $\overline{Q}1$ will fire one-shot 2. One-shot 2 will stay high for 300 μs. At the end of this 300 μs, Q2 will go low. When this happens, $\overline{Q}2$ will provide the rising edge to trigger one-shot 1. Q1 will go high for 100 μs. At the end of 100 μs, Q1 will go low. At this point $\overline{Q}1$ will go high and provide the rising edge to trigger one-shot 2. It appears that we have self-sustaining oscillations. By choosing the appropriate values for the timing components, we can create a custom duty-cycle square-wave. This is a flexible square-wave generator that only uses one IC and four passive components.

9.2.8 The 4047B Monostable/Astable Multivibrator

The 4047B has the ability to perform three separate functions:

1. Retriggerable one-shot
2. Nonretriggerable one-shot
3. Astable multivibrator

The term "astable" refers to a square-wave oscillator (Figure 9.16). Refer to the first half of the function table in Figure 9.16. The 4047B can operate as a square-wave generator in three different modes. Each line in the function table indicates how to connect the 4047B to achieve that type of operation.

CD4047B FUNCTIONAL TERMINAL CONNECTIONS
NOTE: IN ALL CASES EXTERNAL RESISTOR BETWEEN TERMINALS 2 AND 3▲
EXTERNAL CAPACITOR BETWEEN TERMINALS 1 AND 2▲

FUNCTION	TERMINAL CONNECTIONS TO V_{DD}	TO V_{SS}	INPUT TO	OUTPUT PULSE FROM	OUTPUT PERIOD OR PULSE WIDTH
Astable Multivibrator:					
Free Running	4,5,6,14	7,8,9,12	—	10,11,13	t_A (10,11) = 4.40 RC
True Gating	4,6,14	7,8,9,12	5	10,11,13	t_A (13) = 2.20 RC#
Complement Gating	6,14	5,7,8,9,12	4	10,11,13	
Monostable Multivibrator:					
Positive-Edge Trigger	4,14	5,6,7,9,12	8	10,11	
Negative-Edge Trigger	4,8,14	5,7,9,12	6	10,11	t_M (10,11) = 2.48 RC
Retriggerable	4,14	5,6,7,8,9	12	10,11	
External Countdown*	14	5,6,7,8,9,12	—	10,11	

▲ See Text.
\# First positive ½ cycle pulse-width = 2.48 RC, see Note on Page 10.
* Input Pulse to Reset of External Counting Chip External Counting Chip Output To Terminal 4

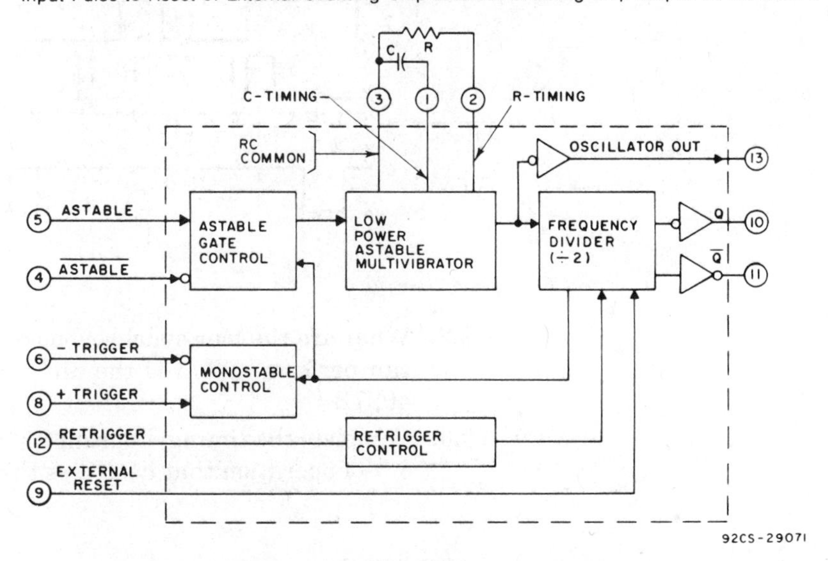

Figure 9.16 4047B one shot/oscillator.

1. "Free running" refers to the normal type of oscillator action. The oscillator that we just constructed, using the 74LS221, is considered to be a free-running type.
2. "True gating" means that the device will only output oscillations when a high level is placed on pin 5, the true astable input pin.
3. "Complement gating" means that the device will only output oscillations when a low level is placed on pin 4, the astable-not input.

The Gate will either enable or disable the output of the oscillator.

The second half of the function table relates to one-shot operation. You are already familiar with the first three types of one-shot operations listed. The last one-shot operation is called *external countdown.*

The external count mode of operation is used to acquire long output pulses. Because one-shot pulse widths are set using *RC* networks, achieving long pulse widths is difficult. To achieve these long *RC* time constants, electrolytic capacitors must be employed. Electrolytic capacitors suffer from leakage problems. The external count mode uses an external digital counter to lengthen the output pulse. In Chapter 10 you will learn how to use MSI counters. The 4047B is an indication of how digital devices will continue to increase in capacity and complexity.

QUESTIONS AND PROBLEMS

9.1. Complete the timing diagram for the 74LS76.

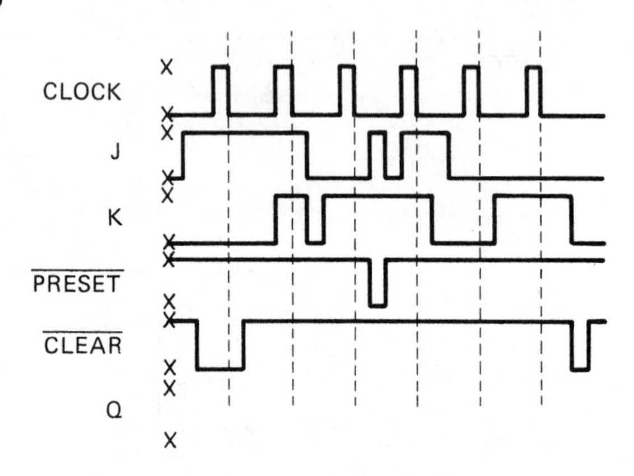

9.2. What are the four synchronous operations that the J-K flip-flop can perform? What is the difference between a 74LS76 and a 4027B?

9.3. Complete the timing diagram for the circuit shown. Write a theory of operation that describes the function of this circuit.

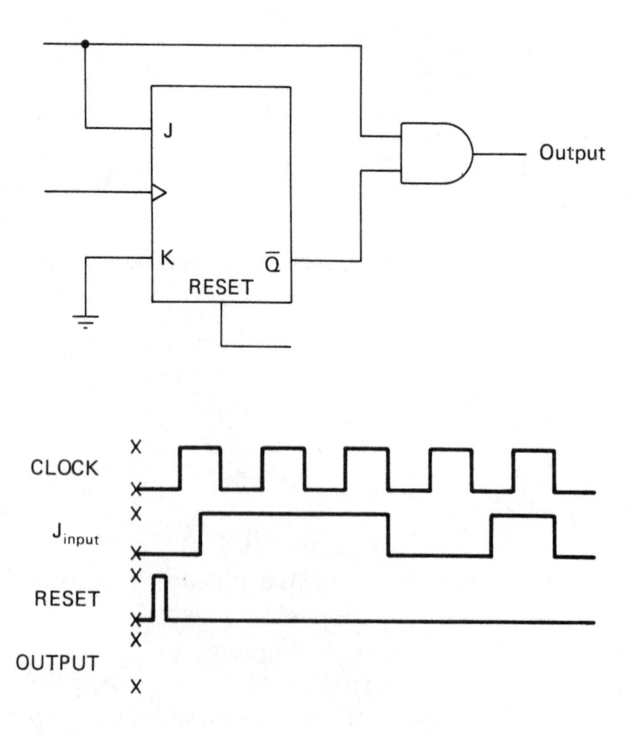

9.4. Create a timing diagram that describes the action of the circuit shown.

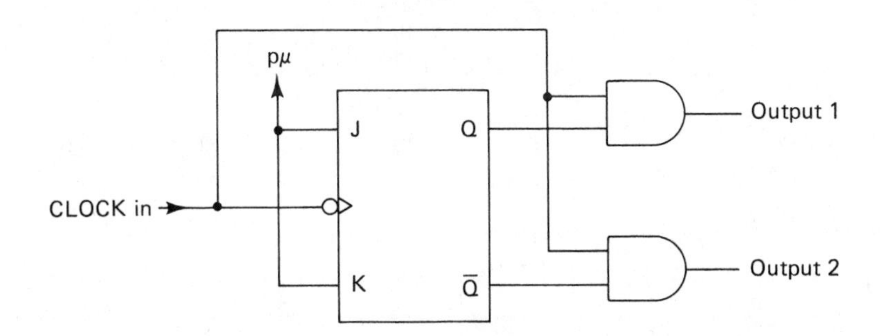

9.5. Create a timing diagram that describes the operation of the following circuit. Be sure to include the following signals: the system clock, the pulse from the debounced switch, Q1, Q2, and $\overline{Q2}$.

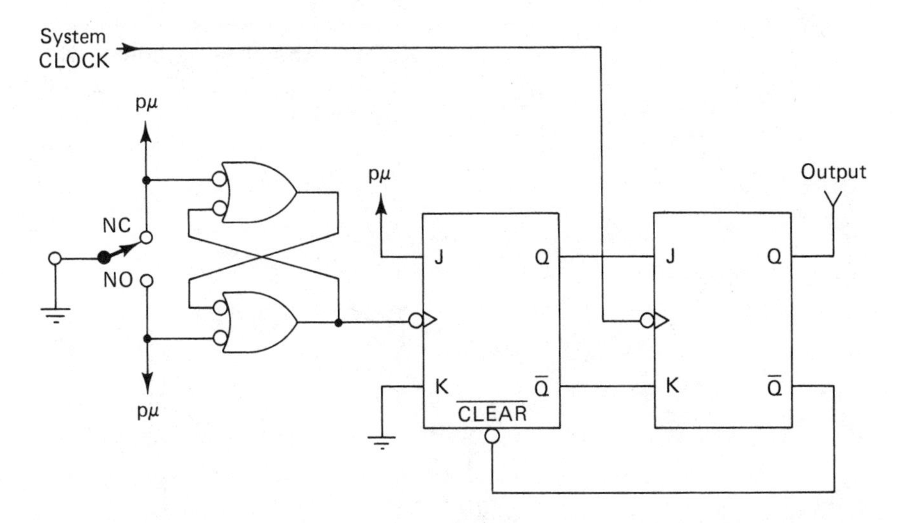

9.6. Answer the following troubleshooting questions concerning Figure 9.6.

(a) What would happen if the clock input for flip-flop 2 was floating?

(b) How would the output be affected if Q2 was shorted to ground?

(c) How would the output be affected if $\overline{Q1}$ was shorted to V_{cc}?

9.7. Use J-K flip-flops to construct a circuit that counts from 0 (0000 binary) to 15 (1111 binary). Be sure to label the least significant and most significant outputs. Also construct a timing diagram to describe the circuit operation.

9.8. Refer to the following circuit. What is the length of the output pulse for this one-shot if the input pulse width is:

(a) Greater than 100 μs?

(b) Less than 100 μs?

Draw a timing diagram that illustrates how this circuit functions. Why is the negative-logic symbol used to represent the OR gate?

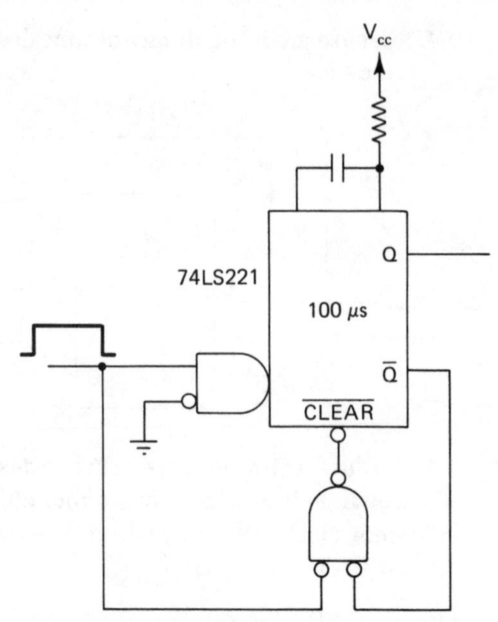

9.9. Refer to the circuit shown, which illustrates a 74LS123 retriggerable one-shot. Create a theory of operation that describes this circuit.

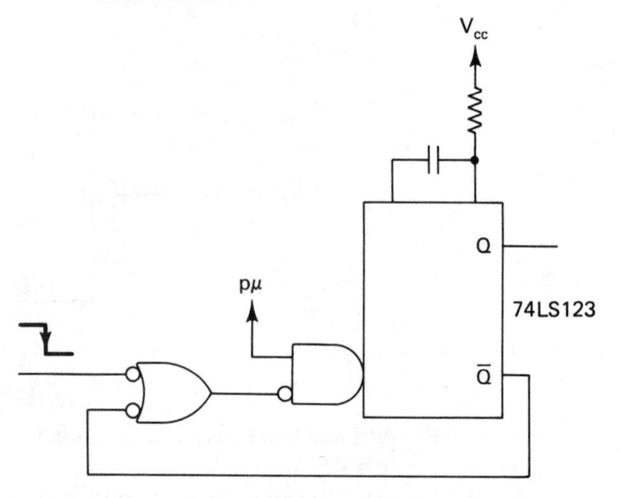

9.10. Draw a timing diagram for the trigger input and Q1 and Q2 outputs of the circuit shown. Label all the events that occur relative to the first rising edge of the trigger.

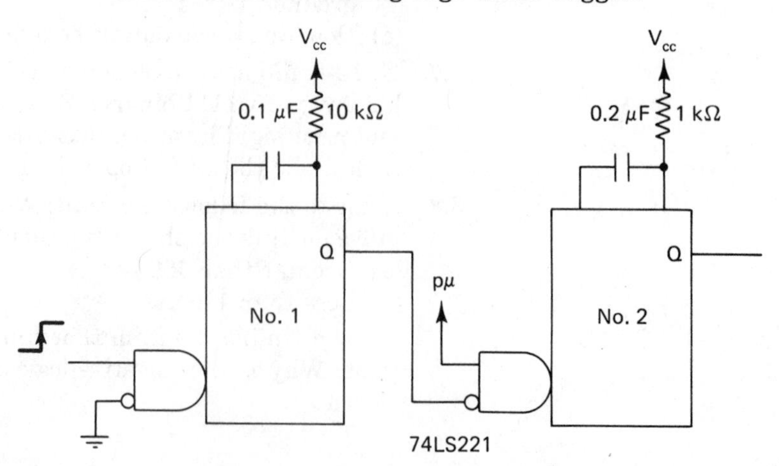

239
240

9.11. Draw a timing diagram that illustrates the operation of the circuit shown.

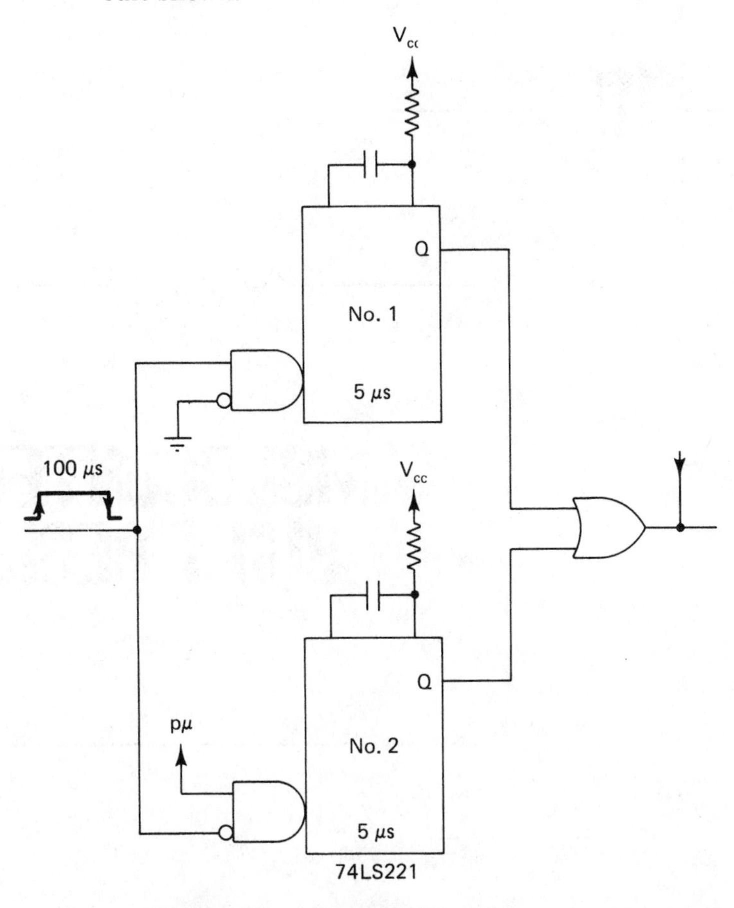

9.12. Using 74LS221 one-shots, design a circuit that delays an incoming pulse by 500 μs. The output pulse should be 1 μs long. Draw a timing diagram to summarize the circuit action.

9.13. If the Q output of a one-shot is logic 0 even though its input is being triggered, what checks would you make to find the problem?

9.14. If a one-shot is designed to output a pulse that is 25 μs long but is outputting a pulse that is 50 μs long, what do you think is the most likely problem?

9.15. Refer to Figure 9.10. How would the circuit be affected if a 220-kΩ resistor was stuffed in place of the 220-Ω resistor?

9.16. How would the oscillator in Figure 9.15 be affected if C1 were open? If R2 were only half of its proper value?

9.17. Redesign Figure 9.15 to use the inverting trigger inputs (pull the noninverting trigger inputs to V_{cc}) instead of the noninverting trigger inputs.

10

MSI COUNTERS AND SHIFT REGISTERS

In Chapter 9 we used three J-K flip-flops to construct a circuit that counted from 0 through 7. Instead of constructing counters from flip-flops, designers use MSI counters. Both the TTL and CMOS logic families provide simple to highly sophisticated MSI counters.

A digital clock is really nothing more than a circuit that counts the rising or falling edges of a 60-Hz reference square wave. The pulse meter in Chapter 8 that illustrated the 74LS75 4-bit latch contained a circuit to count the number of heart beats in a fixed interval.

Another function that counters can perform is to divide a frequency by a particular integer value. A D flip-flop or J-K flip-flop configured to toggle divides the clock frequency by a factor of 2. A divide-by-N-counter is

a circuit that produces one output pulse for every N input clock pulses

Consider a digital clock circuit that divides an input frequency of 60-Hz by a factor of 60. Once every second the output of the circuit will toggle. The toggling output will clock another circuit that counts the number of seconds.

10.1 MODIFYING THE COUNT LENGTH

10.1.1 Modulus and Count-State Diagrams

We have seen that one flip-flop can count from 0 to 1; two flip-flops can count from 0 to 3; three flip-flops can count from 0 to 7. Each time we add a flip-flop the number of possible counts doubles. The modulus of a counter refers to how many counts it will go through before repeating. A counter consisting of four flip-flops will have a modulus of 16, because it counts from 0 to 15 before starting to repeat. An alternative way to describe a counter with a modulus of 16 would be to call it a *mod-16 counter.*

A counter described as a "mod-16 counter" does not necessarily count from 0 to 15. A count-state diagram is used to illustrate the counting sequence of a particular counter (Figure 10.1). The logic symbol indicates a rising edge-triggered clock and outputs Q_a through Q_d. As always, Q_a is the least significant output. When a number is written, the digits are ordered least significant to most significant from right to left. The logic diagram of the counter in Figure 10.1 is derived from the schematic of four separate flip-flops. That is why the least significant bit appears on the left instead of the right. This is true for the logic diagrams of most MSI counters.

The count-state diagram indicates the counting sequence for this particular mod 16 counter. Each circle represents one count state. The count state is indicated with the decimal equivalent output of Q_a through Q_d. The diagram shows that this counter counts from 0 to 15, then begins to repeat.

10.1.2 Modifying the Count

How would we design a mod 12 counter with a count sequence that starts at 0 and proceeds through 11? If we used three flip-flops, the maximum count that could be achieved is 8; if we used four flip-flops, we would jump to a modulus of 16. The answer is

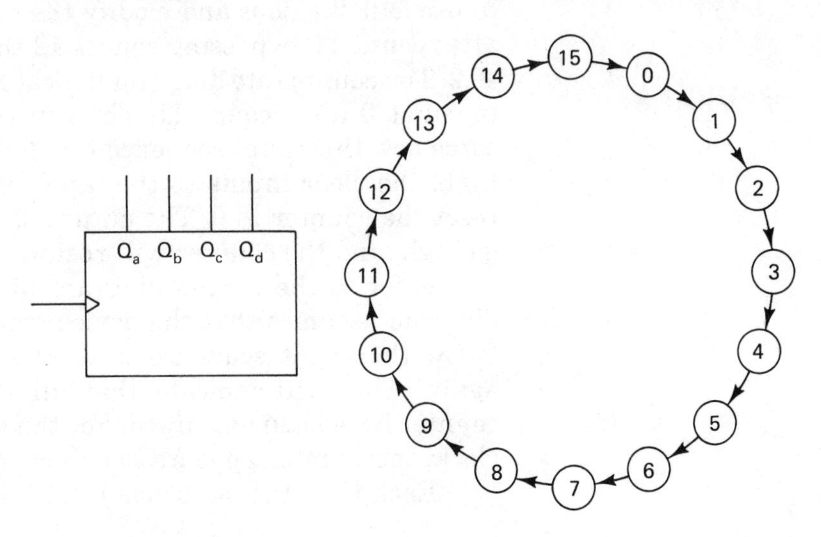

Figure 10.1 Logic symbol and count-state diagram for mod-16 counter.

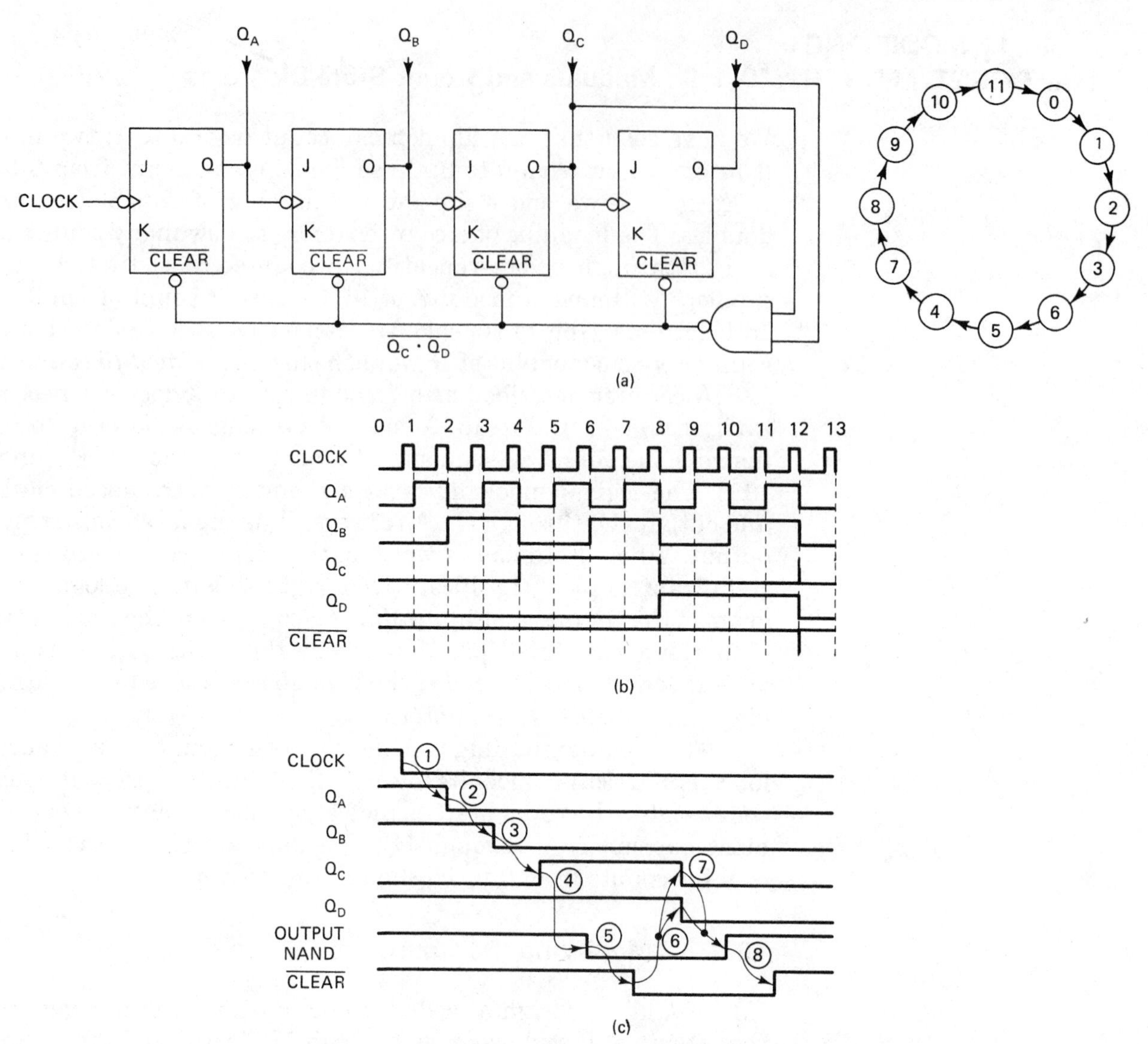

Figure 10.2 Mod-12 counter, timing diagrams, and count-state diagram.

to use four flip-flops and modify the circuit to return to count 0 after count 11, bypassing counts 12 through 15. Refer to Figure 10.2. The count-state diagram indicates that the counter returns to count 0 after count 11. The functional block diagram indicates how this count sequence is achieved. When Qc and Qd are high, the clear inputs to the flip-flops will go active. This will reset the counter to 0. The output of the NAND gate will then go high, and the counter will resume normal operation.

Refer to the timing diagram in Figure 10.2b. The timing diagram assumes that the device starts counting at 0. We will follow the count sequence until the count of 0 happens once again. This will indicate that all the possible states of the counter have been examined. For the first 11 falling edges of the clock, the counter appears to behave normally.

Each time the clock has a falling edge, flip-flop 1 will toggle; each time Q_a outputs a falling edge, flip-flop 2 will toggle; each time Q_b outputs a falling edge, flip-flop 3 will toggle; and each time Q_c outputs a falling edge, flip-flop 4 will toggle.

When count 12 is reached, Q_c and Q_d will go high, causing the output of the NAND gate to go low; this active low is applied to the clear inputs of the flip-flops, forcing Q_a through Q_d to a logic 0 level. The counter will then continue with count 0. The events that cause the counter to reset to count 0 occur in an extremely short period of time. To understand the causes and effects of the actual events that lead to the counter resetting, we must construct a detailed timing diagram.

All digital devices have a propagation delay. After a stimulus is applied, a certain period of time is required before the device will react. Counters of the type that we are now examining are called *ripple counters*. The term "ripple" describes how the output of one device is used to clock the next device. For example, the counter does not go instantly from count 7 to count 8. The falling edge of the clock will cause Q_a to have a falling edge; this will clock flip-flop 2 and cause Q_b to have a falling edge; this will clock flip-flop 3 and cause Q_c to have a falling edge. Finally, this will clock flip-flop 4 and cause Q_d to toggle from low to high.

Consider the following common ripple effect. Four dominoes are stood on end and placed in a line, and the first domino is pushed over; it will take a certain period of time before the last domino will fall.

Refer to timing diagram in Figure 10.2c. Let's follow each event through the timing diagram to understand how the process of modifying the count length actually occurs. Before the falling edge of the clock, the counter is on count 11.

Event 1 The clock has a falling edge; this causes Q_a to toggle from high to low.

Event 2 The falling edge on Q_a will cause Q_b to toggle from high to low.

Event 3 The falling edge on Q_b will cause Q_c to toggle from low to high.

Event 4 The output of the counter is now count 12 (1100). This will cause the output of the NAND gate to go low.

Event 5 When the output of the NAND gate goes low, this will place an active low level on all the clear inputs.

Event 6 Q_a through Q_d are cleared. Q_a and Q_b are already low, and Q_c an d Q_d will go from high to low.

Event 7 The counter has now returned to count 0. This will cause the output of the NAND gate to go high.

Event 8 The output of the NAND gate will place an inactive level on the clear inputs, reenabling the counter.

If each propagation delay was equal to 10 ns, it would take a total of 80 ns for the process of going from count 11 to count 0. It is important that you realize that count 12 actually occurs and is valid for four propagation delays. If the counter's clock frequency is slow compared to the sum of the propagation delays, count 12 will never be seen.

Once again, refer to Figure 10.2a. You should now understand why a positive glitch occurs at Q_c during the reset process. It is these types of glitches that make ripple counters difficult to use at high frequencies.

10.2 MSI RIPPLE COUNTERS

For all the reasons discussed previously, it is not practical to build counters from discrete flip-flops. A full line of MSI counters is available in both TTL and CMOS circuit technologies. The least expensive, general-purpose counters are the MSI ripple counters.

10.2.1 The 74LS90 and 74LS93 Ripple Counters

The 74LS90 is a divide-by-10 counter. Divide-by-10 counters are often called *BCD* or *decade counters,* and the 74LS93 is a divide-by-16, 4-bit binary counter. Because this is an old series of ripple counters, the V_{cc} and ground pins are in non-standard locations. V_{cc} is pin 5 and ground is pin 10. To rectify this problem, the 74LS290 and 74LS293 were introduced. The 74LS290 is the functional equivalent of the 74LS90, but its V_{cc} and ground pins are in the standard positions. The same relationship is true for the 74LS293 and the 74LS93 (see Figure 10.3). The 74LS90 and 74LS93 are inexpensive counters; they each sell for under 50 cents. Refer to the functional block diagram for the 74LS90 in Figure 10.3. When J-K flip-flops are shown with no connection to the J or K inputs, it is assumed that they are both pulled up to V_{cc} and the J-K flip-flop is configured to toggle. Internally, this device is constructed from three J-K flip-flops (configured to toggle) and an S-R flip-flop. Notice the two NAND gates. These two NAND gates are used to modify the count sequence of the 74LS90. When the bottom NAND gate goes active, the counter will reset to count 0; when the top NAND gate goes active, the counter will reset to count 9. There are two negative edge-triggered clock inputs: input A and input B. The first J-K flip-flop is an independent divide-by-2 block, and the last three flip-flops constitute a divide-by-5 block. To achieve divide-by-10 operation, the clock input is placed on input A and Q_a is used to clock input B.

Refer to the reset/count function table for the 74LS90 in Figure 10.3. The first two lines indicate that if both reset-to-0 inputs are at an active high level and either reset-to-9 input is at an inactive low level, the output will be reset to count 0. The third line indicates that if both reset-to-9 inputs are at an ac-

'90A, 'L90, 'LS90
BCD COUNT SEQUENCE
(See Note A)

COUNT	OUTPUT			
	Q_D	Q_C	Q_B	Q_A
0	L	L	L	L
1	L	L	L	H
2	L	L	H	L
3	L	L	H	H
4	L	H	L	L
5	L	H	L	H
6	L	H	H	L
7	L	H	H	H
8	H	L	L	L
9	H	L	L	H

'93A, 'L93, 'LS93
COUNT SEQUENCE
(See Note C)

COUNT	OUTPUT			
	Q_D	Q_C	Q_B	Q_A
0	L	L	L	L
1	L	L	L	H
2	L	L	H	L
3	L	L	H	H
4	L	H	L	L
5	L	H	L	H
6	L	H	H	L
7	L	H	H	H
8	H	L	L	L
9	H	L	L	H
10	H	L	H	L
11	H	L	H	H
12	H	H	L	L
13	H	H	L	H
14	H	H	H	L
15	H	H	H	H

'90A, 'L90, 'LS90
RESET/COUNT FUNCTION TABLE

RESET INPUTS				OUTPUT			
$R_{0(1)}$	$R_{0(2)}$	$R_{9(1)}$	$R_{9(2)}$	Q_D	Q_C	Q_B	Q_A
H	H	L	X	L	L	L	L
H	H	X	L	L	L	L	L
X	X	H	H	H	L	L	H
X	L	X	L	COUNT			
L	X	L	X	COUNT			
L	X	X	L	COUNT			
X	L	L	X	COUNT			

'92A, 'LS92, '93A, 'L93, 'LS93
RESET/COUNT FUNCTION TABLE

RESET INPUTS		OUTPUT			
$R_{0(1)}$	$R_{0(2)}$	Q_D	Q_C	Q_B	Q_A
H	H	L	L	L	L
L	X	COUNT			
X	L	COUNT			

NOTES: A. Output Q_A is connected to input B for BCD count.
B. Output Q_D is connected to input A for bi-quinary count.
C. Output Q_A is connected to input B.
D. H = high level, L = low level, X = irrelevant

functional block diagrams

'90A, 'L90, 'LS90

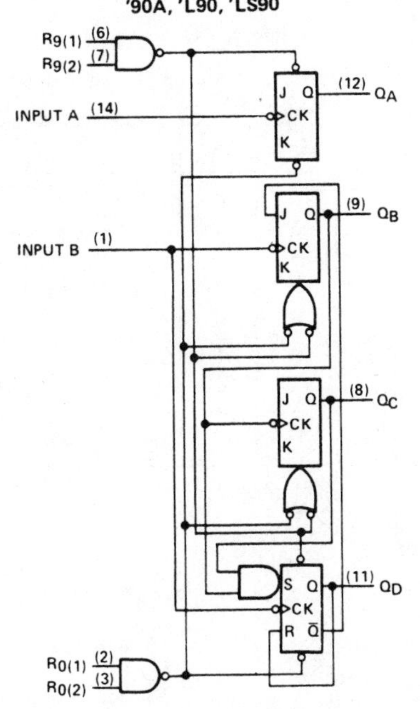

'93A, 'L93, 'LS93

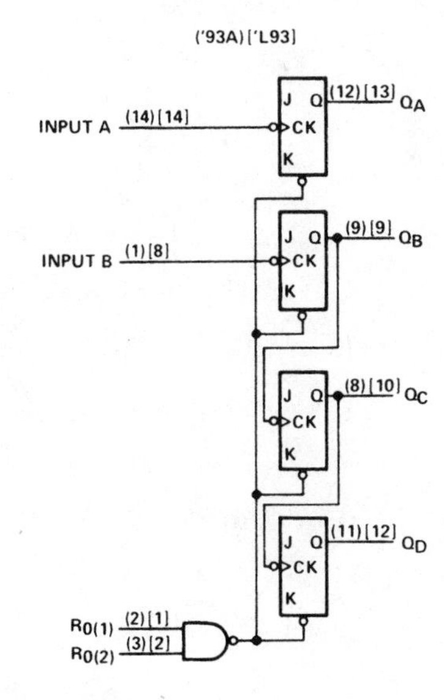

The J and K inputs shown without connection are for reference only and are functionally at a high level.

Figure 10.3 Truth tables and functional diagrams for the 74LS90 and 74LS93.

tive-high level, the output will go to count 9 regardless of the levels at the reset-to-0 inputs. Lines 4 through 7 indicate that if neither NAND gate is at an active-low output level, the counter will be enabled to count.

The BCD count sequence table indicates that if output Q_a is used to clock input B, this device will exhibit a normal BCD count sequence. Figure 10.4 depicts a circuit that uses the reset--to-0 inputs to modify the count length. Notice that Q_a is tied back to clock input B. Q_b is connected to reset-to-0 input 1 and Q_c is connected to reset-to-0 input 2. When Q_b and Q_c are both high the counter will be reset to count 0. Q_b and Q_c will both be high on count 6, but count 6 will never be seen. Just like the counter in Figure 10.2, this last count will only be seen for a couple of propagation delays. The count-state diagram in Figure 10.4 indicates the counting sequence. Creatively using the 74LS90 with a few other simple gates, any count modulus from 1 to 10 can be achieved.

The 74LS93 is the divide-by-16 version of the 74LS90. You will find that most MSI counters come in a BCD version and a 4-bit binary version. The BCD (decade) counter is often used to drive the inputs of a BCD-to-seven-segment decoder/driver, such as the 74LS47. If the outputs of a counter are to only be used

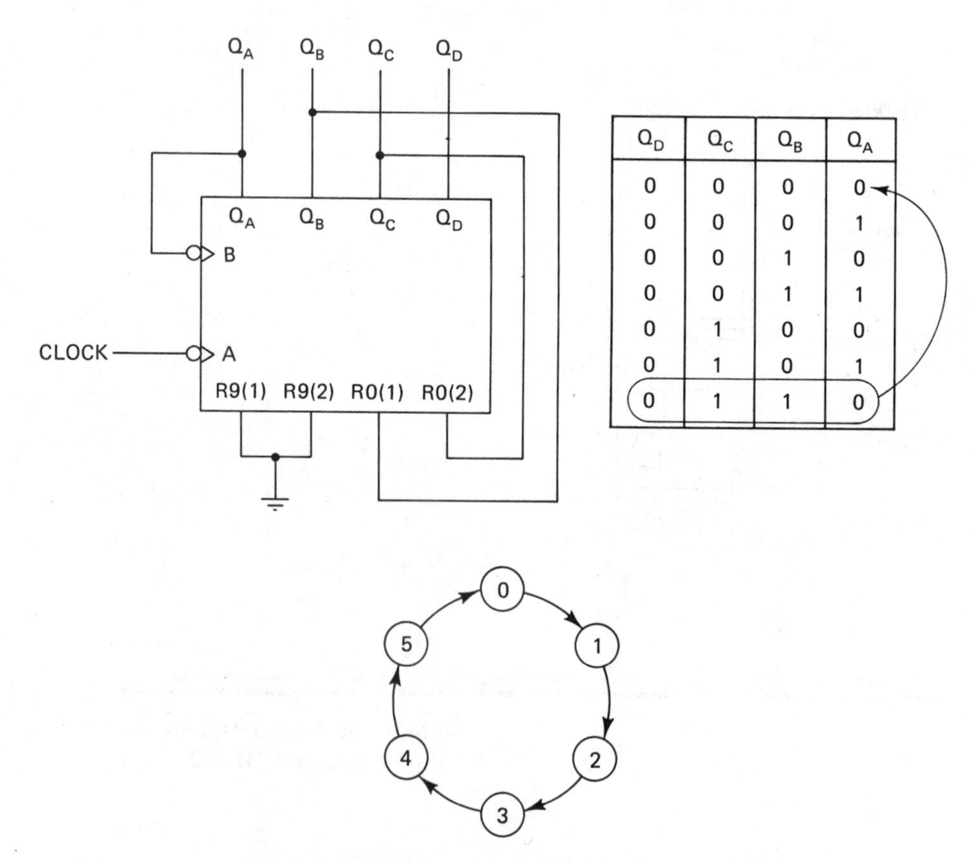

Q_D	Q_C	Q_B	Q_A
0	0	0	0
0	0	0	1
0	0	1	0
0	0	1	1
0	1	0	0
0	1	0	1
0	1	1	0

Figure 10.4 74LS90 with modified count sequence.

within the digital circuitry, a binary counter such as the 74LS93 is used.

The 74LS93 has the same reset-to-0 inputs, but it does not have any equivalent of the reset-to-9 inputs. Study the reset/count function table and the count sequence table for the 74LS93. By using the internal reset to 0 inputs and external gates, the 74LS93 can be configured for any modulus from 1 to 16.

10.2.2 The 4020B, 4024B, and 4040B CMOS Ripple Counter/Dividers

Figure 10.5 shows a series of counters that have 7, 12, and 14 stages. The 14-stage counter only has 12 outputs, so it can be packaged in a standard 16-pin DIP. Each stage is the equivalent of one flip-flop. The 74LS93 can be called a four-stage counter. The clock input is labeled "input pulses." Although it is not indicated by the functional diagrams, these counters are negative edge triggered. A properly drawn schematic would add a negative-edge-trigger symbol to the input pulse pin. The reset is active high and resets all the outputs to a logic 0. External gates can be used to modify the count length of these devices. The greatest possible modulus of each counter can be calculated with the following equation:

$$2^{(\text{number of stages})} = \text{greatest possible modulus}$$

These counters have a large unmodified modulus compared to TTL counters. The circuit in Figure 10.5 is a mod-76 counter. The count state diagram indicates that it counts from 0 to 75. On count 76 the reset input on the counter will go active high and the count will return to 0. When Q3 and Q4 and Q7 are

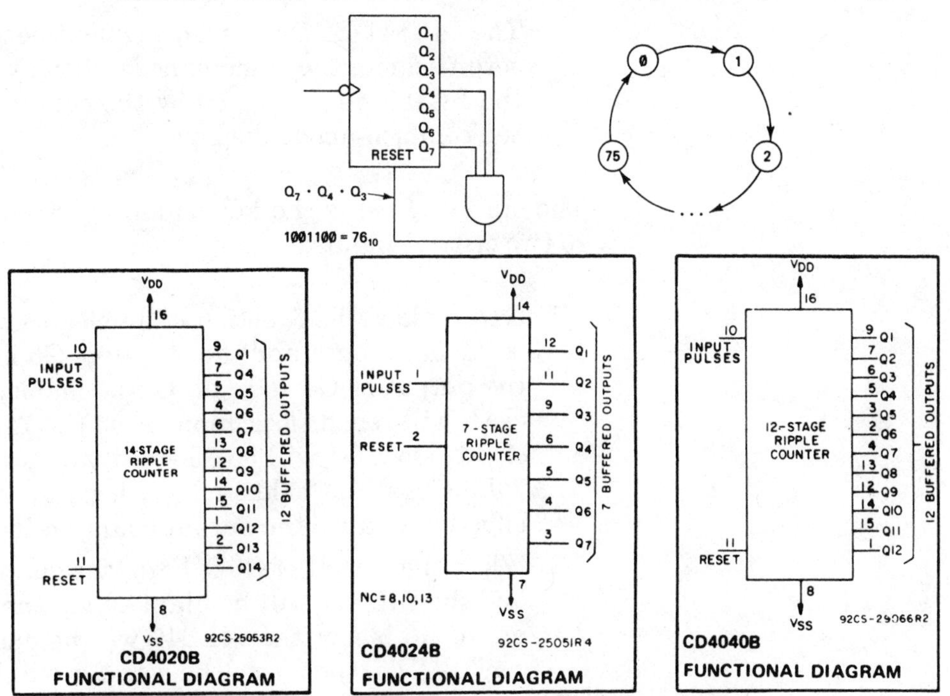

Figure 10.5 4020B, 4024B, and 4040B ripple counters.

high, count 76 has been reached. Notice that there must not be any count state, previous to 76, that will accidentally glitch and reset the counter. Many times all the Q outputs must be sent to external gates to decode the reset pulse. This will guard against the glitches caused by the ripple clocking of the internal flip-flops.

10.2.3 Application of a Ripple Counter

In Chapter 7 we constructed a 10-key keyboard encoding/decoding and display circuit using two 74LS148 priority encoders. We are now going to build the equivalent circuit using a 74LS90 ripple counter, a 74LS151 data selector, and some discrete gates. This circuit is intended to illustrate the operation of a counter in a moderately complex circuit (Figure 10.6). This is the most complex circuit that we have yet examined. Complex circuits should be analyzed with a "top-down" approach. This means that we should avoid getting caught up in the details of the circuit. We will first overview the major blocks that constitute the circuit, then we will examine how each major block functions and how it contributes to the circuit as a whole. The major blocks are:

1. The 74LS151 data selector and gates G3, G4, G5, G6, G7, and G8. These components form a decimal-to-BCD encoder.
2. The 74LS90 decade ripple counter. This counter scans the decimal-to-BCD encoder looking for a closed switch.
3. The 74LS47 BCD-to-seven-segment decoder/driver and the seven-segment common-anode display. This block takes the BCD code generated by the counter and displays it on the common-anode display.

The output of G6 is the key signal in this circuit; it is applied to two critical inputs:

1. G1 provides the function of an electronic switch; the clock is the data input and the output of G6 is the control line. If the output of G6 is high, G1 is enabled and the inverted clock will reach clock input A on the 74LS90; if the output of G6 is low, G1 will be disabled, and the clock input on the 74LS90 will be held high. A logic 0 output from G6 will effectively cause the output count on the 74LS90 to freeze.
2. When the output of G6 is high, the output of G9 will go low and the 74LS47 will be blanked via the BI. When the output of G6 is low, the 74LS47 will be enabled and the "frozen" BCD output from the counter will be displayed.

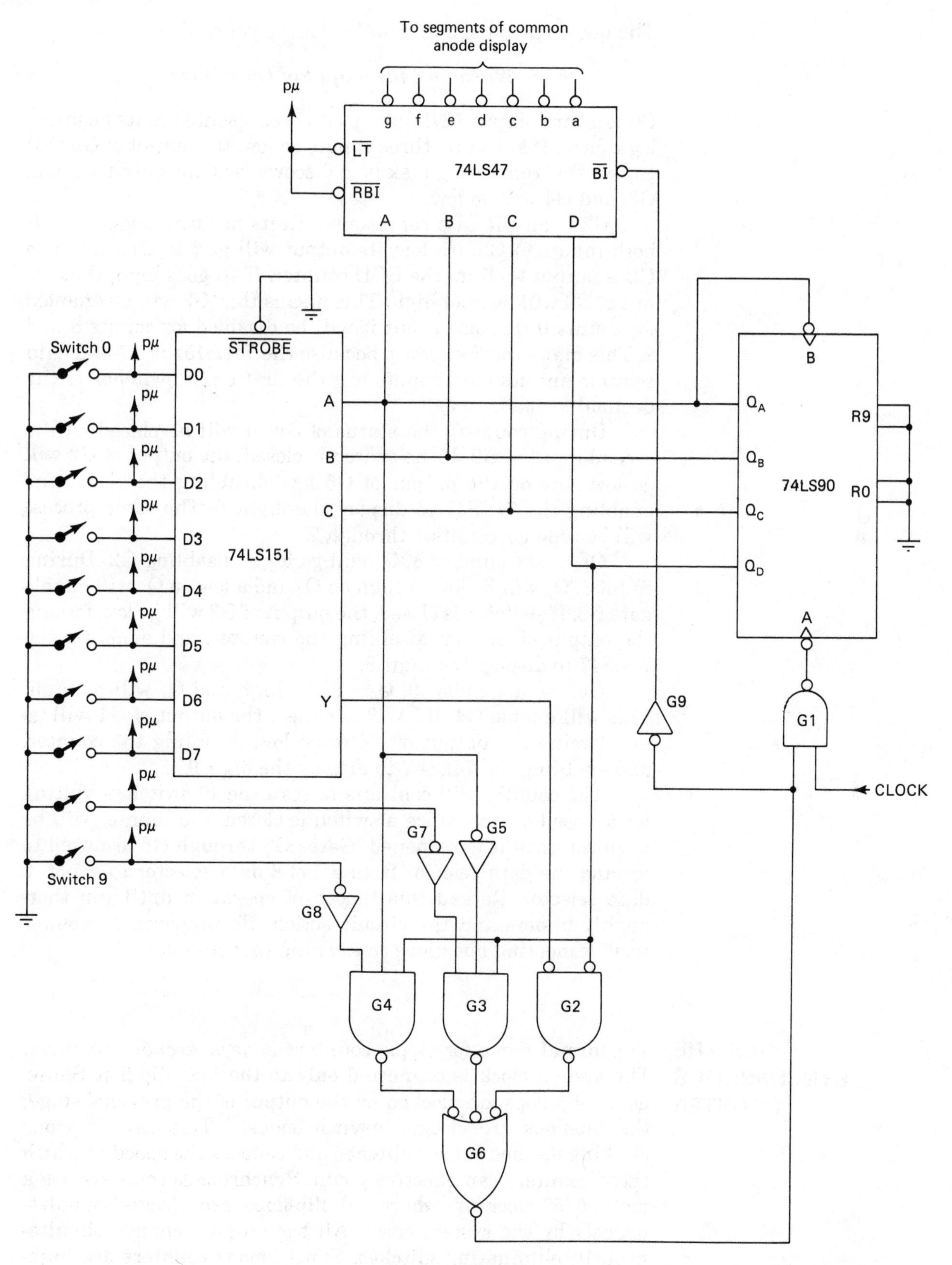

Figure 10.6 Counter-type decimal keypad encoder and display.

The question you should now be asking yourself is:

When will the output of G6 go low?

G6 is a three-input AND gate that is represented in its negative logic form. If any of its three inputs go low, the output of G6 will go low. Our remaining task is to discover how the outputs of G2, G3, and G4 will go low.

G2 is an OR gate represented in its negative-logic form. If both inputs to G2 are low, its output will go low. One input to G2 is output Q_d from the BCD counter. If Q_d goes high, the output of G2 will be held high. This means that G2 will be enabled for counts 0 through 7, but it will be disabled for counts 8 and 9. This makes perfect sense because the 74LS151 is a 1-of-8 data selector and its data inputs are the first eight switches of the decimal keypad.

During count 0, the status of SW 0 will be placed on the second input of G2. If the switch is closed, the output of G2 will go low, forcing the output of G6 low, disabling the clock, and enabling the 74LS47 to display the dight 0. The same process will be true for counts 1 through 7.

On count number 8, Q_d will go high, disabling G2. During count 8, Q_a will be low. A high on Q_d and a low on Q_a will enable gate G3. If switch 8 is closed, the output of G3 will go low, forcing the output of G6 low, disabling the counter, and enabling the 74LS47 to display the digit 8.

On count number 9, Q_d will be high and Q_a will be high. This will enable G4. If SW 9 is closed, the output of G4 will go low, forcing the output of G6 to go low, disabling the counter, and enabling the 74LS47 to display the digit 9.

The counter will continue to scan the 10 switches waiting for a closed switch. Once a switch is closed, the counter will be disabled until it is reopened. Gates G3 through G8 are used to expand the data selector from a 1-of-8 data selector to a 1-of-10 data selector. Reread this theory of operation until you thoroughly understand the circuit action. Be prepared to answer troubleshooting questions concerning this circuit.

10.3 THE SYNCHRONOUS COUNTER

The formal name for ripple counters is *asynchronous counters.* The system clock is connected only to the first flip-flop. Subsequent flip-flops are clocked by the output of the previous stage; the flip-flops are clocked asynchronously. This asynchronous clocking method causes glitches and reduces the speed at which these counters can effectively run. Synchronous counters use a method of clocking where all flip-flops are clocked simultaneously by the system clock. All the outputs change simultaneously, eliminating glitches. Synchronous counters are high-performance devices that also offer many advanced features.

10.3.1 The 74LS160 Family of Synchronous, Presettable Counters

The 74LS160, 74LS161, 74LS162, and 74LS163 all share an identical pinout. The 74LS160 and 74LS162 are identical BCD counters except for the manner in which their clear inputs operate. The 74LS160 has an asynchronous clear. Up to this point, all the clear inputs that we have examined were called direct or asynchronous. When an asynchronous clear goes active, the outputs will instantly react. The 74LS162 has a synchronous clear. The outputs will be affected only after the clear input has gone to its active level and the next active edge of the clock occurs. The clear function will be in sync with the system clock.

The 74LS161 is a 4-bit binary counter with an asynchronous clear. The 74LS163 is identical to the 74LS161 except that it has a synchronous clear. The terms "asynchronous" and "direct" are used interchangeably. Once you have mastered one member of this family of counters, the others are easily learned.

Let's examine the pinout in Figured 10.7 in functional groups.

Clock The clock input to the counter is active on the rising edge.

Clear The clear input is active low. The pinout does not differentiate between an asynchronous and a synchronous clear.

Enables T and P T and P must both be high before the counter is enabled. These enable inputs are used to cascade counters to obtain longer count sequences.

Ripple carry output This output will go high when the counter is on its last count. The last count for a decade counter is 9 and the last count for a 4-bit binary counter is 15. The positive pulse from the ripple carry output will be used to enable the next significant counter in cascaded applications. The 74LS148 priority encoder has an enable output that is used to enable the next downstream encoder. You can think of the ripple

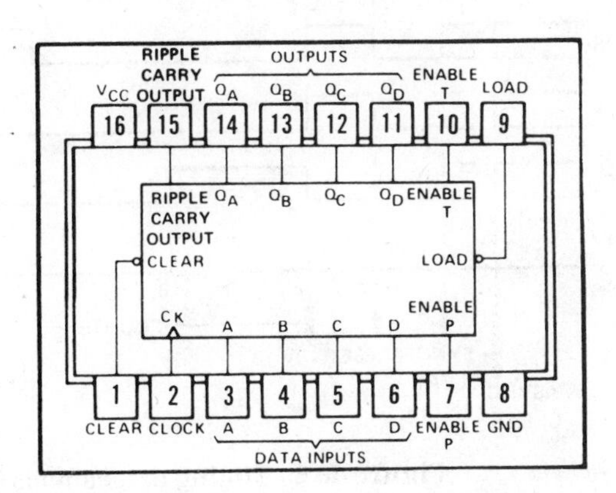

Figure 10.7 Pinout for the 74LS160 series counter.

carry output on the 74LS160 series of counters as working in a similar manner.

Outputs Q_a–Q_d These are the Q outputs from the four internal counting stages.

Data inputs A–D and the load input These counters were described as "presettable." Up to this point, all the counters that we have examined started their count sequence with the count of 0. It is often desirable to start the count sequence at another count. When the load input is taken to its active-low level, the next rising edge of the clock will transfer the levels on the data inputs (A through D) into the count outputs (Q_a through Q_d). This presetting operation is synchronous because it does not occur until the active edge of the clock. Another common name for presettable counters is "programmable" counters.

As always, the best way to understand the operation of a digital device is to examine its timing diagram. The timing diagram in Figure 10.8 is fairly complex. The dashed vertical lines indicate points of interest where we should focus our examination. This

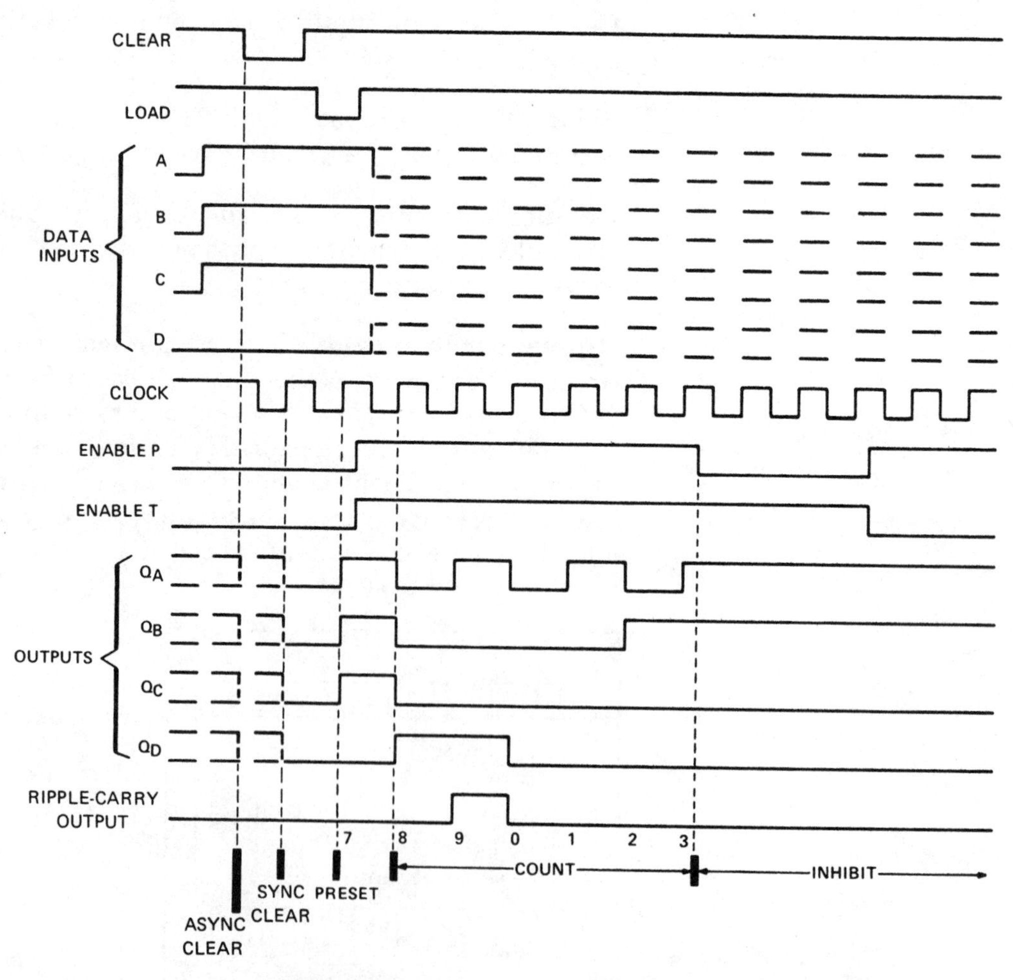

Figure 10.8 Timing diagram for 74LS160 and 74LS162 decade counters.

timing diagram represents the operation of both the 74LS160 and 74LS162 decade counters.

Event 1 The first vertical line illustrates the affect of the async clear on the 74LS160. The clear input is independent of the clock; when the clear input goes active, the outputs Qa through Qd will reset to logic 0. The dashed vertical lines above and below the Qa through Qd outputs represent their unknown states before the active async clear.

Event 2 This vertical line illustrates the affect of the sync clear on the 74LS162. The Q_a through Q_d outputs will not reset until the clear line is active and the clock has an active edge. The remaining events in this timing diagram are equally applicable to both the 74LS160 and 74LS162.

Event 3 This vertical line indicates the preset operation. Data inputs A through C are high and data input D is low. The load input goes active; on the next rising edge of the clock, the data inputs are transferred to the counter's outputs. The counter has been preset to the count of 7.

Event 4 The enable inputs, P and T, go to active-high levels, the clear input is inactive, and the load input is inactive. The counter is now enabled. On the next rising edge of the clock, the counter will sequence from count 7 to count 8. On the next rising edge of the clock, the counter will proceed to count 9. Because count 9 is the highest count of a BCD (decade) counter, the ripple carry will go to an active-high level. The next rising edge will cause the count to return to 0 and the ripple carry output will go to its inactive level. The counter will continue to count until some type of inhibit occurs.

Event 5 The enable P input goes inactive and the counter will be inhibited. The outputs on Q_a through Q_d will be held at the last count (3) until the enable P input returns high, the clear input goes active, or the load input goes active.

The timing diagram for the 74LS161 and 74LS163 is identical to Figure 10.8 except that the ripple carry will go active on count 15.

10.3.2 Cascading the 74LS160 Series

Figure 10.9 illustrates how the 74LS160 series of counters can be connected in cascade to produce longer count lengths. Notice that all four counters in Figure 10.9 are clocked in sync. The first counter is the least significant counter. The enable P and T on the least significant counter must be taken to an active level before any counting will take place. The second counter will be enabled whenever the ripple carry output on the first counter is high. Assume that the four counters are 74LS160 decade counters. When the first counter is at count 9, the ripple carry output will enable the second counter. On the next rising edge

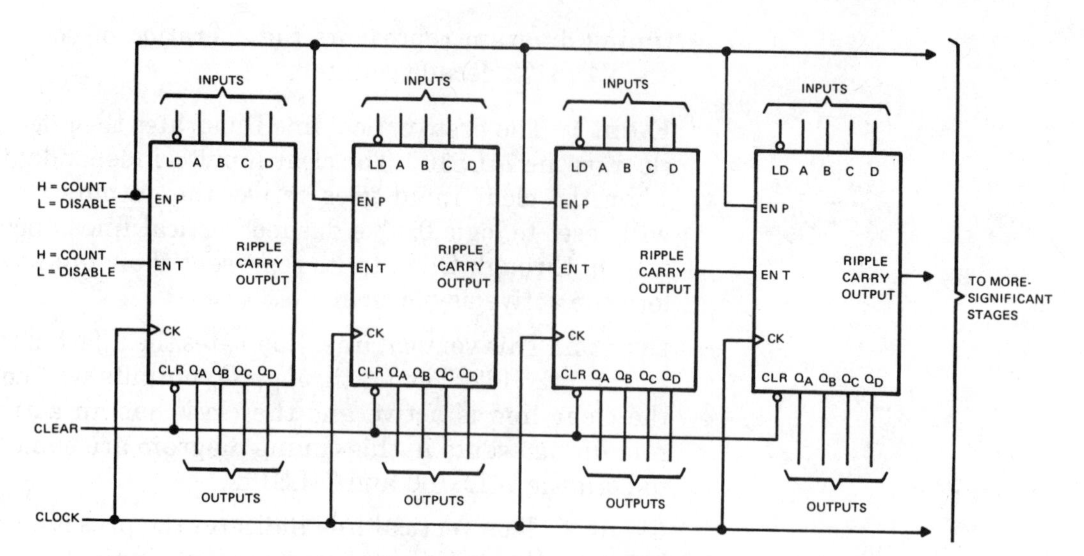

Figure 10.9 Cascading 74LS160 series counters.

of the clock, the first counter will return to 0 and the second counter will advance to 1.

This process of enabling the second counter every 10 clock pulses will continue through count 99. At count 99, both the ripple carry from the first counter and the ripple carry from the second counter are high. This will enable the third counter. On the next rising edge of the clock, counters 1 and 2 will return to 0 and counter 3 will advance to count 1.

This process of enabling the third counter every 100 clock pulses will continue through count 999. At count 999 the ripple carry outputs of counters 1 through 3 will be high. This will enable counter 4. On the next rising edge of the clock, counters 1 through 3 will return to 0 and counter 4 will advance to 1.

This process of enabling the fourth counter every 1000 pulses will continue through the count of 9999. The next rising edge will return all the counters to the count of 0. You should now understand the logic of designing two enable inputs into the 74LS160 series of counters.

10.3.3 An Application Using the Preset Inputs of the 74LS160 Series

Refer to Figure 10.10. U1 and U2 are clocked in synchronous and connected in cascade. As in Figure 10.9, the ripple carry output from U1 (least significant counter) will enable U2, the next significant counter. The 74LS161 is a 4-bit binary counter. When U1 is on count 15, the ripple carry output will enable U2. On the next rising edge of the clock, U1 will return to count 0 and U2 will increment its count by one. Every 16 clock pulses U2 will be enabled.

You may find it strange that the Qa through Qd outputs from either counter are not used. This circuit is specifically de-

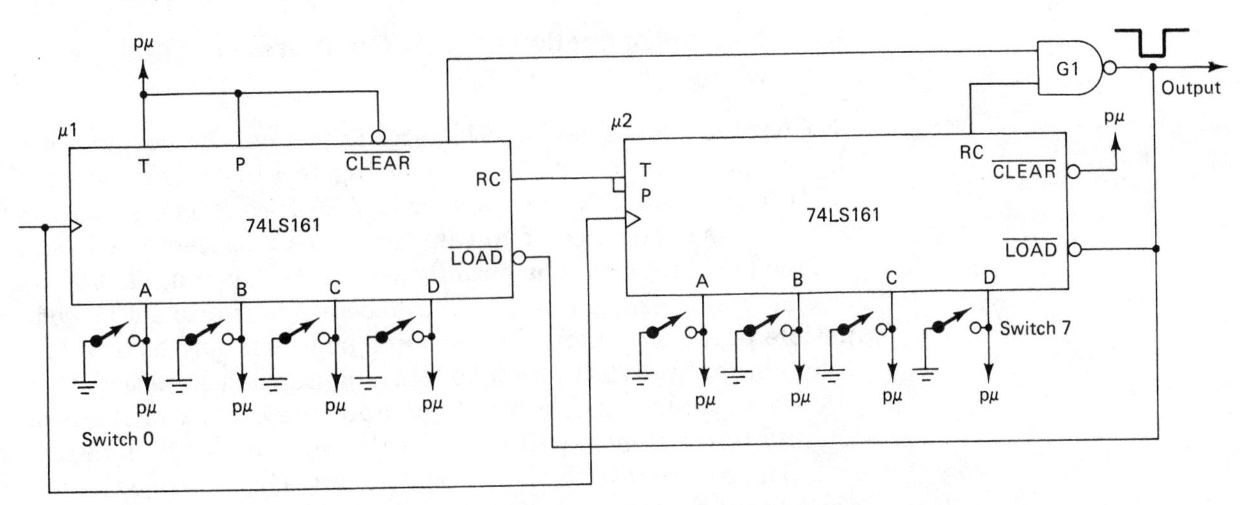

Figure 10.10 Divide-by-2 255 counter.

signed to divide an input frequency by a certain factor. On count 255 (1111 1111 binary) both ripple carries will be high. This will cause the output of the NAND gate to go low. This low is applied to the load inputs on U1 and U2. On the next rising edge of the clock, the count will not return to 0; it will return to the value toggled into SW 0 through SW 7. Each time the count reaches 255, the NAND gate will output a negative pulse, equal to the input clock's period.

We now only need to derive an equation that describes the relationship between the input frequency, the output frequency, and the settings of switches 0 through 7. Consider the case when all eight switches are closed. The preset value loaded into the counter will be equal to 0. This will be the same as clearing the counter. It will take 255 clock pulses before the output of the NAND gate will go active low. Therefore, when all the switches are closed, this circuit will divide the input clock frequency by a factor of 255.

Let's now open switch 0. Each time the counter reaches count 255, the counter will now be preset to the count of 1. It will now only take 254 clock pulses before the terminal count of 255 is reached. The device is now a divide-by-254 counter.

Let's now reclose switch 0 and open switch 1. The counter will now be preset to a value of 2. It will now perform the function of a divide-by-253 counter. We have now experimented enough to derive the following equation:

$$\text{freq}_{\text{out}} = \frac{\text{freq}_{\text{in}}}{255 - (\text{the value of switches 7 through 0})}$$

What happens if all the switches are opened (which constitutes a preset value of 255)? We know that at count 255 both ripple carry outputs will be high. If we preset the counter to a value of 255, the ripple carry outputs will never go low, and the output of the NAND gate will be stuck at a logic 0 level!

10.3.4 An Application of BCD Counters: A Digital Pulse Meter

In Chapter 8 we used 74LS75 latches to save the outputs of a digital pulse meter. We are now going to illustrate the use of BCD counters within the pulse meter (Figure 10.11). There are three boxes in the block diagram that must be accepted without detailed explanation. The box labeled "sensor" is that part of the circuit that senses a heart beat and converts it into a TTL-compatible pulse. This could be accomplished as simply as having a light shine through the earlobe into a phototransistor. Each time the heart contracts, the blood flow into the earlobe will cut off the light and a pulse will be generated by the phototransistor.

The box next to the heart beat sensor will multiply the sensor's input frequency by a factor of 10. This meter will display the pulse in units of beats per minute. Updating the display once every minute would not be an effective method of showing the true pulse. If we multiply the number of pulses (counted in the 6-second interval) by a factor of 10, the display can be updated every 6 seconds. This will result in a more accurate display. The multiplication process is accomplished by a device called a phase-locked loop. Phase-locked loops are extremely important circuits, in both the digital and analog worlds. Do not concern yourself with the operation of the frequency multiplier. It will simply allow us to update the display 10 times more often, greatly improving the operation of our circuit.

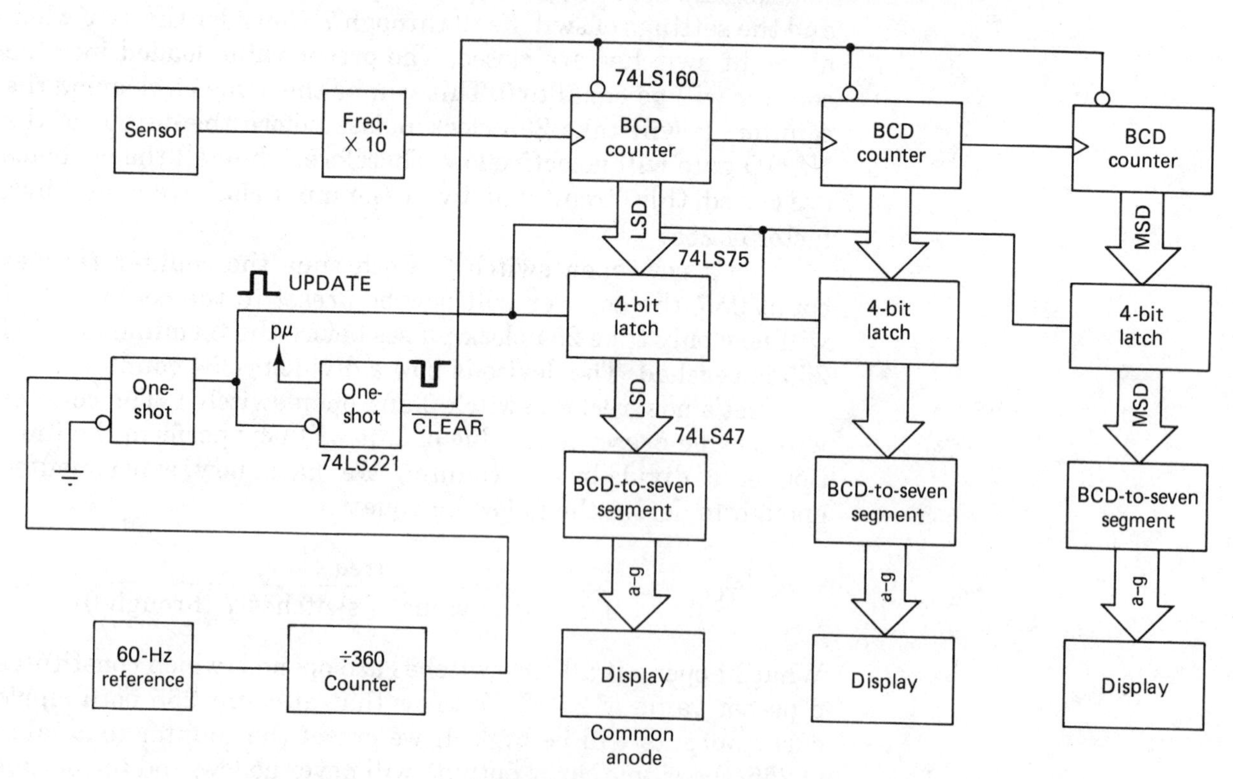

Figure 10.11 Block diagram of improved digital pulse meter.

The third box that will not be examined in detail is the 60-Hz reference. Because the power-line frequency is 60 Hz, it is often used to establish a reference for time-keeping operations. This 60-Hz reference is easily derived using a transformer, half-wave rectifier, and a Schmitt trigger. A *Schmitt trigger* is a device that uses the concept of hysteresis to convert slowly rising and falling waveforms into digital square waves. The Schmitt trigger will be examined in Chapter 13.

The three counters are connected in cascade to achieve a maximum count of 999. Every 6 seconds an update strobe will latch the present count into the three 4-bit latches. These latches will drive BCD-to-seven-segment decoder/drivers.

A divide-by-360 counter will divide the 60-Hz reference frequency into $\frac{1}{6}$ Hz. Every 6 seconds the positive edge from the divide-by-360 counter will fire the first one-shot. This one-shot will deliver a positive update strobe to latch the present count into the latches. The falling edge of the update strobe will fire the second one-shot. This negative pulse will clear the counters back to zero. After each 6-second sample, the process of updating the displays and clearing the counters will occur. Figure 10.12 depicts the actual schematic for the digital pulse meter.

U1 through U3 are 74LS160 decade counters with direct clear. They are connected in cascade as explained in Figure 10.9. U4 through U6 are 74LS75 4-bit latches. They will provide the temporary display memory for the 6 seconds between update pulses. U7 through U9 are 74LS47 BCD-to-decimal decoder/drivers. Notice that they are connected to blank leading zeros. U10 is the update one-shot. The pulse width of 100 μs is not critical. Any pulse width from 1 μs to 0.1 s would function correctly. U11 provides the clear pulse to reset the counters. Its pulse width is also not critical. Counters U12, U13, and U14 provide the divide-by-360 function; U12 is a divide-by-10 counter, U13 is a divide-by-12 counter, and U14 is a divide-by-3 counter. Notice that when counters are connected in cascade, the total factor of division is equal to the product of each counter's factor of division.

Take a moment to examine the schematic in Figure 10.12 in great detail. Try to understand how each device contributes to the system function.

10.3.5 An Application of Counters: A Seven-Segment Display Multiplexing Circuit

It is often desirable to control many seven-segment displays with a single BCD-to-seven-segment decoder/driver. The segment outputs (a through g) from the sole BCD-to-seven-segment decoder are connected to the segments of each display. One and only one display will ever be enabled simultaneously.

Consider the case where one BCD-to-seven-segment decoder will drive eight displays. The proper BCD inputs for the

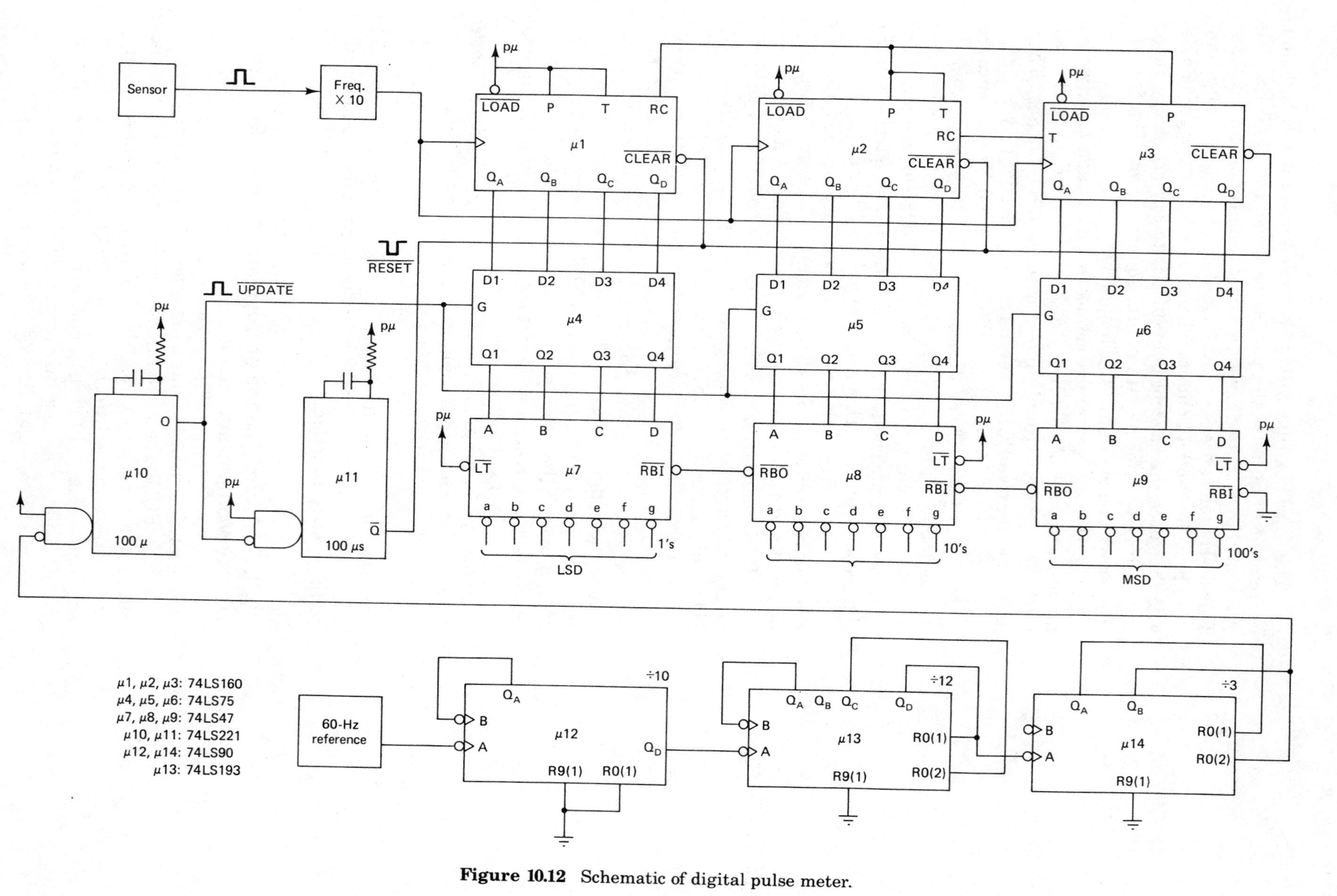

Figure 10.12 Schematic of digital pulse meter.

first display must be placed on the inputs of the BCD-to-seven-segment decoder. Display 1 must be enabled. The proper digit will then illuminate on the first display. After a short period of time, the first display will be disabled and the proper BCD code for the second display will be placed on the inputs of the BCD-to-seven-segment decoder. The second display will be enabled, and the proper digit will then illuminate on display 2. This process will be repeated for displays 3 through 8. After display eight has been enabled and displayed the proper digit, the process should then repeat with display 1.

What digital devices will be required to build this display multiplexing circuit? Obviously, we will need some multiplexers. We will have to multiplex eight groups of four BCD codes onto the inputs of a BCD-to-seven-segment decoder/driver. A device will be required that can select one of eight displays to enable. That sounds like a 3-line-to-8-line decoder. Finally, a device will be required to provide the code that selects the proper BCD codes to multiplex onto the inputs of the BCD decoder and also provide the correct 3-bit code to the 3-line-to-8-line decoder. That function can be provided by a mod-8 counter.

We have already used a decade counter to scan the inputs of a decimal keypad. We will now use a mod-8 counter to scan and select the proper BCD code and the proper display (Figure 10.13). U7, the 74LS90 ripple counter, is configured as a mod-8 counter. When Q_d goes high at count 8, the reset to 0 NAND gate will go active and the counter will reset to count 0. (Remember that count 8 is so short that it will never be seen.) The outputs of the mod-8 counter will drive the select inputs of the 74LS151 8-line-to-1-line multiplexers and the 74LS138 3-line-to-8-line decoder. The A select bit for each BCD code is applied to U4. The B select bit for each BCD code is applied to U3, as are the C and D select bits applied to U2 and U1, respectively. The 3-bit output from the counter will select a particular BCD code; it will be reconstructed on the Y outputs of U1 through U4 and placed on the inputs of U5.

Notice that the BCD-to-seven-segment decoder is a 74LS48. A 74LS48 differs from the 74LS47 in two ways: the segment outputs are active high and current-limiting resistors are internally integrated into the IC. Because the segment outputs are active high, the 74LS48 is designed to drive common-cathode displays. The counter will also drive the select inputs on the 74LS138. The 74LS138 will choose one of eight displays to enable. The active-low output of the 74LS138 will drive a 7407 noninverting buffer. The 7407 can sink approximately 45 mA of current. The output of each 7407 will be connected to the common cathode of one display. The output of the selected 7407 will go active low, providing the common ground for the selected display.

Carefully examine the circuit in Figure 10.13. Concentrate on understanding the timing relationship between the BCD codes and the selected display.

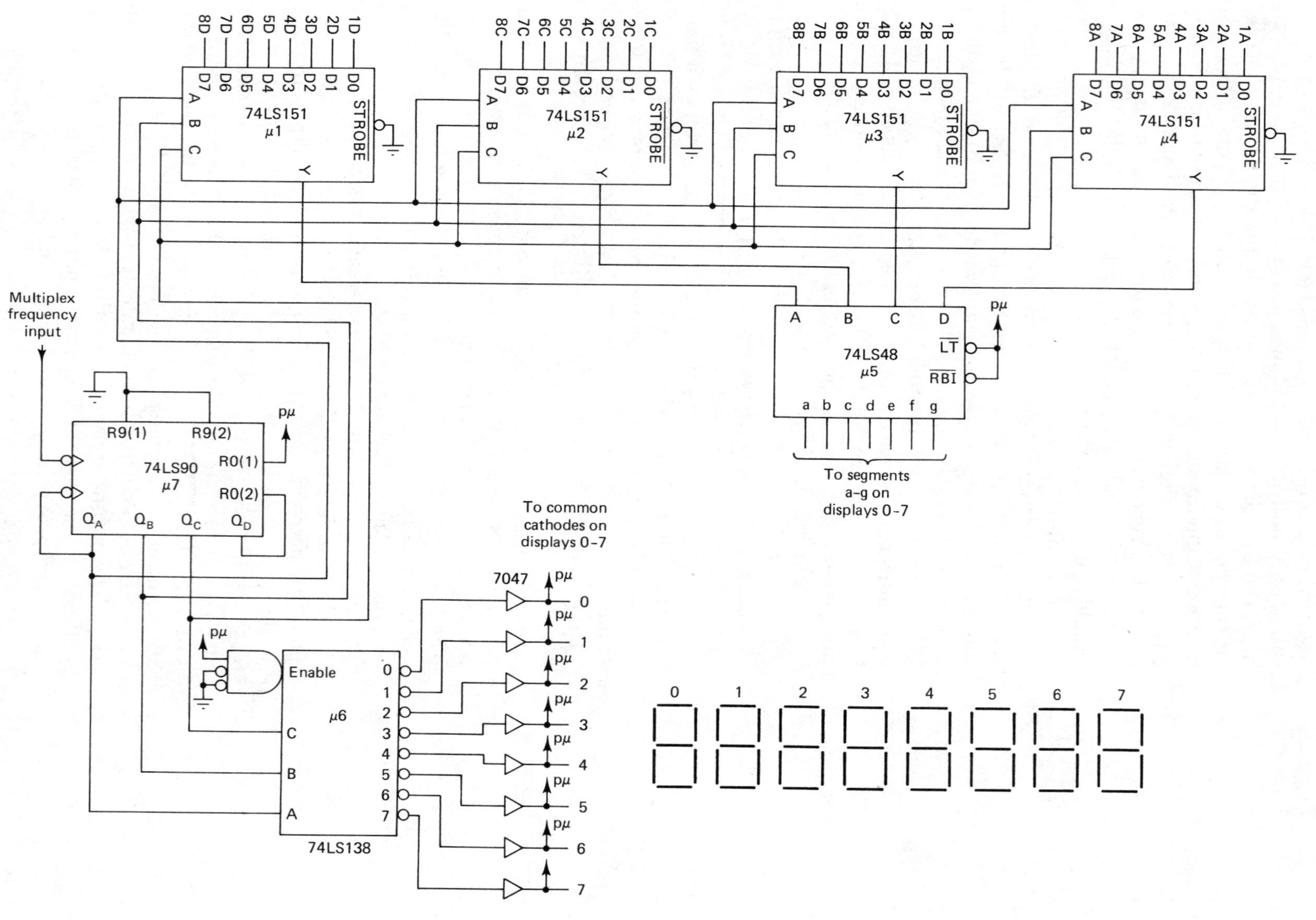

Figure 10.13 Seven-segment multiplexing circuit.

10.4 UP/DOWN COUNTERS

There are many applications that require a counter to count in both directions. Consider a digital display on a videotape recorder that indicates the present location of the tape. As the tape reel moves forward, the display (indicating inches of tape) will continue to increase. If the tape direction is reversed, the position indicator will start to decrease.

10.4.1 The 74LS190 and 74LS191 Synchronous Up/Down Counters

The 74LS190 and 74LS191 are programmable, reversible up/down decade and 4-bit binary counters. The pinout for the 74LS190 and 74LS191 appears in Figure 10.14. Let's examine the pinout in Figure 10.14 according to function.

Outputs Q_a–Q_d The 74LS190 is a decade counter and the 74LS191 is a 4-bit binary counter. Other than this distinction, they are exactly alike. Q_a through Q_d, as we have seen in every MSI counter, are the count output pins.

Data inputs A–D and load These counters are presettable, just like the 74LS160 series. There is one major difference between the presetting operation of the 74LS160 series and the 74LS190/LS191. The 74LS160 counters have a synchronous preset; the data inputs appear on the Q outputs only after the load input has gone active and a rising edge appears at the clock input. The 74LS190 and 74LS191 have a direct (asynchronous) preset. When the load input goes active, the Q outputs will instantly change to agree with the data inputs, independently of the clock!

The 74LS190 or 74LS191 do not have a clear pin. To clear the counter, the data inputs are taken to logic 0 and the load input is pulled low.

Clock The 74LS190 and 74LS191 are positive edge-clocked devices.

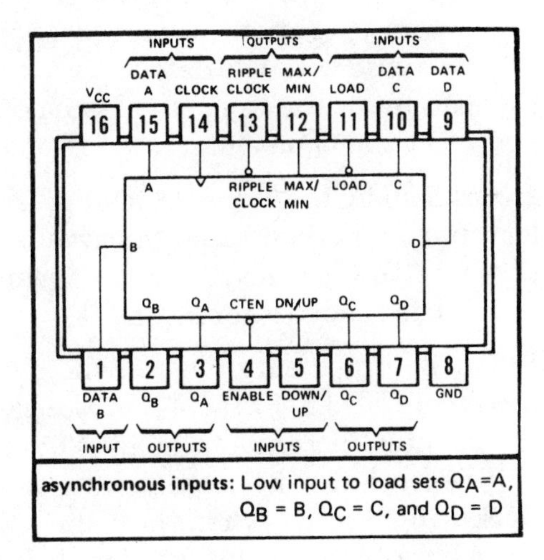

Figure 10.14 Pinout for 74LS190 and 74LS191 up/down counters.

DN/UP—down/up The logic level on this pin will determine the direction of the count. A high on this pin will cause the counter to count down; a low on this pin will cause the counter to count up. The DN/UP pin can be driven by a static level pulled up to V_{cc} or down to ground, or the DN/UP pin can be driven dynamically by the output of a digital device.

G—enable input The 74LS160 series of counters have two active-high enable inputs: P and T. As you should remember, P and T are used to cascade counters to achieve greater count lengths. The enable on the 74LS190 series is a single active-low input. The 74LS190 series uses a different technique to cascade counters. Counting is inhibited until the G input is taken to a logic 0 level.

Ripple clock The ripple clock output is an active-low equivalent of the ripple carry output on the 74LS160 series of counters. It will produce a low-level output equal to the low-level portion of the clock input whenever the counter overflows or underflows. A carry on a BCD counter occurs when the counter sequences from a count of 9 to a count of 0. The formal term describing a carry is *overflow.* When a BCD counter is down-counting and reaches the count of 0, the next count that it will advance to is the count of 9. In simple math this is known as a borrow; in digital electronics this event is called an *underflow.* One method of cascading counters is to use the rising edge of the ripple clock output to clock the next counter stage.

Max/Min This output produces a high-level pulse equal in period to the clock input whenever the counter overflows or underflows. Max/Min is used to signal to other devices that the counter is about to overflow or underflow.

Figure 10.15 is a timing diagram illustrating the operation of the 74LS191 4-bit binary counter. Each vertical line in Figure 10.15 indicates an important event.

Event 1 The load input is taken to an active-low level. Because the preset operation is asynchronous, the Q outputs will instantly agree with the data inputs, regardless of the clock. While the load input is active, any rising edges on the clock input will not affect the Q outputs.

Event 2 The counter has been preset to the count of 13 and the load input is taken to an inactive level. The enable input is active and the low level on the down/up pin indicates that the counter is ready to count up.

Event 3 The rising edge of the clock occurs and the counter increments from 13 to 14. On the next rising edge of the clock the counter increments to 15. Notice that the max/min output will be high for the total length of the count of 15. The ripple clock will go active when the counter is on count 15 and the clock has a high-to-low transition. On the next rising edge of the clock,

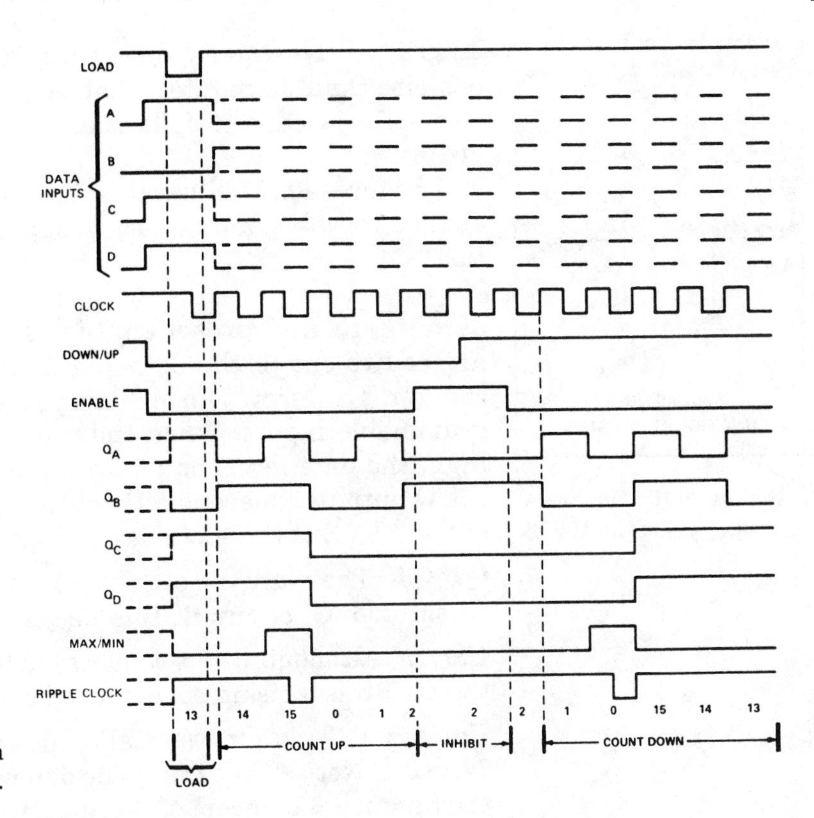

Figure 10.15 Timing diagram for the 74LS191.

the counter overflows to count 0. The max/min and ripple clock outputs will return to their inactive states. Notice how the rising edge of the ripple clock can be used to clock the next counter stage.

Event 4 The enable input goes inactive. The counter will not respond to any rising edges on the clock input. The counts stays frozen at 2.

Event 5 During the period of time that the counter was inhibited by a high signal on the enable input, the logic level on the DN/UP pin toggled from low to high. The counter is now enabled to count in the down direction.

Event 6 The active edge of the clock decrements the count from 2 to 1. The next rising edge of the clock decrements the count to 0. The max/min and ripple clock outputs will go active to indicate an underflow operation. Notice that an overflow and underflow condition cannot be differentiated by only observing the max/min or ripple clock outputs.

The 74LS190 operates exactly as the 74LS191 except that it overflows on count 9 instead of count 15.

10.4.2 The 4029B Presettable Up/Down Counter

The 4029B is a highly versatile counter. It can count up or count down, operate as a 4-bit binary or BCD counter, and be cascaded in a synchronous or asynchronous mode. As we have seen with previous CMOS devices, the terminology used in describing the

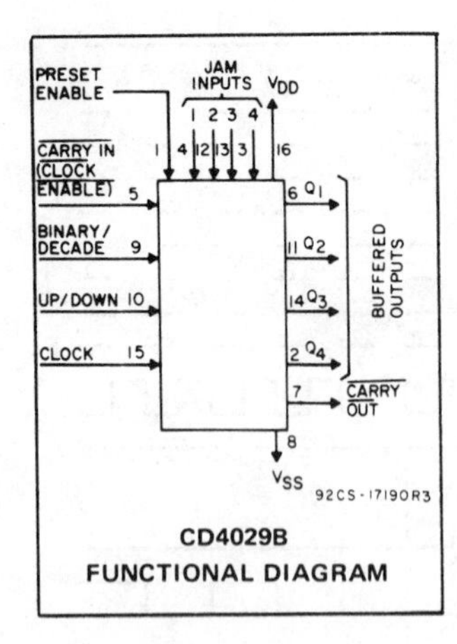

Figure 10.16 Functional diagram of the 4029B.

4029B will be slightly different than the terminology used to describe the TTL counters that we have examined. We will draw on our knowledge of TTL counters to help us understand the 4029B.

Figure 10.16 illustrates the functional diagram of the 4029B. We will examine each group of pins according to its function.

Jam inputs and preset enable The jam inputs on the 4029B are equivalent to the data inputs of any TTL programmable counter. The term "jam" is used to emphasize the nature of programmable inputs. When the preset enable input goes active high, the data levels on the jam inputs will be "jammed" into the Q outputs independently of the clock. The preset operation of the 4029B is therefore asynchronous.

Q1–Q4 These are the counting outputs. They are equivalent to Q_a through Q_d of any BCD or decade TTL counter.

Clock Although it is not indicated on the functional diagram, the 4029B is a rising edge-clocked counter.

Up/Down This pin was called down/up on the 74LS190 and 74LS191. Notice that the order of down and up are reversed in the functional diagram of the 4029B. A high on the up/down pin will cause the 4029B to perform as an up counter; a low on this pin will cause it to perform as a down counter.

Binary/Decade The logic level on this pin will define the unmodified modulus of the 4029B. A high will define a modulus of 16 and a low will define a modulus of 10. In actual applications, the binary/decade pin is usually tied to V_{cc} or ground; it is not driven by the changing output of another digital device.

Carry out This pin will go active low for the period of one clock pulse to indicate that the counter has reached an overflow or underflow state. It is used like the ripple clock on the 74LS190 and 74LS191 counters. The positive edge of the carry output can be used to clock the next counter stage for counters connected in a ripple clocked fashion.

Carry in (clock enable) This is the active-low enable input of the 4029B. It is called by two different names because it can be used to enable the counter in different situations. It is often used like the P and T enable inputs of the 74LS160 counters. When 4029B's are clocked in synchronous, the carry-in input will be driven by the carry-out output of the previous counter stage. When the upstream counter has reached an overflow or underflow condition, the next counter will be enabled. Figure 10.17 illustrates the two different methods of cascading 4029Bs.

The parallel clocking method is exactly like the cascading method that was illustrated for the 74LS160 series of counters. The ripple-counting method employs external OR gates that function as electronic switches. If the carry-out output of the pre-

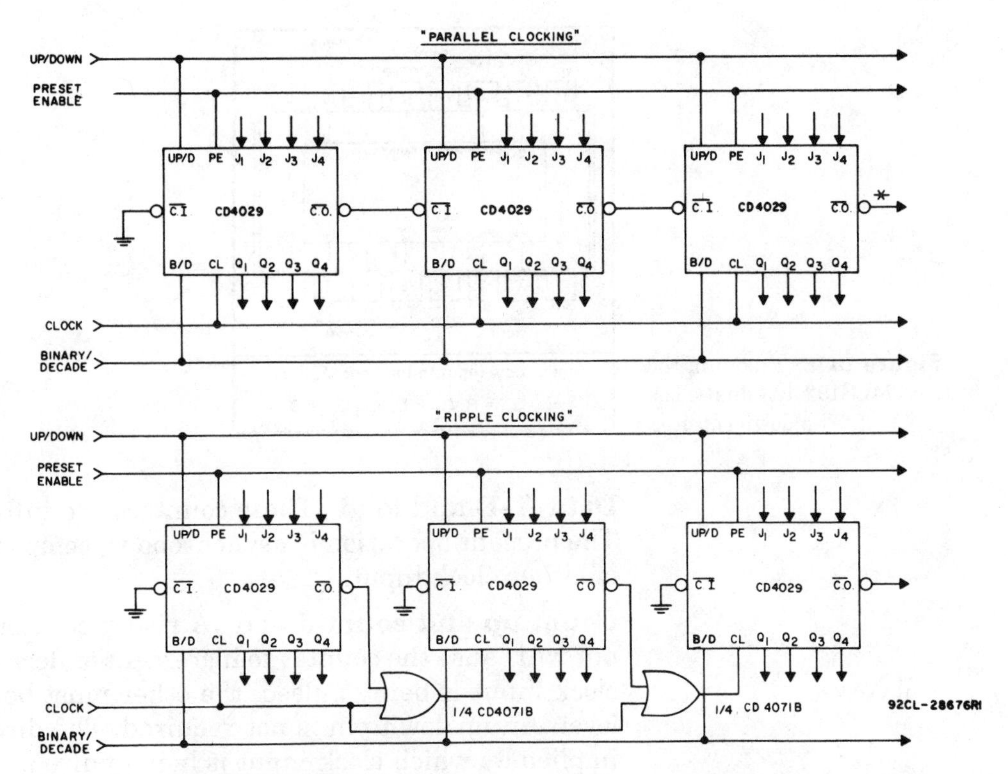

Ripple Clocking Mode:
The Up/Down control can be changed at any count. The only restriction on changing the Up/Down control is that the clock input to the first counting stage must be high.

* $\overline{\text{CARRY OUT}}$ lines at the 2nd, 3rd, etc., stages may have a negative-going glitch pulse resulting from differential delays of different CD4029B IC's. These negative-going glitches do not affect proper CD4029B operation. However, if the $\overline{\text{CARRY OUT}}$ signals are used to trigger other edge-sensitive logic devices, such as FF's or counters, the $\overline{\text{CARRY OUT}}$ signals should be gated with the clock signal using a 2-input OR gate such as CD4071B.

For cascading counters operating in a fixed up-count or down-count mode, the OR gates are not required between stages, and $\overline{\text{CO}}$ is connected directly to the CL input of the next stage with $\overline{\text{CI}}$ grounded.

Figure 10.17 Synchronous and ripple counting modes of cascaded 4029Bs.

vious stage is inactive high, the output of the OR gate will stay high, blocking the clock pulses. When the ripple carry output goes active low, the OR gate will pass the clock onto the next counting stage.

The timing diagram of the 4029B is similar to the timing diagram of the 74LS190/191. The only differences are in the active levels of the inputs and outputs.

10.4.3 The 74LS192 and 74LS193 Dual-Clock Counters

The 74LS192/193 is different than other up/down counters because it has two clock inputs. One clock input is used to count up and the other is used to count down. Figure 10.18 illustrates the pinout for the 74LS192/193. We will examine the pinout in Figure 10.18 in functional groups.

Q_a–Q_d These are the four counting outputs. The 74LS192 is a decade counter and the 74LS193 is a 4-bit binary counter.

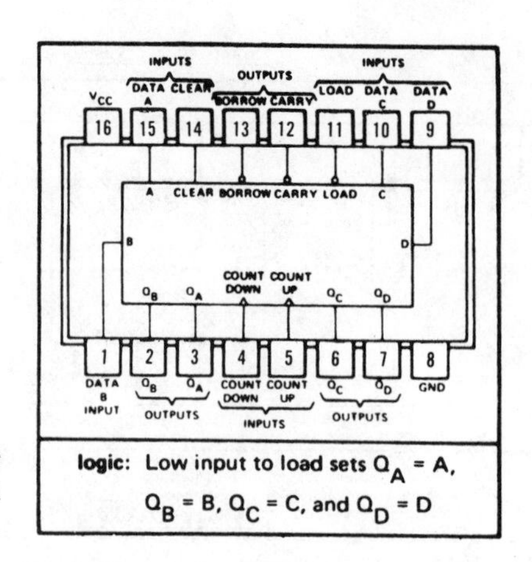

Figure 10.18 Pinout of the 74LS192/193 dual-clock up/down counter.

Data A–D and load These counters are fully programmable. The present operation is asynchronous, completely independent of either clock input.

Count up and count down A rising edge on either clock input will cause the counter to increment or decrement. While one clock input is being pulsed, the other must be held at a logic 1 level. An up/down pin is not required. The direction of count is implied by which clock input is being pulsed.

Borrow and carry The up/down counters that we have previously studied have had single clocks and single outputs that react to both overflow and underflow conditions. Because the 74LS192/193s have dual clocks, independent outputs are provided for overflow and underflow indicators. The carry output will go active to indicate an overflow condition, and the borrow output will go active to indicate an underflow condition.

The counters can be cascaded by simply connecting the carry output and the borrow output to the count-up and count-down inputs of the next stage.

Clear The clear input is active high and asynchronous. When the clear input is taken to its active level, outputs Q_a through Q_d will all go to a logic 0 level, independent of either clock.

Figure 10.19 illustrates a timing diagram describing the operation of the 74LS193 binary up/down counter. We will examine the events that occur at each vertical line in Figure 10.19.

Event 1 The asynchronous clear input goes to an active high level. The outputs, Q_a through Q_d, reset to a count of 0.

Event 2 The clear input returns to an inactive level. The count will remain at 0 until a load operation occurs or a clock input is pulsed with a rising edge.

Event 3 The load input goes to its active level. The logic levels on the data inputs are transferred to the data outputs. The counter is preset to a count of 13.

Event 4 The load input returns to an inactive level. The counter is now enabled to start counting at 13.

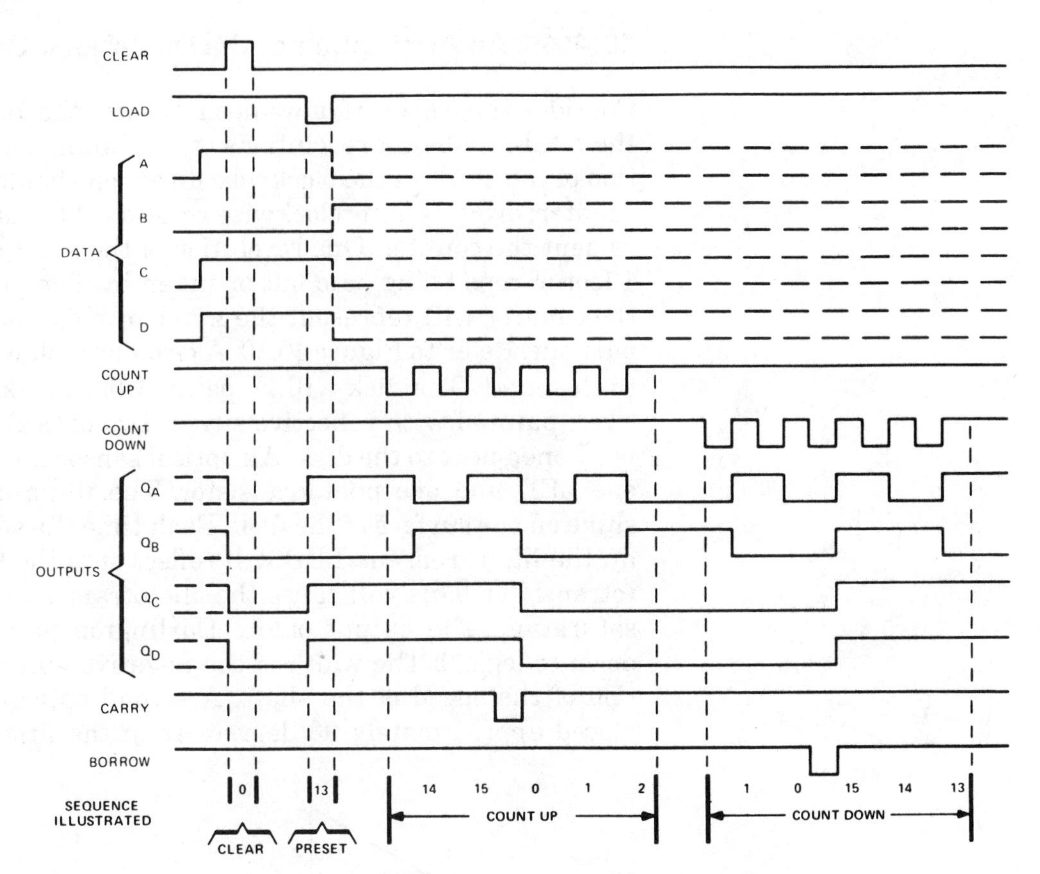

Figure 10.19 Timing diagram for 74LS193.

Event 5 The clear and load inputs are inactive and the count-down input is at a logic 1 level. The rising edge on the count-up input will cause the counter to increment to a count of 14.

The next rising edge on the count-up input will cause the counter to increment to 15. The carry output will go low, for a period equal to the negative portion of the count up-clock, to indicate that the counter is at an overflow state. Notice that the rising edge of the carry output can be used to clock the count-up input of the next counter in cascade. The 74LS193 continues to count up to a count of 2.

Event 6 Both clocks are taken to logic 1 levels. The counter will not increment or decrement during this period. The outputs on Qa through Qd will remain stable at the count of 2.

Event 7 On the rising edge of the count-down clock, the counter will decrement to a count of 1. On the next rising edge on the count-down inputs, the counter will decrement to a count of 0. The borrow input goes active for a period equal to the negative portion of the count-down clock to indicate that the counter is in an underflow state. Notice that the rising edge on the borrow output can be used to clock the count down input of the next significant counter in cascade.

Event 8 The counter will count down from 2 to 13 between events 7 and 8.

10.4.4 An Application of the Dual-Clock Counter

Consider the case where we desire to use the 74LS192 to count the total number of revolutions of a spinning shaft. Every rotation of the shaft in the clockwise direction should increment the counter; every counterclockwise rotation of the shaft should decrement the counter. One revolution of the shaft could represent 1 foot of rope being paid out or taken in. The present output of the counter will represent the number of feet of rope presently paid out. Refer to Figure 10.20. A circular disk will be connected to the shaft. The disk will be painted flat black, except for one circle painted with reflective silver. Two optical sensors will be positioned next to the disk. An optical sensor is constructed from one LED and one phototransistor. The illuminated LED will shine on the surface of the disk. Each time the silver spot passes by, the light from the LED will reflect onto the base of the phototransistor. This will cause the phototransistor to conduct into saturation. The output of the Darlington pair will be pulled down to logic 0. The width of the negative pulse will be a function of the speed of the shaft. A second optical sensor will be placed approximately 90 degrees from the first optical sensor.

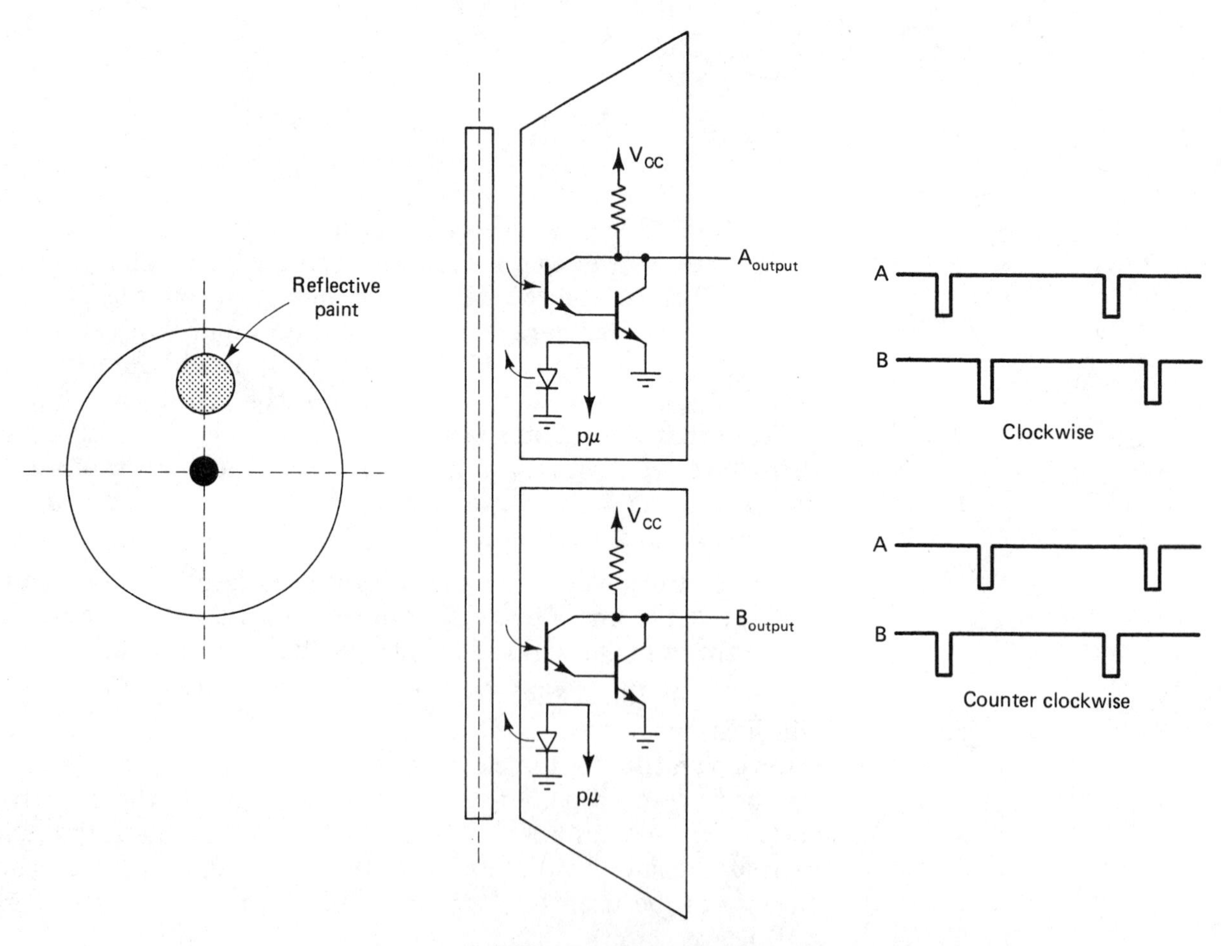

Figure 10.20 Optical sensors for disk rotation.

The output waveforms for clockwise and counterclockwise rotation is illustrated in Figure 10.20.

When the shaft is traveling in a clockwise direction, the negative pulse on output A will lead the negative pulse on output B. When the shaft is traveling in a counterclockwise direction, the pulse on B will lead the pulse on A. The preceding statements are true, regardless of the speed of the shaft. We must design a circuit that translates phase information into a clock pulse.

If the pulse on A is leading the pulse on B, then a rising edge should be applied to count up.

If the pulse on B is leading the pulse on A, then a rising edge should be applied to count down.

The circuit operation will be simple. It must remember which pulse comes first, and wait for the other pulse before it provides the rising edge to the proper clock input. Refer to Figure 10.21. Assume that the Q outputs from U1 through U4 are initially low. We will first examine the case when the pulse on input A is leading the pulse on input B (indicating a count-up operation).

The rising edge of the negative pulse will clock D flip-flop U1, and it will also be applied to the positive input on one-shot U4. U1 will set whenever a rising edge occurs on its clock because its D input is pulled up to V_{cc}. The $\overline{Q}$ output of U1 will go low, enabling one-shot U3 to fire when it receives a rising edge from input B. The positive edge on U4 will not fire the one-shot because the $\overline{Q}$ output of U2 is initially high, disabling U4.

The rising edge of the lagging negative pulse on input B will fire one-shot U3. The $\overline{Q}$ pulse from U3 will clock the count up input of a 74LS192 counter and it will also clear both D flip-flops, via G1.

The circuit is now back to its initial state where all Q outputs are cleared. If the shaft continues rotating in a clockwise direction, the positive edges of the pulses on $\overline{Q}$ of U3 will continue to increment the counter.

Let's assume that the shaft has reversed direction and is now rotating counterclockwise. The pulses at input B are now leading the pulses at input A. The positive edge on input B will set U2. $\overline{Q}$ on U2 will go low and enable U4. When the positive edge occurs on input B, U4 will be fired. The negative pulse on $\overline{Q}$ of U4 will clock the count down input of the counter and also clear the D flip-flops via G1.

In summary, let's define the function of each element in the phase to clock pulse translation circuit. D flip-flops U1 and U2 are used as simple memory elements. The state of the $\overline{Q}$ output from U1 or U2 will indicate whether input A or input B is leading. When the $\overline{Q}$ level from either D flip-flop goes to its active low output level, the associated one-shot will be enabled.

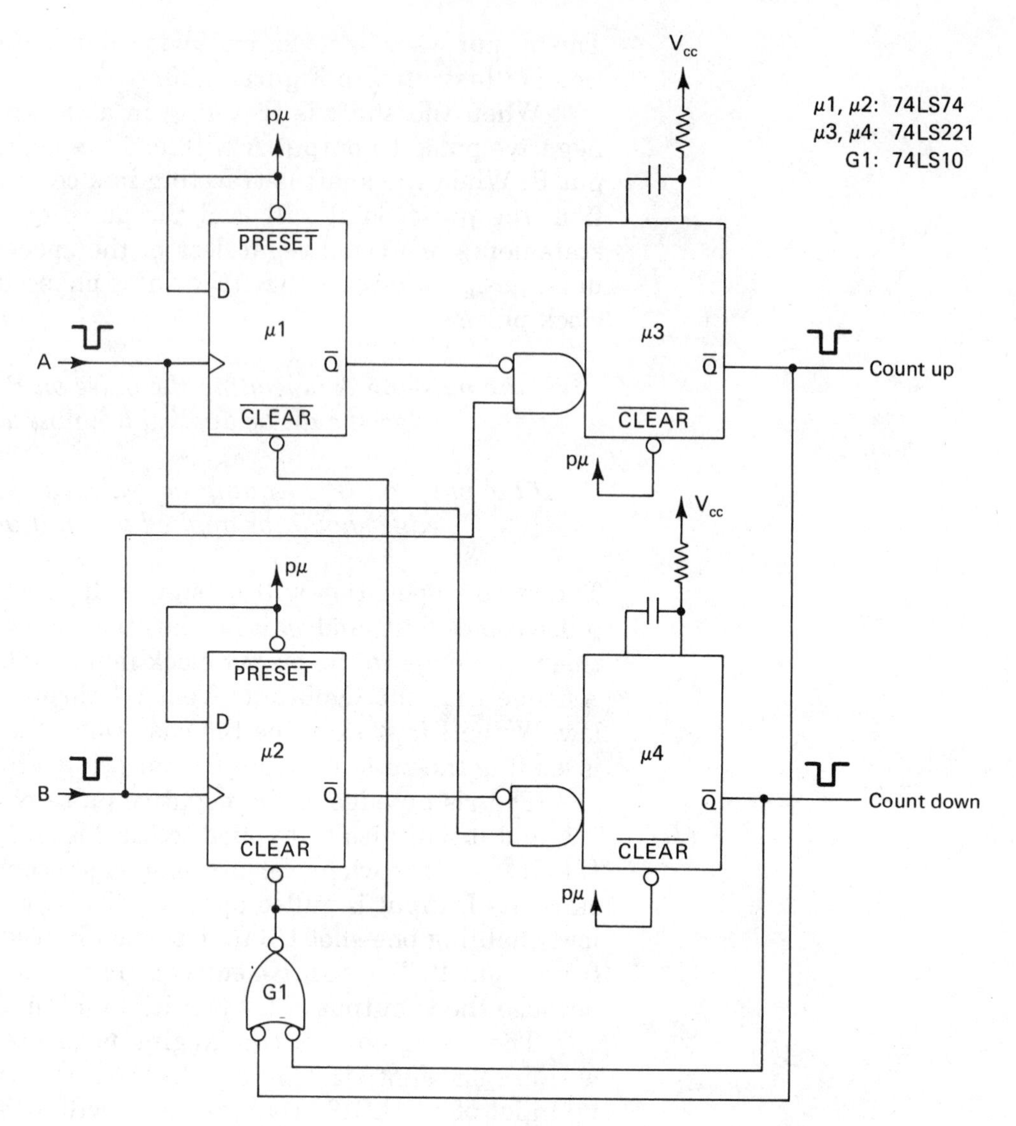

Figure 10.21 Phase-to-clock pulse translation circuit.

The one-shots U3 and U4 are used to provide the rising edge required to clock the up/down counter and also to provide the logic 0 level required to clear the memory elements, U1 and U2.

AND gate G1 is represented as a negative-logic active-low input, active-low output OR gate. If $\overline{Q}$ from U3 goes low or $\overline{Q}$ from U4 goes low, the D flip-flops will be cleared.

10.4.5 Special-Purpose Counters

We have examined the basic counters in the TTL and CMOS logic families. There are many specialized counters available. The CMOS 4017B and 4022B are counters with decoded outputs. The 4022B is a BCD counter, but instead of having four outputs Q1 through Q4, it has 10 outputs 0 through 9. The 4022B is called a fully decoded counter. Each possible count state (0 through 9) has an associated output. A decoded BCD counter is nothing more than a decade counter followed by a BCD-to-decimal decoder.

We have seen many applications of counters that require the use of 4-bit latches and BCD-to-seven-segment decoder/drivers. The 4026B and 74LS143 include a BCD counter, 4-bit latch, and BCD-to-seven-segment decoder all integrated into one IC. Your solid foundation in counter theory should enable you to successfully read and interpret the data sheets describing any counter-type device.

10.5 SHIFT REGISTERS

The basic building blocks of counters are flip-flops that are configured to toggle. Another application of flip-flops is shift registers. Shift registers are used for a wide variety of applications, such as serial-to-parallel conversion, parallel-to-serial conversion, sequencing circuits, and time delays.

10.5.1 The 74LS164 Serial-In, Parallel-Out Shift Register

The best way to learn the concepts of a PISO (parallel in, serial out) shift register is to examine the pinout, function table, and timing diagram of a typical unit. The 74LS164 is a popular device that inputs serial data and outputs 8 bits of parallel data. In the computer world data comes in two different forms: serial and parallel. Serial data is nothing more than a stream of single bits of digital information: logic 1's and logic 0's. Parallel data is a group of related bits. The most common size of parallel data is 8 bits. A parallel group of 8 bits is know as a *byte.* A byte of information represents a decimal number between 0 and 255 ($2^8 = 256$). This byte of information can represent a number, a letter, or a command. It is not important that you fully understand what a byte of data represents yet. That concept will be fully discussed in your class on microprocessors. What you should now understand is the difference between serial and parallel data (see Figure 10.22). You have seen many different pinouts. Although you do not know how a parallel output shift register works, a good deal of information can still be derived from analyzing the pinout in Figure 10.22.

Clock The 74LS164 has a positive edge-triggered clock. Whatever function the 74LS164 provides, it will perform its function on the rising edge of the clock.

Clear Every clear input that we have seen will reset the Q outputs when it is taken to an active level. The bubble on the clear input indicates that it is an active-low input. When the clear input goes active, all the Q outputs will be driven to logic 0.

Qa–Qh Q output pins have always represented outputs of sequential devices. These eight Q's must be the byte of parallel data that we have previously discussed.

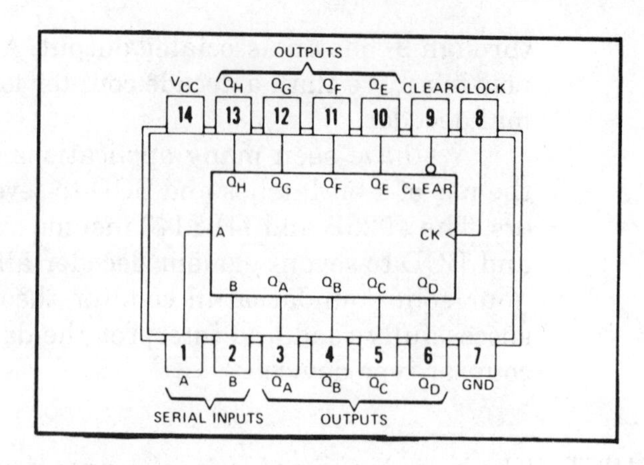

FUNCTION TABLE

INPUTS				OUTPUTS		
CLEAR	CLOCK	A	B	Q_A	Q_B . . .	Q_H
L	X	X	X	L	L	L
H	L	X	X	Q_{A0}	Q_{B0}	Q_{H0}
H	↑	H	H	H	Q_{An}	Q_{Gn}
H	↑	L	X	L	Q_{An}	Q_{Gn}
H	↑	X	L	L	Q_{An}	Q_{Gn}

H = high level (steady state), L = low level (steady state)
X = irrelevant (any input, including transitions)
↑ = transition from low to high level.
Q_{A0}, Q_{B0}, Q_{H0} = the level of Q_A, Q_B, or Q_H, respectively, before the indicated steady-state input conditions were established.
Q_{An}, Q_{Gn} = the level of Q_A or Q_G before the most-recent ↑ transition of the clock; indicates a one-bit shift.

schematics of inputs and outputs

Figure 10.22 Pinout and function table for 74LS164 parallel shift register.

A and B These pins are described as serial inputs. Because the 74LS164 is a serial-to-parallel shift register, the levels applied to inputs A and B must constitute serial data.

Refer to the function table in Figure 10.22. We will examine the truth table, line by line, to derive the manner in which the 74LS164 functions.

Line 1 When the clear input is at an active-low level, all other inputs become don't cares and the Q outputs will all be cleared to logic 0s. This is a typical direct (asynchronous) clear that we have seen in flip-flops and counters. (In the remainder of the function table the clear input will be held at an inactive level.)

Line 2 The clock is low. This line illustrates the situation where the clock in not active. The outputs, Q_a through Q_h, will maintain their memory states.

Line 3 The clock is active with a rising edge and the serial data inputs are both logic 1s. A logic 1 will be shifted into the Q_a output. The contents of the Q_a output, previous to the rising edge of the clock, will be transferred into the Qb output. The contents of the Qb output, previous to the rising edge of the clock, will be transferred into the Q_c output . . . and so on. The only question is what happens to the previous contents of the Qh output? Each of the eight stages in the shift register are responsible for remembering one bit of information. Because it is the last stage, the previous contents of Q_h are lost.

Lines 4 and 5 These lines demonstrate that the A and B serial inputs are ANDed together to form the serial bit of information that is transferred into Q_a on the rising edge of the clock. If input A is low or input B is low, the result of the AND function will be a logic 0; that logic 0 will be the serial data transferred into Q_a. Only if both inputs A and B are high will the serial input data be high.

We have discovered that the 74LS164 is composed of eight memory cells. On the rising edge of the clock the ANDed contents of the A and B data inputs is transferred into the first memory cell, Q_a. All the contents of the memory cells, previous to the rising edge of the clock, are transferred to the next significant memory cell. The contents of the last memory cell, Q_h, is lost during the shift operation. Figure 10.23 is a timing diagram that illustrates the operation of the 74LS164. The first vertical line

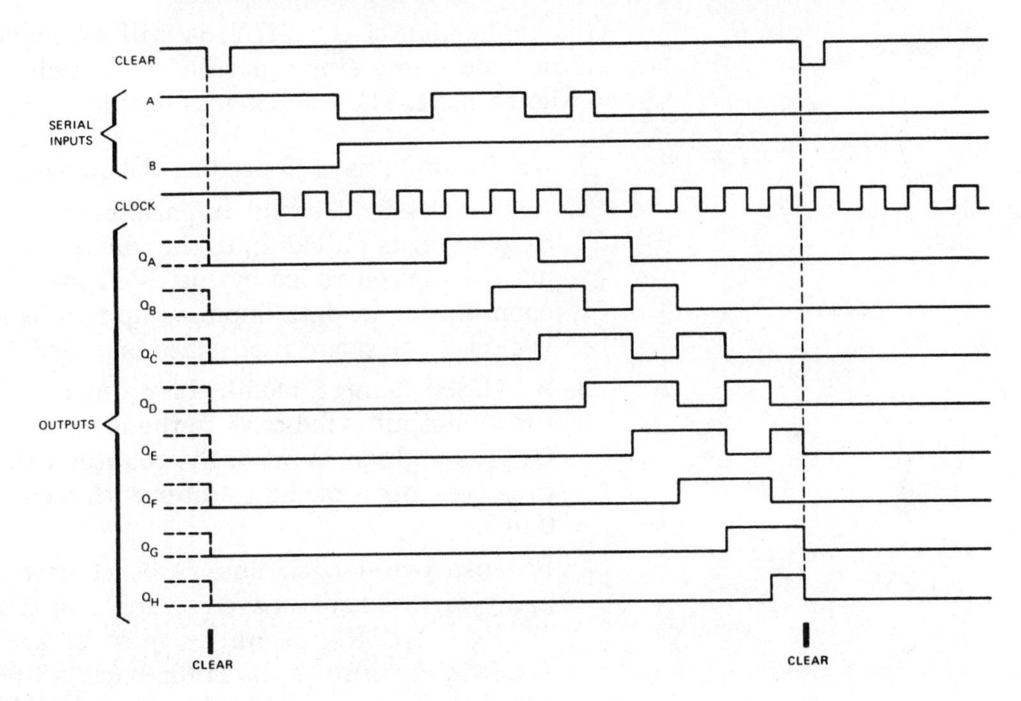

functional block diagram

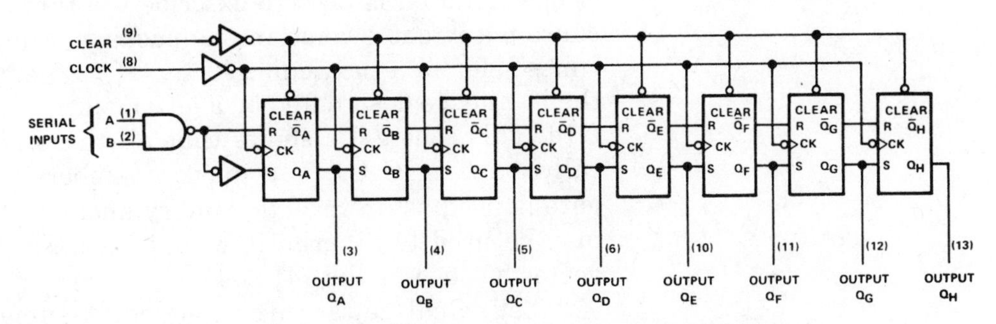

Figure 10.23 Timing diagram for 74LS164.

indicates when the clear input goes active. All the Q outputs will clear to logic 0. After the clear input goes high, the 74LS164 will be ready to operate. During the first three rising edges of the clock, the ANDed value of inputs A and B is logic 0. Shifting logic 0's into a cleared shift register has no effect on the outputs.

On the fourth rising edge of the clock, the ANDed value of A and B is a logic 1. This logic 1 is shifted into the Q_a output. The logic 1 in the Q_a output will continue to shift through the shift register until it is lost by being shifted out of Qh on the ninth rising edge or when the clear input goes active.

Examine the rest of the timing diagram and assure yourself that you understand the operation of the 74LS164.

10.5.2 An Application of the SIPO (Serial In, Parallel Out) Shift Register

This application of the 74LS164 will assemble 8 bits of serial information into a single parallel byte. Refer to the schematic in Figure 10.24. The components in this system are:

1. An S-R latch used to produce debounced clock pulses.
2. A 74LS164 8-bit serial in/parallel out shift register. The clear input is pulled up to the inactive level. The serial input data is controlled by the SPST switch. If the switch is open, the input data is pulled up to a logic 1; if the switch is closed, the serial input data is pulled down to a logic 0.
3. A 74LS90 counter, modified for a modulus of 8. Notice that the Q_d output is fed back to the reset to zero inputs. When Q_d goes high on count 8, the counter will reset to count 0, thus creating a mod-8 counter with a count sequence from 0 to 7.
4. Because byte-length quantities of data are so common, many digital devices are available in groups of eight per IC. The 74LS374 is an example of such a device. The 74LS374 contains eight D flip-flops, all positive edge triggered by a common clock input, and packaged in a 20-pin IC. The pin OE is the output enable input. When it is at an active-low level, the eight flip-flops are enabled.

A new notation is used to describe the three NAND gates. Because many gates come in one package, schematics will sometimes refer to a particular IC as G1, G2, etc.; the gate within the IC may be designated by a letter, such as G1a, G1b, G1c, and G1d. This schematic shows that all three NAND gates physically exist in IC G1. You may remember that when a NAND gate is used as an inverter, the symbol of a two-input inverter may be used to differentiate that inverter from an actual inverter IC such as the 74LS04.

Let's analyze the circuit action. Assume that the counter

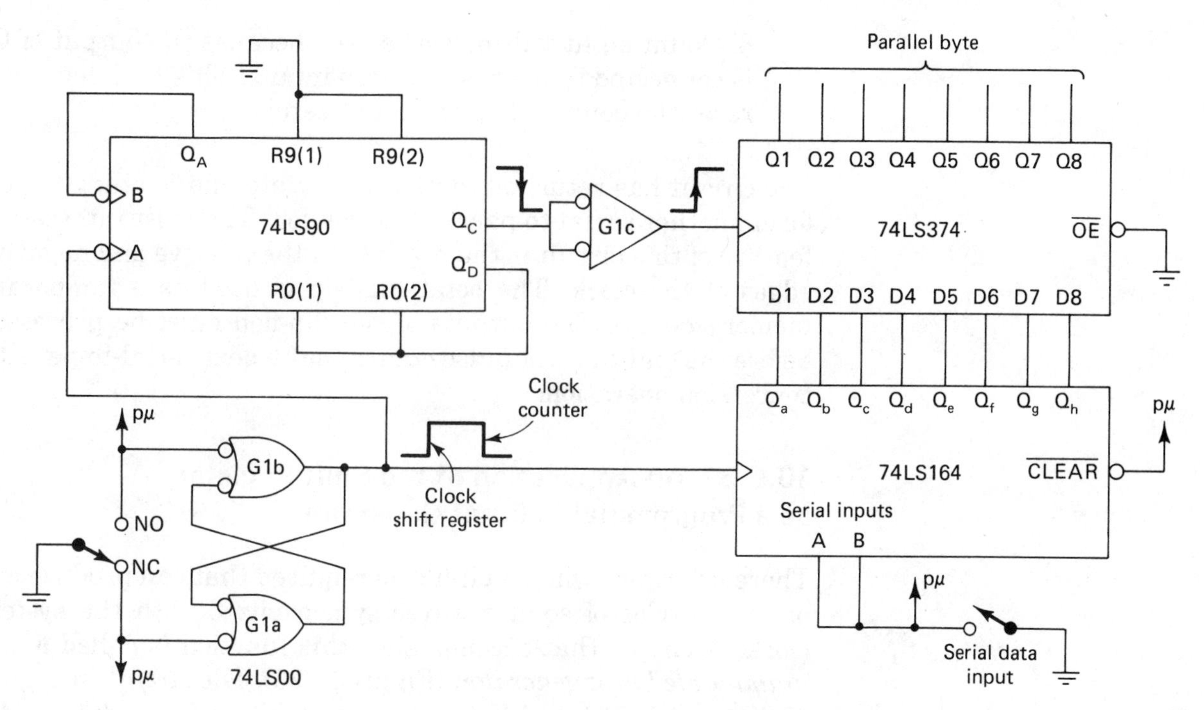

Figure 10.24 Serial-to-parallel converter.

is initially cleared to a count of 0. The most significant bit of the 8-bit word will be toggled into the serial input switch. After the data is set, the SPDT switch will be toggled. The rising edge from G1b will clock the first data bit into Q_a of the shift register. The falling edge from G1b will increment the counter from 0 to 1. This action will be repeated for the last 7 bits of the byte. On each rising edge the data will be shifted one place to the right in the shift register, and the counter will increment by one.

Picture the circuit after the first 7 bits have been entered: Q_g through Q_a contain the 7 bits, where the bit in Q_g is the first entered and the most significant. The counter is on the count of 7. The last (least significant) data bit will be toggled into the serial data switch. On the rising edge from G1b, the last data bit will be shifted into the shift register. The eight serial bits of data have now been reformed into a parallel byte residing on Q_h through Q_a. On the falling edge from G1b, the counter will be incremented from count 7 to count 8. Two important events will happen:

1. When the counter increments from 7 to 8, Q_c will have a falling edge. This falling edge will be inverted by G1c. The rising edge from G1c will latch the 8 bits of data on Q_h through Q_a into the octal flip-flop, where they will be temporarily saved.
2. Because the 74LS90 is a ripple counter, the action of Q_c falling from high to low will clock the Q_d stage of the counter. This will cause Q_d to go from low to high, for count number

8. Count eight will never be seen because the output of Q_d is connected to the reset to zero inputs. This high level will reset the counter back to count zero!

The circuit has returned to its initial state and is ready to perform another serial-to-parallel conversion. The most interesting feature of this circuit is the use of both the positive and negative edges of the clock. The octal flip-flop is used as a temporary memory device. The outputs of the flip-flop must be processed before the eighth clock pulse occurs on the next serial-to-parallel conversion operation.

10.5.3 An Application of the Shift Register as a Programmable Burst Generator

There are times when a circuit is required that can produce one or more cycles of square waves synchronized with the system clock. A circuit that accomplishes this function is called a *programmable burst generator.* (Figure 10.25). The circuit in Figure 10.25 consists of two J-K flip-flops, an 8-bit shift register, and a 1-of-8 data selector. The first J-K flip-flop is configured as a simple S-R flip-flop. When flip-flop 1 is cleared, a flip-flop 2 will be configured to perform a reset function; when flip-flop 1 is set, flip-flop 2 will be configured to perform a set function.

G1 acts as an electronic switch. When Q2 is low, the output of G1 will be high; when Q2 is set, G1a will invert and pass the system clock. G2 will reinvert the system clock into the original phase.

The shift register and the data selector work together to provide the circuit's programmability. The required number of cycles in the burst will be placed on the select inputs of the data selector. The input data for the shift register is tied high. Each rising edge from G1 will shift the logic 1 data bit one cell farther to the right. When the logic 1 reaches the selected data input, the complemented output, W, of the data selector will go low. This will clear both flip-flops. The logic 0 on the Q2 output will block any further transitions of the system clock.

A busy circuit like this can be fully understood only by examining a timing diagram. The timing diagram in Figure 10.25 describes the sequence of events that occur after a start pulse is applied to the circuit. In general, the number of cycles in the burst will be equal to the select code on the 74LS151 plus 1. If the select code is 0, the number of bursts will be equal to one; if the select code is 7, the number of bursts will be equal to eight. We will assume that the select inputs on the data selector are equal to 3. This will produce a four-cycle burst.

Although the system clock is always running, the output will be stuck low until Q2 is set. We will follow the cause and effects of the events illustrated in the timing diagram.

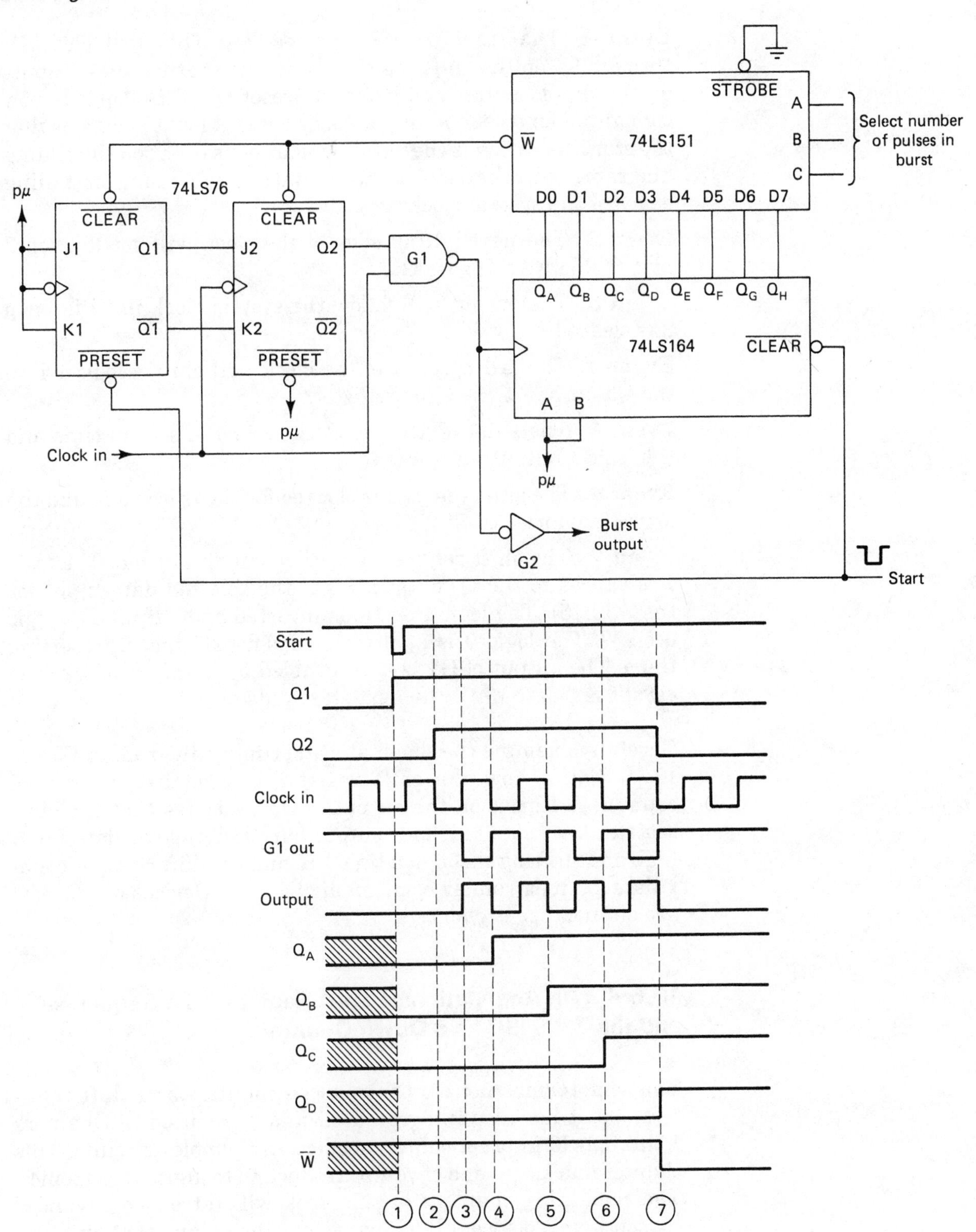

Figure 10.25 Programmable burst generator.

Event 1 The start input is pulsed low. This will clear Q_a through Q_h on the shift register (assuring that the clear inputs on the flip-flops are inactive) and preset Q1. Flip-flop 2 is now configured for a set operation. An important fact to notice is that the start operation is the only asynchronous event on the timing diagram. All other events occur during the rising or falling edges of the system clock.

Event 2 The next falling edge of the clock will set flip-flop 2 and enable NAND gate G1.

Event 3 The output will follow the system clock until flip-flop 2 is cleared.

Event 4 The falling edge of the clock will shift a logic 1 into the Qa output of the shift register.

Event 5 The shift register is clocked for the second time and the logic 1 output advances to Q_b.

Event 6 The shift register is clocked for the third time and the logic 1 advances to Q_c.

Event 7 The shift register is clocked one final time. The logic 1 advances to the Q_d output. D3 is the selected data input for the 74LS151. The logic 1 on D3 is inverted and output as a logic 0 on W. The logic 0 is applied to flips-flops 1 and 2, resetting them. The output of G1 is now disabled by a logic 0 on Q2. The circuit is frozen until the next start pulse.

Carefully examine the schematic and timing diagram in Figure 10.25. Notice how the shift register is used like a decoded counter. A logic 1 on the output of Qd indicates that the shift register has "counted" four output bursts. Tying the data input high and shifting through a logic 1 to make a shift register count pulses is an extremely common application. Devices of this sort are often called *sequencers.*

10.5.4 An Application of Shift Register as a Sequencer and the 74LS190 as a Down Counter

This chapter has not only introduced counters and shift registers, but it has also integrated devices introduced in Chapters 1 through 10 in practical and moderately complex circuits. This chapter marks the end of your introduction to digital electronics. The remaining chapters in this book will introduce advanced concepts and devices. It is extremely important that you feel comfortable with the material in the previous chapters before continuing onto Chapter 11. The most efficient way to learn advanced topics is first to build a solid foundation with the fundamental concepts of combinational and sequential logic.

The last application in this chapter is the front-end circuitry in a photocopy machine. Refer to the block diagram il-

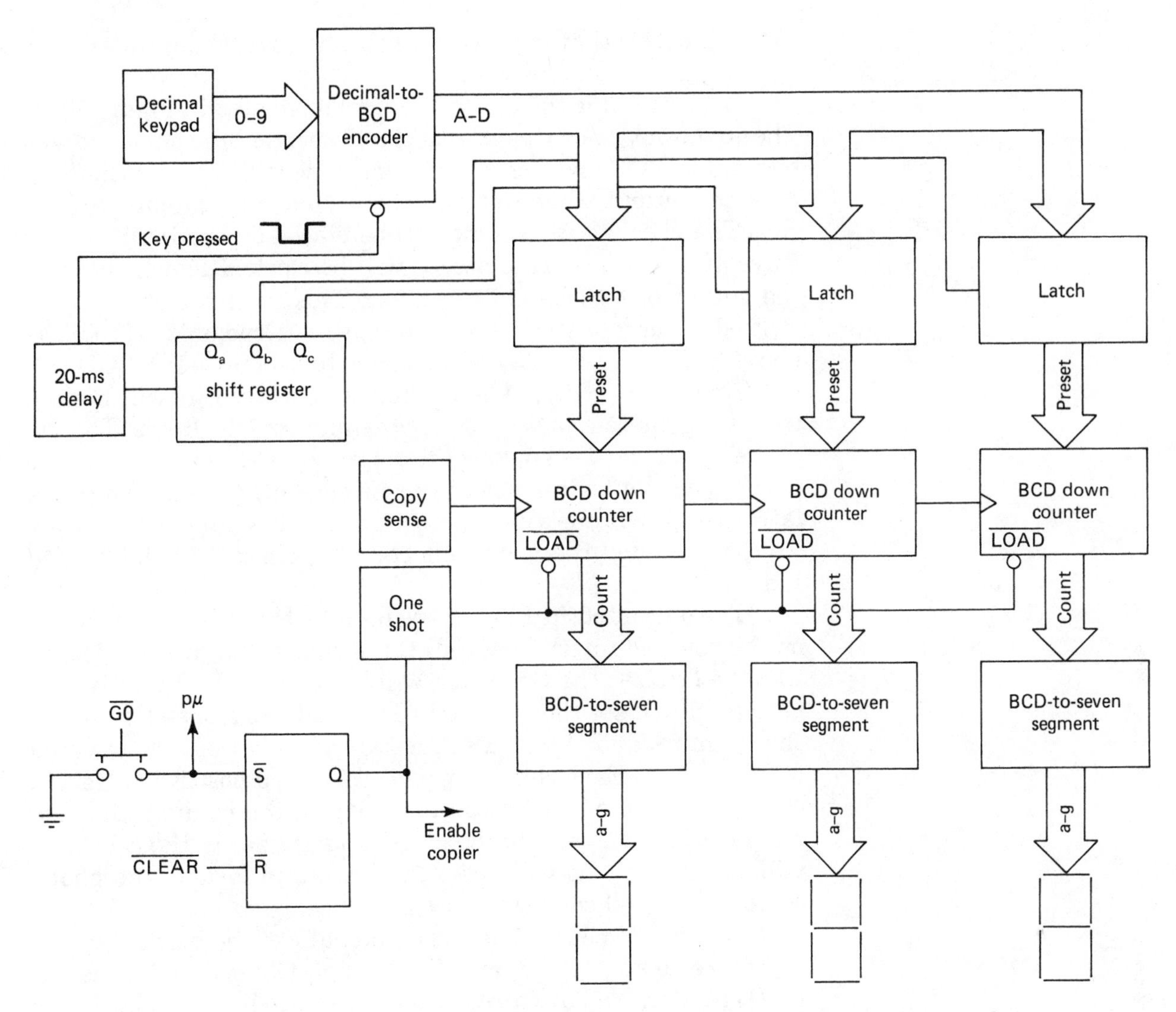

Figure 10.26 Block diagram of photocopier control circuitry.

lustrated in Figure 10.26. The operation of the photocopier will entail the following steps:

1. The number of copies required (from 1 through 999) will be entered via a 10-key keypad. The hundred's digit will be entered first, then the ten's digit, and finally the one's digit.
2. The go switch, which is a momentary normally open push-button switch, will be depressed to start the copy operation.
3. The three-digit display will count down the number of copies left to produce. When the display counts down to zero, the copy process is finished.

The operation of this circuit may seem awkward to you. Remember that it is used to illustrate the operation of a down counter and a sequencing shift register. A practical circuit would entail so much detail that the intended emphasis on the counters and shift register would be lost.

Let's examine the block diagram illustrated in Figure 10.26. The 10 output lines from the decimal keypad are connected to a decimal-to-BCD encoder. A negative pulse will occur on the "key-pressed" output whenever any of the 10 decimal inputs are active. The encoded BCD digit will be placed on the inputs of the three 4-bit latches. The latches will be clocked sequentially by the output of a SIPO shift register. This shift register will be clocked, after a 20-ms delay, by the key pressed output from the encoder. The 20-ms delay will be used to debounce the BCD output from the encoder. When a key is depressed, it will bounce from a logic 0 to a logic 1 for approximately 1 to 10 ms. The 20-ms delay will be used to give the depressed key a chance to settle down to a solid logic 0, before the encoded digit is clocked into a latch. By providing a 20-ms delay of the key-depressed signal, we have avoided the need to debounce each key in the keypad individually.

The output of the latches are used to drive the preset inputs on the down counters. After all three encoded digits are clocked into the latches, the Go button will be depressed. The active-low Go signal will set the S-R latch. This will fire a one-shot to load the counters and it will also enable the photocopier's internal copy circuitry. Each time a copy is produced a positive pulse will clock the three cascaded counters. When the counters have finally counted down to 0, the clear input signal to the S-R latch will go active. This will reset the latch and disable the photocopier's internal copy circuitry.

Figure 10.27 is a completed schematic of the block diagram in Figure 10.26. U1, U2, and G1 through G3 form a decimal-to-BCD encoder. We have already used this circuitry as a keyboard encoder in Figure 7.13. You should thoroughly understand its operation. The encoded digit of the key pressed will be applied to the inputs of latches U3 through U5. Notice that unlike previous latches, U3 through U5 are rising edge triggered instead of level transparent. U3 through U5 are 74LS379 quad D-type flip-flops. This circuit requires a memory latch that will react only to the rising edge of a clock.

Assume that U14, the SIPO shift register, is initially cleared. Notice that the serial input data is pulled up to a logic 1. Each time U14 is clocked by the rising edge of the 20-ms delay one-shot, a rising edge will shift through the Q outputs. When the first key is depressed (hundred's place), the rising edge from $\overline{Q}$ of the one-shot will clock the shift register. This will transfer a logic 1 into Qa. The rising edge from Qa will latch the contents of the encoded digit into U5. The next key closure will cause the encoded digit to be latched into U4. The final encoded digit will be latched into U3.

When the Go button is depressed, U15 will set. This rising edge on the Q output will fire U13. The negative pulse from U13 is used to load the encoded digits into the Q outputs of the

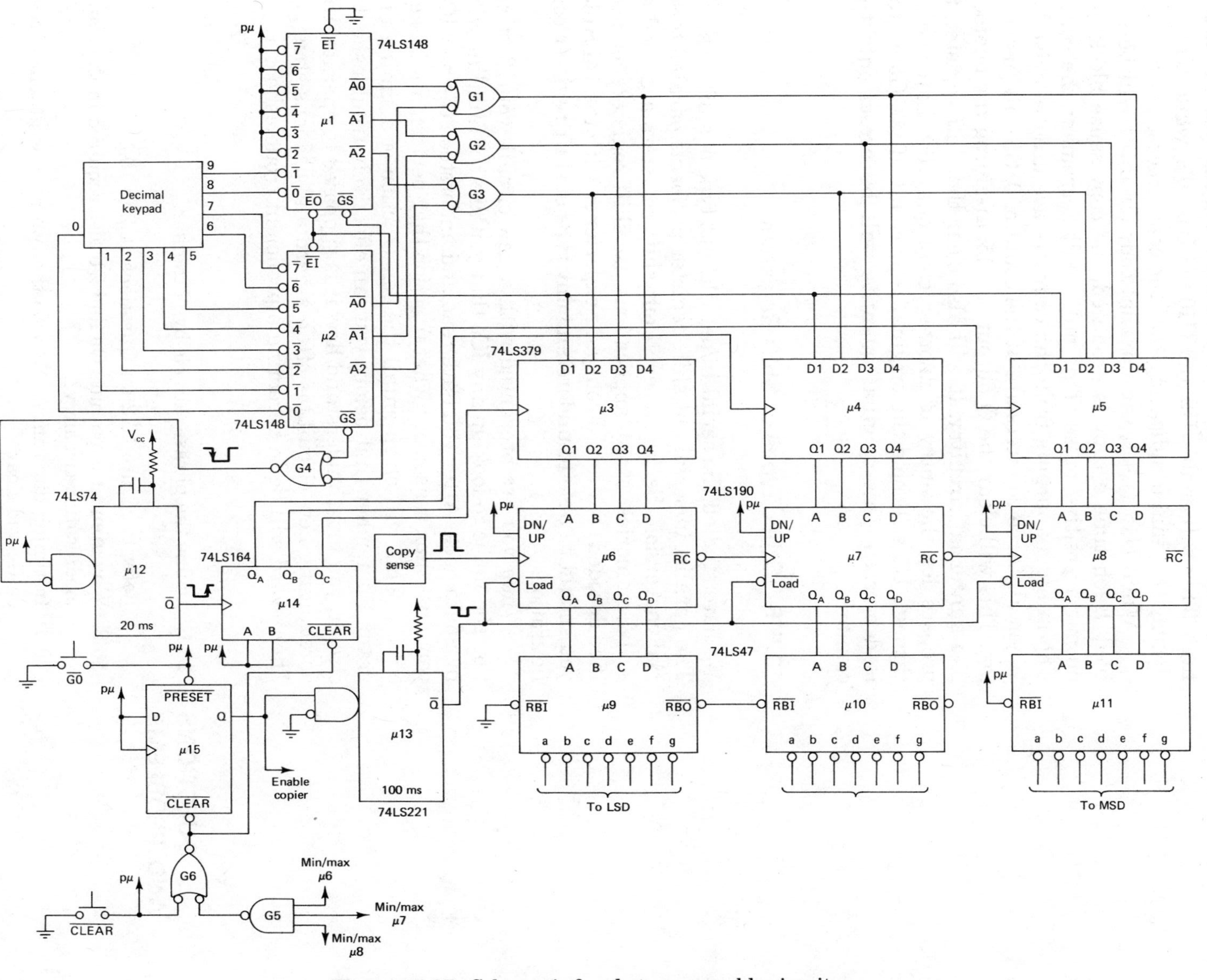

Figure 10.27 Schematic for photocopy enable circuitry.

counters. These outputs will drive the 74LS47 BCD-to-seven-segment decoders. The outputs from the 74LS47s will drive the seven-segment displays. Notice that leading edge blanking is used on U9 and U10, but the ripple blanking input of U11 is tied high. This will enable U9 and U10 to blank 0's, while U11 will display the final 0 indicating that the copy process is over.

When U15 is set, the internal copy circuitry will be enabled. Each time a copy is produced, the copy sense block will output a positive pulse. This pulse will decrement the count. The min/max outputs from the counters are used to drive G5. When all three counters are at the count of 0, G5 will go active low. This will clear the Q output of U15, disabling the photocopier's internal circuitry. It will also clear the shift register, to prepare for the entry of another copy quantity. Latches U3 through U5 still hold the original copy count. If the Go button is depressed, this original quantity will be preset into the counters.

10.5.5 Other Types of Shift Registers

We have studied the serial in/parallel out type of shift register. The parallel in/serial out shift register performs parallel-to-serial conversion. Serial in/serial out shift registers are used as serial memories. Many applications use shift registers as modified decoded counters. From your experience with the 74LS164, you should be able to understand any type of shift register configuration.

When you are presented with a new circuit, your first action should be to look up any ICs that you are not familiar with in a data book. After you have read and understood the new ICs, you must then reason out what function they perform in the new circuit. This will often entail the construction of a timing diagram. Remember that before you can effectively troubleshoot a circuit, you must understand how it is supposed to operate. Understanding new circuits is often a hard and arduous struggle. The reward will be in becoming a proficient troubleshooter.

QUESTIONS AND PROBLEMS

10.1. Define the phrase "divide-by-*N* counter."

Refer to Figure 10.2 for Questions 10.2 to 10.5.

10.2. Why aren't the outputs Qa and Qb also required in the count modification circuitry?

10.3. In timing diagram A: Why do the two narrow glitches occur on Qc and clear?

10.4. If count 12 actually occurs, why isn't it included in the count-state diagram?

10.5. Assume a propagation delay of 10 ns for all flip-flops and gates. How wide is the clear input pulse in timing diagram B?

10.6. Create a count-state diagram and write a theory of operation that describes the following counter circuit.

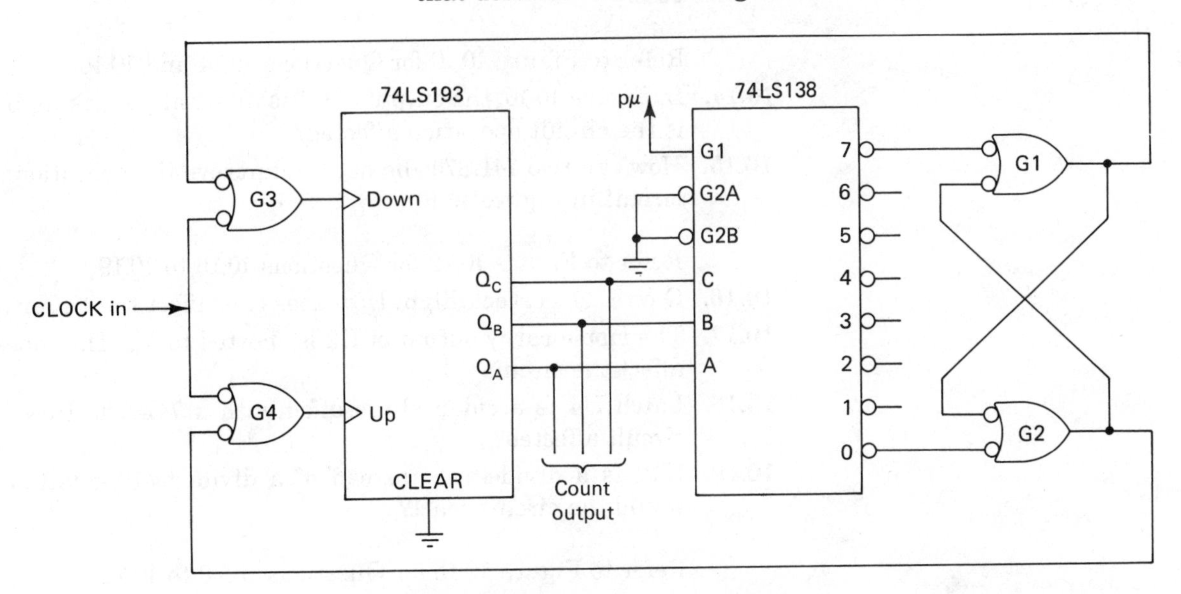

10.7. By what factor does the following circuit divide the input frequency?

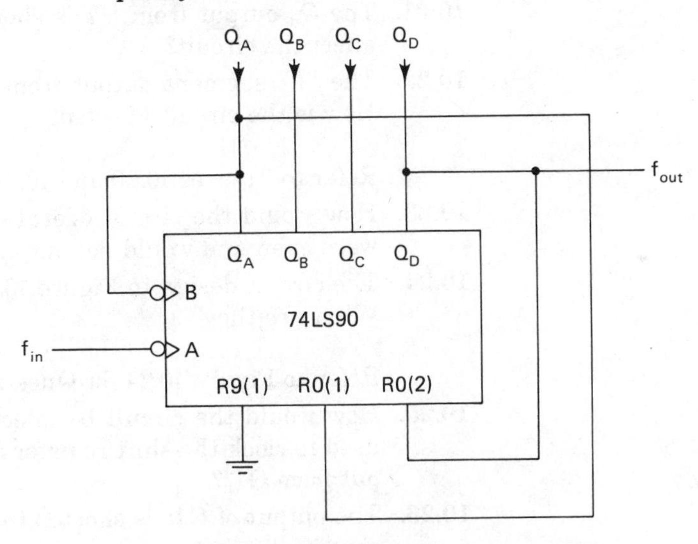

10.8. What is the major difference between ripple and synchronous counters?

10.9. Use a 74LS93 to create a mod-14 counter.

Refer to Figure 10.6 for Questions 10.10 to 10.13.

10.10. The display appears to flicker displaying the digit eight, with segments of varying intensity. What could be the possible problem?

10.11. How would the circuit operation be affected if the trace running from the output of G6 to G1 were open? Shorted to ground?

10.12. How would the circuit be affected if the output pin of G8 were bent underneath the IC during the stuffing operation?

10.13. How does the clock frequency affect the operation of the circuit?

Refer to Figure 10.10 for Questions 10.14 and 10.15.

10.14. In Figure 10.10, the output of G1 is internally stuck high. How is the circuit operation affected?

10.15. How can two 74LS75s be used to improve the operation of the circuit in Figure 10.10?

Refer to Figure 10.12 for Questions 10.16 to 10.19.

10.16. $\overline{Q}$ on U11 is stuck high. How does this affect the circuit?

10.17. The ripple carry output of U2 is shorted to V_{cc}. How does this affect the circuit?

10.18. Latch U4 is accidentally stuffed with a 74LS76. How is the circuit affected?

10.19. U12 is a divide-by-5 instead of a divide-by-10 counter. How would the circuit react?

Refer to Figure 10.13 for Questions 10.20 to 10.22.

10.20. Displays 4 through 7 are always dark. What checks would you make to discover the problem?

10.21. The Q_d output from U7 is shorted to ground. How does this affect the circuit?

10.22. The "a" segment output from the 74LS48 is shorted to V_{cc}. How is the circuit affected?

Refer to Figures 10.20 and 10.21 for Questions 10.23 and 10.24.

10.23. How would the circuit operation be affected if a photosensor were open and would not output a negative pulse?

10.24. The circuit design in Figure 10.21 has two major design flaws. What are they?

Refer to Figure 10.24 for Questions 10.25 to 10.28.

10.25. How would the circuit be affected if the output of G1a were used to clock the shift register and counter instead of the output from $G1_b$?

10.26. The output of G1c is shorted to the output of G1b. How is the circuit affected?

10.27. If the outputs of Q1 through Q8 of the octal D flip-flop are always logic 0's, what checks would you perform?

10.28. How would the circuit operation be affected if the clear input of the shift register was floating?

Refer to Figure 10.25 for Questions 10.29 to 10.32.

10.29. The output always follows the system clock. What checks would you perform?

10.30. The start input is stuck low. How does this affect the circuit operation?

10.31. Q1 is stuck low. How does this affect the circuit operation?

10.32. The strobe input on the data selector is floating. How is the circuit affected?

Refer to Figure 10.27 for Questions 10.33 to 13.6.

10.33. The pulse width out of one-shot U12 is 20 μs instead of 20 ms. How does this affect the circuit?

10.34. The output of G1 is stuck high. How is the circuit affected?

10.35. The output of G4 is shorted to the Q input of U2. How is the circuit affected?

10.36. The ripple carry output of U6 is open. How does this affect the circuit?

11

THREE BUS ARCHITECTURE

We have stated that SSI and MSI digital devices are the "glue" that is used to interconnect complex LSI ICs. LSI ICs are said to be intelligent devices. Here the term "intelligent" refers to an LSI IC's ability to execute a series of programmed instructions and alter the sequence in which the instructions are executed according to established criteria. The microprocessor is the most common example of an intelligent device.

This book is a text on digital electronics, not microprocessors. An in-depth knowledge of digital electronics is a prerequisite to the study of microprocessors. The problem we now face is that in order to fully understand and appreciate the remaining chapters in this book, the student must have a fundamental idea of how microprocessors operate. It appears to be a catch-22; before you can study microprocessors you must first understand digital electronics; but before you can pursue a study of advanced digital devices you must first understand the fundamentals of microprocessors.

Any endeavor entails a certain amount of compromise. This chapter is a quick peek at the microprocessor and its associated circuitry. You will be exposed to just enough information to make your study of the remaining chapters in this book a significant experience.

11.1 TALKERS AND LISTENERS

11.1.1 Three-State Logic: The High-Impedance Output

We have observed two different TTL output structures: the totem pole and the open collector. If totem-pole outputs are tied together, indeterminate logic levels will result. When the output of common collector devices are tied together, any device can take the common node to a logic 0. An open-collector connection functions like the negative logic interpretation of an AND gate; a low on any input will pull the output low. There is a third type of output structure. This structure embodies characteristics of both totem poles and open collectors (see Figure 11.1). The circuit in Figure 11.1 makes it possible for one BCD-to-seven-segment decoder and seven-segment display to be shared between four different input sources. (Notice that this is not a multiplexing circuit as was shown in Chapter 10. This circuit has only one seven-segment display.) A 2-bit select code will be applied to the 2-line-to-4-line decoder. The output of this decoder will enable one of four BCD digits to appear on the inputs of the BCD-to-seven-segment decoder. By changing the select code, any of the four BCD digits can be displayed.

You should be puzzled by U1 through U4. The four output lines of each of these devices are tied to the similar output lines on the other three devices. You are aware that totem-pole outputs cannot be tied together. Therefore, these four devices cannot have totem-pole output structures. Furthermore, these devices cannot be open collector, because there are no pull-up resistors illustrated in the schematic.

U1 through U4 are examples of the third type of output structure—the three-state output. A three-state device has three possible outputs: logic 0, logic 1, and high impedance. It is the high-impedance output that distinguishes the three-state device from totem poles and open collectors. You must assume from the schematic in Figure 11.1 that the outputs of three-state devices can be tied together with no undesirable effects.

Refer to the truth table and schematic symbol of the three-state buffer in Figure 11.1. The schematic symbol illustrates a normal buffer with an extra line. This line is the enable input. Because it is bubbled, the enable input is active low. Lines 1 and 2 of the truth table show that when the enable is at an active-low level, the buffer passes the input level to the output. In the third line of the truth table, the enable input goes to the inactive level. The input is now designated as don't care and the output is labeled "high-z." High-z refers to the third possible level of a three-state device—the high-impedance output level. When a three-state device is at a high-z level, its output is effectively open or disconnected!

The model of the three-state device in Figure 11.1 shows how the output is disconnected. The enable input controls the position of a SPST switch. When the enable input is active, the

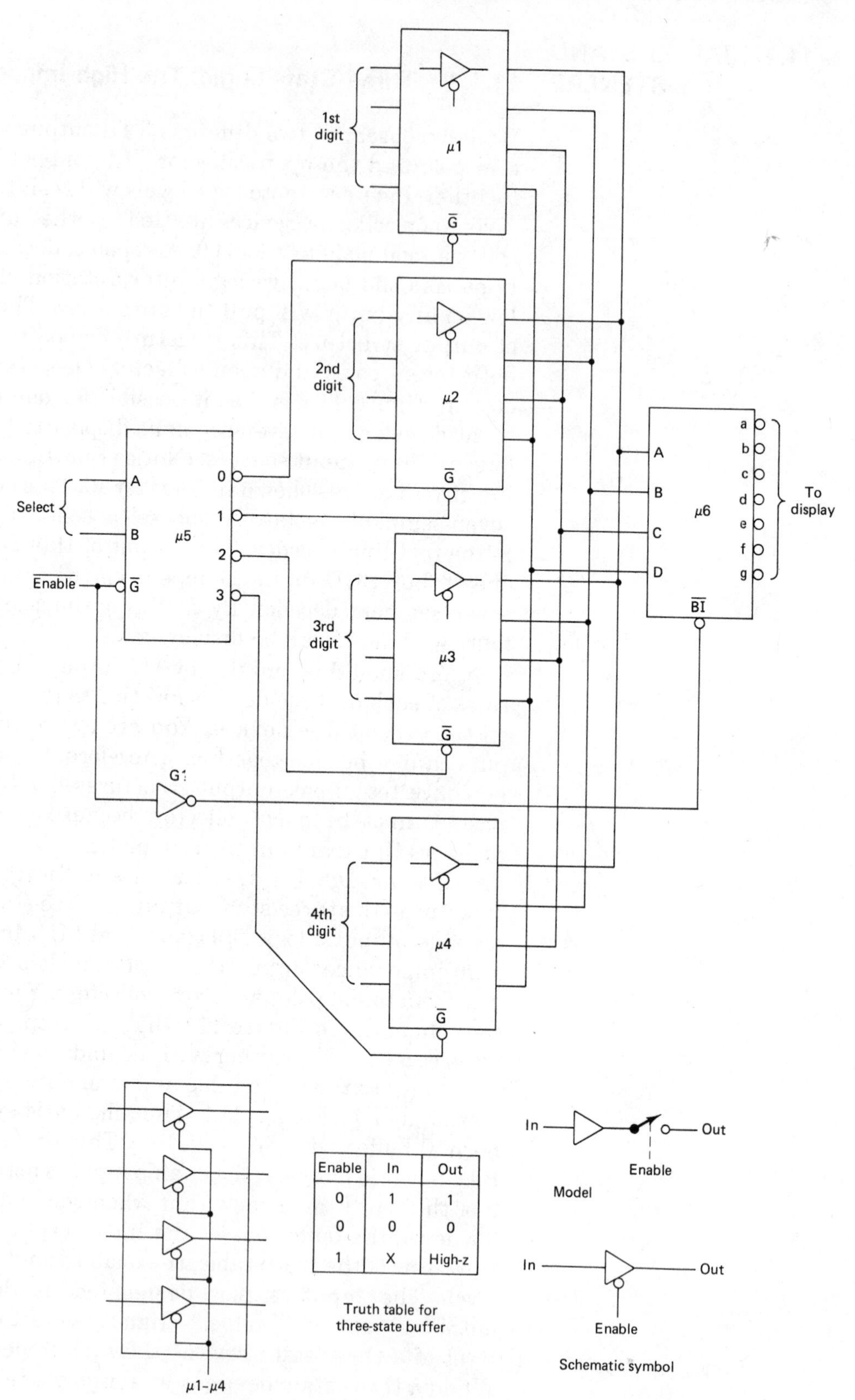

Figure 11.1 Three-state buffer: schematic symbol, model, and truth table.

switch is closed and data pass in a normal fashion. When the enable input goes inactive, the SPST will open. The output of the buffer is now disconnected from the output of the switch.

With this newfound knowledge in hand, let's examine how the circuit in Figure 11.1 actually shares the display. U1 through U4 each contain four three-state noninverting buffers. The four buffers in each device are controlled by a common enable input. When the enable input of a particular device is active, all four buffers within that device operate normally. When the enable input of a particular device is inactive, all four buffers within that device go to a high-z output level. The 2-line-to-4-line decoder will place an active-low level on one enable input, while placing inactive-high levels on the other three enable inputs. The device that is selected by the 2-bit select code will pass a BCD code through its four buffers; the outputs of the other three devices will be effectively disconnected.

There are two important cases that we must consider before the concept of three-state devices can be fully understood:

1. What happens if more than one three-state device sharing the same output lines is enabled simultaneously?
2. What happens if all the three-state devices sharing a common node are disabled?

Case 1 A three-state device can be thought of as a modified totem pole. The outputs of three-state devices sharing the same node should never be enabled simultaneously. This action would result in indeterminate logic levels. The open collector is the only output structure where common devices can be simultaneously active.

Case 2 Assume that the enable input to the 2-line-to-4-line decoder has gone to an inactive level. All the decoded outputs, 0 through 3, will be taken to the inactive-high level. None of the three-state buffers will be enabled; all the inputs on the common node are effectively disconnected. If you placed an oscilloscope probe on any BCD input of the BCD-to-seven-segment decoder, what voltage level would you observe? If all the outputs of the three-state buffers are disconnected, the entire node is floating. You would observe 1.4 to 1.8 V, a normal TTL floating input level! It would appear as if all the traces running from the three-state buffers were cut to produce opens.

When you are troubleshooting a PCB that contains three-state devices, a floating input level will not necessarily indicate a broken trace, open feedthrough, or any other previous problem that would result in a floating input level. There will be an added possibility that the three-state devices sharing a particular node are all disabled. That in itself may or may not indicate a circuit malfunction. There will be many legitimate instances when all the three-state devices sharing a particular node will be disabled.

In the circuit illustrated in Figure 11.1, if the enable input of the 2-line-to-4-line decoder goes to an inactive-high level, the blanking input of the BCD-to-seven-segment decoder will go active low to blank the seven-segment display. When the BCD inputs to the BCD-to-seven-segment decoder are floating, they will appear to be logic 1's. You may remember that the 74LS47 will automatically blank the display when the BCD inputs are all equal to logic 1's. Therefore, it may appear that there is no need to blank the BCD-to-seven-segment decoder when the 2-line-to-4-line decoder is disabled. Noise glitches can momentarily pull the levels on floating inputs to logic 0. This would induce flashes on the seven-segment display. Blanking the display when the 2-line-to-4-line decoder is disabled ensures that noise glitches will not affect the seven-segment display.

Each of the quad three-state buffers can be classified as a *talker.* A talker is a device that is capable of outputting a signal. On the other hand, the BCD-to-seven-segment decoder can be classified as a *listener.* It will receive a BCD input from one talker. As in any well-mannered conversation, only one talker can be active simultaneously. If more than one talker is active simultaneously, the conversation will be garbled and the listener will not receive the correct information. Three-state output structures allow many talkers to share the same listener. If the three-state devices are enabled in the proper manner (only one active at a time), the listener can receive information from many different sources.

Three-state devices will allow us to create a circuit structure called a *bus.* A bus is nothing more than a group of signals grouped together according to a common function. In Figure 11.1 the traces that are connected to the BCD input of the BCD-to-seven-segment decoder can be called a bus. The common function of these signals is that they all carry a BCD digit to the BCD-to-seven-segment decoder. Using only conventional totem-pole or open-collector outputs, the circuitry required to create the bus in Figure 11.1 would be extensive and complex. The use of three-state devices and decoders makes the bus circuit almost trivial.

11.2 A SURVEY OF THREE-STATE DEVICES

Any digital device can be manufactured with a three-state output structure. Popular devices used to create buses are inverting and noninverting buffers, transparent latches, and D flip-flops.

11.2.1 The 74LS240 Octal Three-State Buffer

A three-state buffer is used to isolate the logic levels of a bus talker from the bus itself. Three-state buffers usually have the ability to sink and source much more current than the typical digital device. The 74LS240 is an example of a typical three-

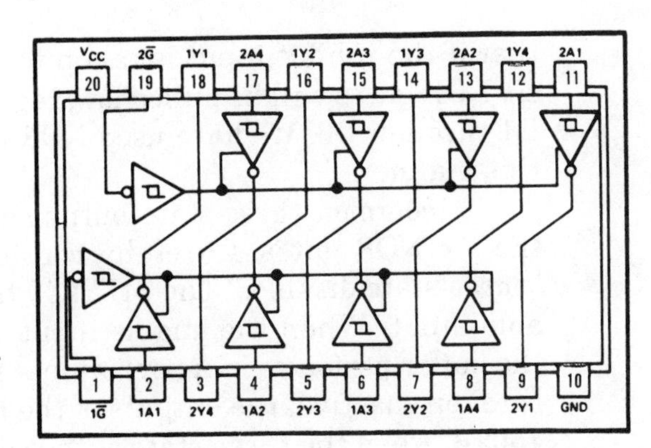

Figure 11.2 Pinout of 74LS240 octal buffer/line driver with three-state output.

state inverting buffer (Figure 11.2). The first detail that you may notice in the pinout of the 74LS240 is the elongated square in the center of each buffer. This is the symbol representing the property of hysteresis. The hysteresis symbol designates a special type of input structure that is used to reduce the effect of noise. We will cover the concept of hysteresis in Chapter 13. Until then you should treat these devices as normal three-state inverters.

The 74LS240 is an octal buffer/line driver with three-state outputs. Octal means that there are eight buffer/line drivers in this 20-pin package. You will discover that quad and octal devices are extremely popular. In Chapter 10 we discussed the idea that the quantity of 8 bits, called a byte, is the most common width of parallel data; many three-state ICs are designed to accommodate byte-wide quantities.

TRUTH TABLE

DISABLE	INHIBIT	Dn	Qn
0	0	0	1
0	0	1	0
0	1	X	0
1	X	X	Z

Logic 0 = Low

Z = High Impedance

X = Don't Care

Logic 1 = High

The term buffer/line driver refers to the 74LS240's ability to sink and source great amounts of current. The 74LS240 can sink 24 mA and source 15 mA. A line driver is a device that is used to drive signals through long transmission lines. This is where the 74LS240's good current and noise-reduction abilities are used.

Each of the eight inverting buffers have a three-state output structure. The first four buffers are controlled by the active-low enable on pin 1. The second group of four buffers is controlled by the enable input on pin 19. Used separately, each group of four buffers can control a 4-bit quantity such as a BCD digit, as was done in Figure 11.1. If pins 1 and 19 are tied together, the 74LS240 can be used to control a byte of data.

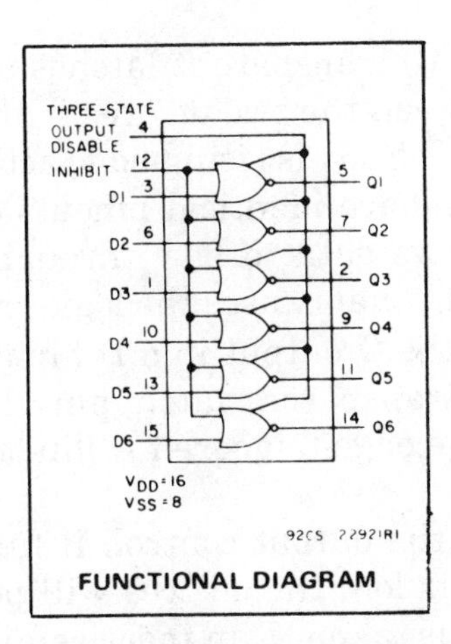

Figure 11.3 4502B strobed hex inverter/buffer with three-state outputs.

11.2.2 The 4502B Strobed Hex Inverter/Buffer with Three-State Outputs

CMOS devices are also available with three-state output structures. Refer to the functional diagram in Figure 11.3. The package contains six (hex) NOR gates. One input of each NOR gate is tied together at the inhibit input pin. Remembering that the dynamic input level of an NOR gate is a logic 1, you should realize that if the inhibit input is taken to a logic 1, the outputs of the six NOR gates will be pulled low, regardless of the data on inputs D1 through D6. We have used NOR gates as digital switches in this manner.

A common three-state output disable is also connected to the six NOR gates. Carefully consider the name of this pin—"three-state disable." The 74LS240 three-state buffer has an enable input. When the enable input is taken to an active level, the buffer performs in a normal manner; when the enable input is taken inactive, the outputs of the buffers go to high-*z*. On the 4502B, when the three-state output disable pin is at an inactive-low level, the output of the NOR gates will appear on the Q outputs; when the three-state output disable is taken to an active-high level, the Q outputs will go to high-*z*.

Refer to the truth table in Figure 11.3. The first three lines illustrate the conditions that occur when the three-state output disable is at an inactive-low level. If the inhibit input is at a nondynamic low level (lines 1 and 2), the NOR gate will function as a common inverter. In line 3, the inhibit input is at a dynamic high level; all the outputs will go to logic 0, regardless of the levels on the data inputs. In the fourth line, the three-state output disable input is at an active-high level. The inhibit and data inputs are now don't-care conditions, because the Q outputs will go to high-*z*.

The 4502B has two ways to block data on the inputs from appearing on the Q outputs. The inhibit input can be taken to a dynamic level, forcing all the Q outputs low, or the disable input can be taken to a dynamic level, forcing all the outputs to high-*z*. the term "strobe" in the description of the 4502B can refer to either the inhibit input or the three-state disable input.

11.2.3 The 74LS363/364 Three-State Transparent Latches and D Flip-Flops

The 74LS363/364 are byte-wide (octal) transparent latches and D flip-flops. The only difference between the two devices is that the 74LS363 is level active and the 74LS364 is rising-edge active (see Figure 11.4). The 74LS363/364 share identical pinouts except for pin 11. The input that allows data to flow through a latch is called an enable and the input that causes the logic level on the D input to be transferred to the Q output in a D flip-flop is called a clock. The 74LS363 is a transparent latch; pin 11 is the enable. The 74LS364 is a positive edge-triggered D flip-flop; its pin 11 is the clock.

Pin 1 of these devices is called the output control. It functions as the output enable. If pin 1 is low, the devices will perform their normal functions; if pin 1 is taken to an inactive-high level, the eight Q outputs will go to high-*z*. The 74LS363/364 are 20-pin packages. Because of these pin limitations, the $\overline{Q}$ outputs are not available.

'LS363
FUNCTION TABLE

OUTPUT CONTROL	ENABLE G	D	OUTPUT
L	H	H	H
L	H	L	L
L	L	X	Q_0
H	X	X	Z

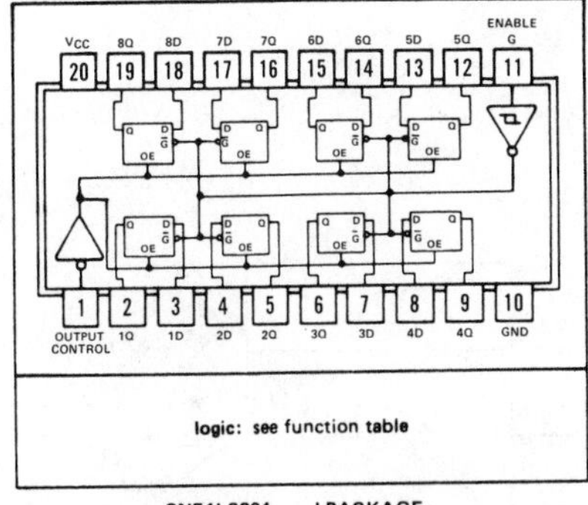

SN54LS364 . . . J PACKAGE
SN74LS364 . . . J OR N PACKAGE
(TOP VIEW)

'LS364
FUNCTION TABLE

OUTPUT CONTROL	CLOCK	D	OUTPUT
L	↑	H	H
L	↑	L	L
L	L	X	Q_0
H	X	X	Z

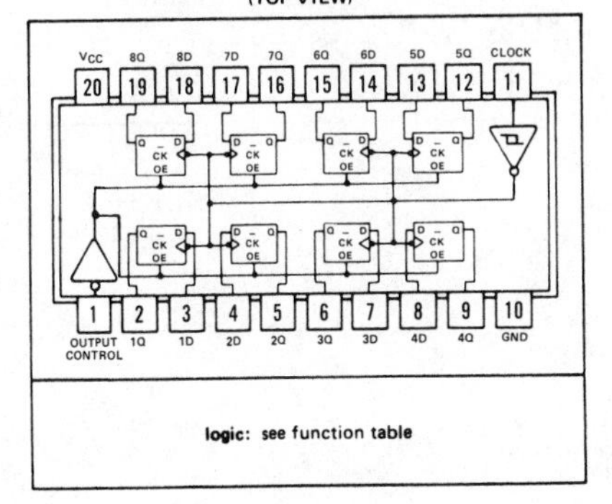

Figure 11.4 Pinouts and function tables for the 74LS363/364.

Refer to the function tables in Figure 11.4. The function tables are identical except for the columns of enable and clock. The 74LS363 has an active-high enable and the 74LS364 is rising edge clocked. When the output control pin is at a logic 0, the octal latches and flip-flops will behave in a normal fashion. The fourth line in the function table indicates the case when the output control is at a logic 1. The eight Q outputs will go to high-*z*; the enable/clock and data inputs become don't cares.

The 74LS363/364 have one other interesting characteristic. The guaranteed V_{OH} (logic 1 output voltage) is 3.45 V instead of the normal 2.4 V. This enables the 74LS363/364 to interface with MOS devices that require a higher logic 1 level than the standard 2.4 V. The 74LS373/374 that we used in the serial-to-parallel converter in Chapter 10 is the same as the 74LS363/364 except that it has a standard 2.4-V V_{OH} specification.

11.2.4 The 4076B 4-Bit D Flip-Flops with Three-State Outputs

The 4076B consists of four D flip-flops, data input disable, and data output disable circuitry. When the output enable control of the 74LS363/364 is high, the Q outputs go to high-*z* and no further information can be clocked into the latches or flip-flops. The 4076B is designed to enable the clocking of new data, even when the outputs are at high-*z*.

Refer to the functional diagram in Figure 11.5. The data input disable circuitry is represented by four switches that control the data inputs to the quad flip-flops. The output disable

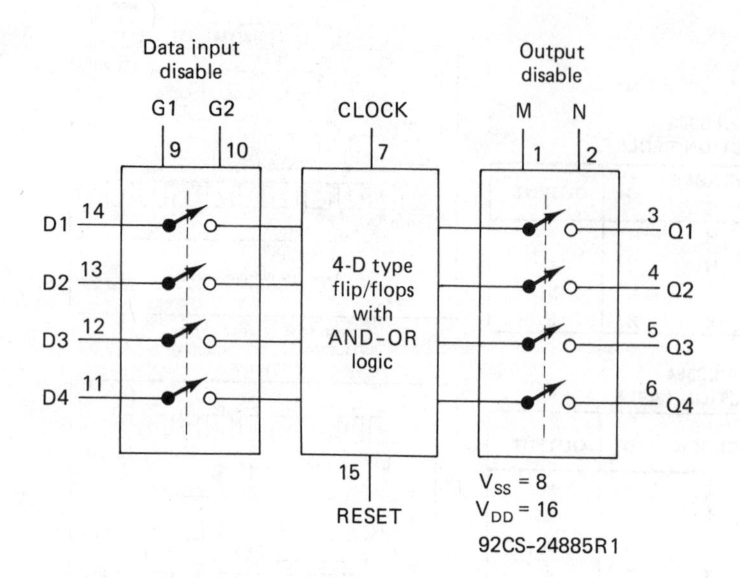

Functional diagram

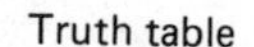
Truth table

RESET	CLOCK	Data input disable G1	G2	Data D	Next state output Q	
1	X	X	X	X	0	
0	0	X	X	X	Q	NC
0	$\uparrow$	1	X	X	Q	NC
0	$\uparrow$	X	1	X	Q	NC
0	$\uparrow$	0	0	1	1	
0	$\uparrow$	0	0	0	0	
0	1	X	X	X	Q	NC
0	$\downarrow$	X	X	X	Q	NC

When either output disable M or N is high, the outputs are disabled (high impedance state), however sequential operation of the flip flops is not affected

1 = high level X = don't care

0 = low level NC = no change

Figure 11.5 Functional diagram and truth table for the 4076B.

circuitry controls whether the Q outputs receive data from the outputs of the flip-flops or go to a high-*z* output.

Refer to the truth table in Figure 11.5. The note following the table explains how the output disable functions. Output disable inputs, M and N, are ORed together to from an active-high internal disable signal. If input M is high or input N is high, the Q outputs will go to high-*z*. However, unlike the 74LS363, the internal operation of the quad D flip-flops is not affected when the outputs go to high-*z*.

Let's analyze the truth table on a line-by-line basis. We will be examining the internal outputs of the D flip-flops and the Q outputs of the IC.

Line 1 The asynchronous clear is at an active high level. All other conditions become don't cares and the internal Q outputs will be cleared to logic 0s. (For the remainder of the truth table the clear input will be at an inactive level.)

Line 2 The clock is low. The flip-flops are in a memory state, their Q outputs remain unchanged.

Lines 3 and 4 These two lines demonstrate that the data input disable signals, G1 and G2, are internally ORed to produce an active-high data input disable signal. If G1 is high or G2 is high, the symbolic switches in the functional diagram will be open, and the internal Q outputs will remain in a memory state.

Lines 5 and 6 These lines illustrate the normal operation of a positive-edge-triggered D flip-flop. Both the data input disable inputs are at inactive-low levels. On a rising edge, the high or low data will be clocked into the internal Q output.

Line 7 When the clock is at a high level, the Q output will remain unchanged.

Line 8 As in lines 2 and 7, the falling edge of the clock will have no effect on the internal Q outputs.

11.3 BIDIRECTIONAL BUS DRIVERS

Every pin that we have seen on digital ICs could be designated as either an input or an output. In computer systems, data must travel in both directions. The data pins on a microprocessor are both inputs and outputs! The data buses require buffers to drive them—buffers whose data pins are capable of both receiving data (as inputs) and transmitting data (as outputs) (see Figure 11.6).

11.3.1 A Discrete Bus Transceiver.

This circuit is constructed from two three-state buffers and a conventional inverter. The input of one three-state buffer is tied to the output of the other three-state buffer. A visual inspection of the circuit does not reveal whether point A or point B is an input or an output. In fact, both points A and B are connected to inputs and outputs. Refer to the simple two-line truth table in Figure 11.6. There are two cases that we must investigate: when the control input is a logic 0 and when the control input is a logic 1.

Control input is a logic 0 The logic 0 will enable three-state buffer 1. The inverter will place a logic 1 on the control input of three-state buffer 2. This logic 1 will effectively disconnect the output of three-state buffer 2 from the node at point A. Data can

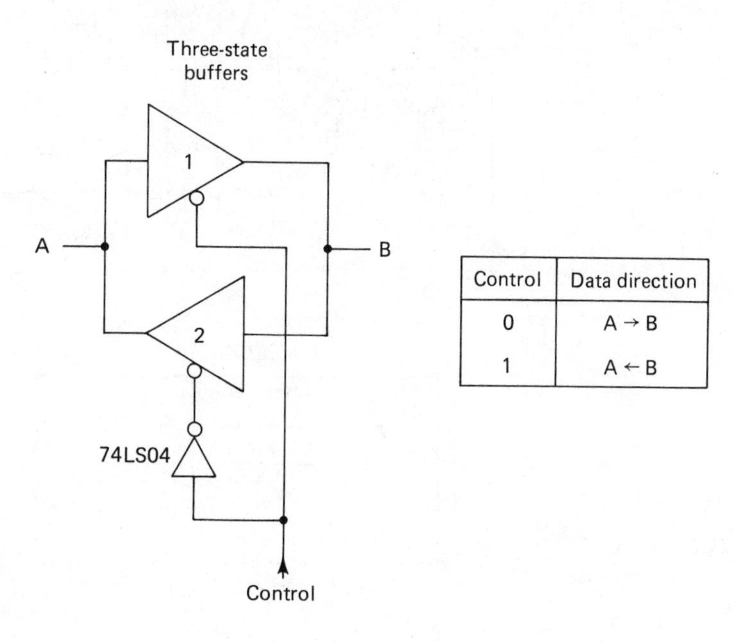

Control	Data direction
0	A → B
1	A ← B

Figure 11.6 Discrete bus transceiver.

now move from point A to point B via three-state buffer 1. There is no conflict at node A because three-state buffer 2 is at high-z.

Control input is a logic 1 Three-state buffer 1 will go to high-z, effectively removing its output from node B. Three-state buffer 2 will be enabled via the inverter. The data can now move from point B to point A via three-state buffer 2. There is no conflict at node B because three-state buffer 1 is at high-z.

The circuit in Figure 11.6 is called a *bus transceiver*. The term "transceiver" indicates that data can travel in either direction. (The terms "bus transceiver" and "bidirectional bus driver" are used interchangeably.) It is extremely important to realize that data do not travel in both directions simultaneously. In a correctly operating circuit, the inverter ensures that three-state buffers 1 and 2 will never be enabled simultaneously. The control input should be thought of as a direction input. If the logic level on the direction input is low, data flow from A to B; if it is high, data flow from B to A.

11.3.2 The 74LS245 Octal Bus Transceiver with Three-State Outputs.

Building bus transceivers from discrete gates is impractical. The 74LS245 is an example of a byte-wide bus transceiver contained in a 20 pin package (see Figure 11.7). The 74LS245 functions like the discrete bus transceiver in Figure 11.6 except that it

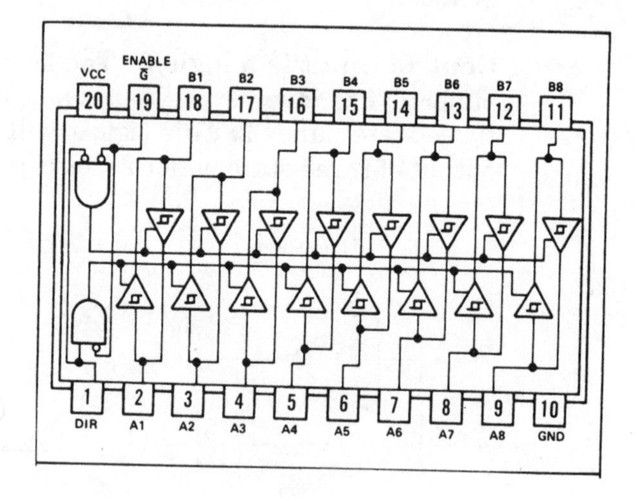

FUNCTION TABLE

ENABLE $\overline{G}$	DIRECTION CONTROL DIR	OPERATION
L	L	B data to A bus
L	H	A data to B bus
H	X	Isolation

H = high level, L = low level, X = irrelevant

Figure 11.7 74LS245 octal bus transceiver.

has one added mode—the ability to bring both the A and B pins to high-*z*.

Refer to the function table in Figure 11.7. When the enable input is low, the 74LS245 functions like a normal bus transceiver; the direction of data movement is controlled by the direction control input. When the direction control input is a logic 0, the data flow from B to A. When the direction control input is at a logic 1, the data flow from A to B.

In the third line of the truth table, the enable input goes to an inactive-high level. Both the A side and the B side will go to high impedance. The 74LS245 is effectively removed from the circuit.

11.4 THE MASTER OF THE SYSTEM BUS: THE MICROPROCESSOR

11.4.1 Block Diagram of a Computer.

Digital systems are composed of many complex bus structures. They require a device that can coordinate and synchronize all the circuit action in the system. This "master of the system bus" must be capable of retrieving and executing stored commands. It must be an "intelligent" device that can make logical decisions when confronted with choices. An LSI device called the *microprocessor* is used for such applications.

The term *computer* is extremely difficult to define. It is hard to think of an area or product where microprocessors are not used. Automobile ignitions, kitchen appliances, watches, calculators, and an infinite number of other objects are controlled by microprocessors. Having a device controlled by a microprocessor does not make that device a computer. Microprocessors are used to give devices simple limited local intelligence. Consider a programmable microwave oven. You can instruct the oven to start a defrost cycle at a certain time of the day for a predetermined length of time. Then the oven can be instructed to cook a roast until it reaches a temperature of 275 degrees. This sequence of events that the microprocessor must execute constitutes a *program*. The term "computer" is usually reserved to describe a general-purpose computing device that is capable of performing a wide variety of tasks.

All computers spend the majority of their time reading instructions and data from memory and writing the result of calculations back into memory. The next chapter is about digital memory systems. Before you can thoroughly understand how digital memories function, you must have a basic idea of what microprocessors really are and how they read and write to and from memory. Do not expect to acquire a general understanding of computer concepts from this introduction to microprocessors. You will devote a whole quarter or semester to that pursuit.

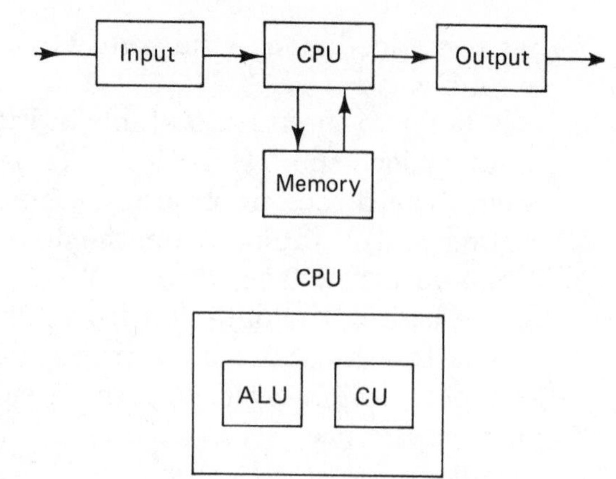

Figure 11.8 Block diagram of computer.

Consider the traditional block diagram of a computer (Figure 11.8). The block diagram of the computer consists of four separate units: input circuitry, central processing unit (CPU), memory, and output circuitry. The input and output circuitry is used to interface the computer with the outside world. A typical input device is a keyboard; a typical output device is a video display (CRT). Our interest lies in the two other blocks: the CPU and the memory. Notice that arrows point in both directions between the CPU and the memory. That indicates that the CPU can read and write into the memory.

The CPU is the heart and brains of the computer system. The CPU is divided further into two separate blocks: the arithmetic/logic unit (ALU) and the control unit (CU). The ALU performs simple arithmetic (such as addition and subtraction) and logic functions (such as NOT, AND, OR, XOR, and shifting). The CU coordinates all the activity in the computer system. A microprocessor is nothing more than a CPU implemented in a single IC package. Microprocessors can think, but they can't hear, they can't talk, and they can't remember anything. That is why input/output circuitry and memory devices are required to form a complete computer system.

The microprocessor's programmed instructions are stored in memory. Together with the programmed instructions is stored data. If we program a microwave oven to cook for 30 minutes at 12:30, cook is an instruction while 30 minutes and 12:30 are pieces of data. When you first enter the cooking program, the microprocessor takes the values from the keyboard on the oven and writes them into memory for future retrieval.

Our objective is to understand how the microprocessor reads and writes data to and from the memory.

11.4.2 Three-Bus Architecture.

The pins of a microprocessor can be broken in three groups of buses: address, data, and control/status (Figure 11.9).

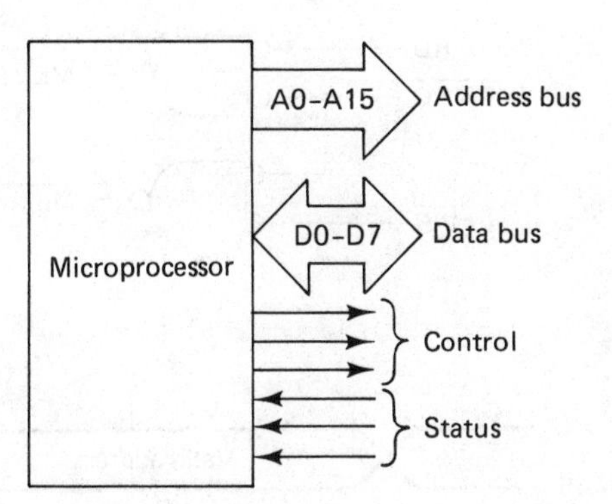

Figure 11.9 Three-bus system of a microprocessor.

Address bus If you want to send a letter, you must provide the mail carrier with a unique address. This unique address will guide the carrier to the correct residence. When a microprocessor is required to read from or write to memory, it must also provide a unique address. A typical microprocessor has an address bus that is 16 bits wide (A0 through A15). With 16 address lines the microprocessor can generate 2^{16} unique addresses; 2^{16} is equal to $65,536_{10}$. To avoid suffering under the influence of ungainly numbers the concept of 1K of memory was developed. 2^{10} is equal to 1024; 1024 bytes of memory is called 1K of memory. Therefore, 16 address lines are said to produce 64K unique memory locations. It is understood that 64K actually means 65,536.

The address bus is output only. It does not make any sense for a microprocessor to input a memory address. Remember that the microprocessor is the master of the bus; it will provide the address of the memory location that will be accessed.

Data bus The data bus is the path in which data enters and exits the microprocessor. Because memory locations will be read from and written to, the data bus must be bidirectional. The data bus in this example is 8-bits (one-byte) wide. Each memory location will store 8 bits of data. Microprocessors are available with data buses of 8 and 16 bits.

Control/status bus A control line controls the interaction between the microprocessor and external devices (such as memories). Control lines are output only. Status lines are input only. The microprocessor uses status inputs to monitor the state of particular lines and sense the occurrence of important external events. A typical microprocessor has a control/status bus that is 6 to 12 bits wide.

Because of our interest in the relationship between the microprocessor and memories, we will focus our attention on three control lines that a typical microprocessor might use to interface with memories: $\overline{\text{MEMRQ}}$, $\overline{\text{RD}}$, and $\overline{\text{WR}}$. (Figure 11.10). When the microprocessor needs to access memory, the $\overline{\text{MEMRQ}}$ (memory request) control line will go active low. The logic levels on the $\overline{\text{RD}}$ and $\overline{\text{WR}}$ lines will indicate whether the microproces-

(a)

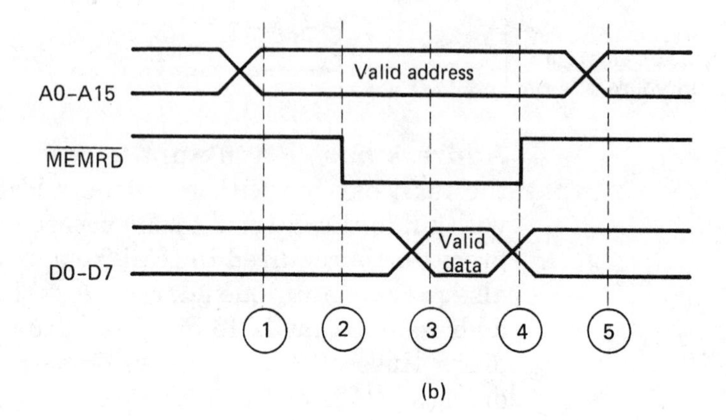

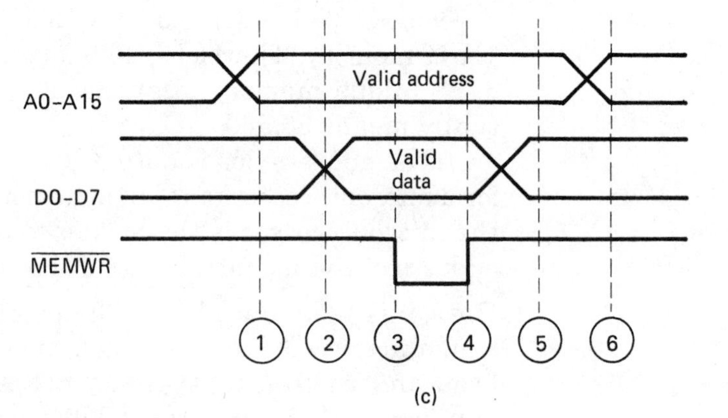

Figure 11.10 Gate (a) and control lines for memory read (b) and write (c).

sor is going to read from or write into memory. If the $\overline{RD}$ line is low and the $\overline{MEMRQ}$ line is low, the microprocessor is going to read from memory. If the $\overline{WR}$ line is low and the $\overline{MEMRQ}$ line is low, the microprocessor is going to write to memory. The $\overline{RD}$ and $\overline{WR}$ lines will never both be at active-low levels simultaneously. The simple two-OR-gate circuit in Figure 11.10a decodes the $\overline{MEMRQ}$, $\overline{RD}$, and $\overline{WR}$ control lines into $\overline{MEMRD}$ (memory read) and $\overline{MEMWR}$ (memory write commands.

Refer to the timing diagram in Figure 11.10b that describes a memory-read operation. The data output of the memory device is a three-state output structure. This structure will stay at high-*z* unless the microprocessor initiates a memory-read or memory-write operation.

Event 1 The microprocessor will present the address of the memory location that it wishes to read.

Event 2 The $\overline{MEMRD}$ signal will go low. This indicates to the memory IC that is should enable its data bus drivers and retrieve the byte of information stored at the indicated address.

Event 3 After a specified interval, the memory will place the data byte onto the data bus. This data byte will be read into an internal storage location in the microprocessor.

Event 4 The microprocessor will take the $\overline{\text{MEMRD}}$ signal back to the inactive level. When this occurs the memory will remove the data byte from the data bus by going to a high-z output.

Event 5 The microprocessor will then remove the address from the address bus. The transfer from the memory device to the microprocessor is now complete.

Refer to the timing diagram in Figure 11.10b that describes the memory-write operation.

Event 1 The microprocessor will first output the address of the memory location where the data byte is to be stored.

Event 2 The microprocessor then places the data byte to be stored in memory onto the data bus.

Event 3 The $\overline{\text{MEMWR}}$ signal is brought to an active-low level. The memory device will now transfer the byte of information on the data bus into the storage location being pointed to by the contents of the address bus.

Event 4 The $\overline{\text{MEMWR}}$ signal must be held active for a minimum specified interval. The $\overline{\text{MEMWR}}$ will then be returned to an inactive level.

Event 5 The microprocessor will remove the byte of data from the data bus.

Event 6 The microprocessor will finally remove the address from the address bus.

These examples of memory read and write operations have been greatly simplified. There are many factors and specifications that must be considered in practical memory circuits. These specifications will be thoroughly explored in the next chapter.

11.5 THE HEXADECIMAL NUMBER SYSTEM

11.5.1 Introduction to Hexadecimal Notation

Consider the following typical address:

1011 0001 1111 0101

Attempting to describe an address that is 16 bits wide can be extremely confusing. In groups of 16 (or even eight) strings of ones and zeros have a tendency to blur together. In Chapter 1 you learned the binary number system. The binary number system was used to describe the actual outputs of digital circuits. Because of the long data bytes and addresses in microprocessor circuitry, another number system is required that efficiently compacts these long strings of 0s and 1s into an easily readable form.

This new number system should meet the following criteria:

1. It must have the capacity to handle long groups of bits in a readily readable form.
2. Conversion between binary and this new number system must be extremely simple.

There is a number system that uniquely meets these criteria. The new number system is called *hexadecimal.* The prefix "hex" is the Greek word for the number six. The word "decimal" is equal to 10. The hexadecimal number system has a base of 16 (10 + 6). Although a number system with a base of 16 may seem awkward or clumsy, you will find that hexadecimal numbers are extremely easy to understand and manipulate. If a technician is going to work with microprocessor or memory circuit boards, a fluent command of hexadecimal is required.

A number system with a base of 16, by definition, must have 16 unique counting symbols. The first 10 are borrowed from the decimal number system and the last six are taken from the alphabet.

0, 1, 2, 3, 4, 5, 6, 7, 8, 9, A, B, C, D, E, F

What is the simple relationship between binary and hexadecimal numbers? You know that 4 binary bits can represent 16 unique numbers. Because the hexadecimal number system has 16 symbols, any group of 4 binary bits can be represented with one hexadecimal symbol.

Binary	*Hexadecimal*
0000	0
0001	1
0010	2
0011	3
0100	4
0101	5
0110	6
0111	7
1000	8
1001	9
1010	A
1011	B
1100	C
1101	D
1110	E
1111	F

Notice that the first 10 lines appear to be a standard binary-to-BCD conversion. Instead of having the binary combinations of 1010 to 1111 declared as illegal, they are now represented by the first six letters in the alphabet.

Reconsider the 16-bit address that we examined at the beginning of this section.

1011 0001 1111 0101

B 1 F 5

The string of 16 bits can be represented by four hexadecimal symbols. To convert any string of binary bits into a hexadecimal number, start with the least significant bit and form groups of 4 bits. If the last group has less than 4 bits, pad the missing bits with zeros. Convert each group of 4 bits into the corresponding hexadecimal symbol. The binary-to-hexadecimal conversion is now complete. To avoid confusion, the letter "H" is used to identify a hexadecimal number. For example,

1011 0011 = B3H

10 1001 1111 = 29FH

1100 0101 1010 1011 = C5ABH

Converting from hexadecimal back to binary is also a simple task. Each hexadecimal symbol should be reconverted back into 4 bits.

11.5.2 Hexadecimal-to-K's Conversion

It is often required to translate an address in hexadecimal to a value in K's of memory. You already know that 1K of memory is equal to 1024 bytes. The following table is a convenient means of converting back and forth between hexadecimal and K's.

Refer to Figure 11.11. The first column denotes the powers of 2 (from highest to lowest) for a 16-bit address. The second column indicates the decimal equivalent of each power of 2. The third column indicates the K value of each decimal number in the second column. Finally, the last column indicates the hexadecimal equivalent for each of the previous three columns.

Let's assume that you need to know the hexadecimal equivalent of a memory location at 13K.

Step 1 Break 13K into values that are listed in Figure 11.11:

13K = 8K + 4K + 1K

Step 2 Sum the hexadecimal equivalents of the individual values derived in step 1.

(13K) = 2000H + 1000 + 0400 = 3400H

Powers of 2	Decimal	Value in K's	Hexadecimal
2^{15}	32,768	32	8000
2^{14}	16,384	16	4000
2^{13}	8,192	8	2000
2^{12}	4,096	4	1000
2^{11}	2,048	2	0800
2^{10}	1,024	1	0400
2^{9}	512	1/2	0200
2^{8}	256	1/4	0100
2^{7}	128	1/8	0080
2^{6}	64	1/16	0040
2^{5}	32	1/32	0020
2^{4}	16	1/64	0010
2^{3}	8	1/128	0008
2^{2}	4	1/256	0004
2^{1}	2	1/512	0002
2^{0}	1	1/1024	0001

Figure 11.11 Hexadecimal-to-K's conversion table

The hexadecimal equivalent of 13K is 3400H. In Chapter 12 we will examine the topic of memory mapping. An important part of the memory mapping process is converting between hexadecimal and Ks.

11.5.3 Hexadecimal-to-Decimal Conversion

Consider the place values of a four-symbol hexadecimal number:

Place Value

X	X	X	X
$16^3 = 4096$	$16^2 = 256$	$16^1 = 16$	$16^0 = 1$

To convert a hexadecimal number to its decimal equivalent you must first convert the hexadecimal symbol in each column to its decimal equivalent. Multiply the decimal equivalent of each symbol by the place value of the column. Finally, sum up all the products from the multiplication and the result is the decimal equivalent of the hexadecimal number.

As an example, let us convert A3F9H to decimal.

Step 1 The decimal equivalent of each column is

10 3 15 9

Step 2 Multiply each of these column values by the proper place value:

$10 \times 4096 = 40960$ $3 \times 256 = 768$ $15 \times 16 = 240$ $9 \times 1 = 9$

Step 3 Sum the products from step 2.

$$
\begin{array}{r}
40960 \\
+ \quad 768 \\
+ \quad 240 \\
+ \quad \underline{\quad 9} \\
41977_{10}
\end{array}
$$

This procedure is identical with the binary-to-decimal conversion procedure. The only difference is the place values.

11.5.4 Decimal-to-Hexadecimal Conversion

This procedure will also be identical to the decimal-to-binary procedure that we studied in Chapter 1.

We will use Figure 11.11 as an aid in the conversion process. The first step in the conversion process is to find the largest decimal number in column 2 of Figure 11.11 that is smaller than the number to be converted. This value in column 2 will be subtracted from the decimal number that we are converting. The hexadecimal equivalent, in column 4, will become the first value in a hexadecimal sum. We will continue to subtract decimal numbers from the intermediate value to be converted and sum their hexadecimal equivalents until the intermediate value is equal to 0. Refer to Figure 11.12 on page 272 for an example of a decimal-to-hexadecimal conversion.

It requires six subtractions to reduce the decimal number to 0. Notice how the hexadecimal column is summed to produce the converted hexadecimal number. The second part of Figure 11.12 is a reconversion from the hexadecimal answer to decimal. The reconversion checks that our hexadecimal answer is equivalent to the decimal number that we wanted to convert.

11.5.5 Adding Hexadecimal Numbers

The hexadecimal addition during the decimal-to-hexadecimal conversion will not generate any carries. A hexadecimal carry is generated when an addition causes a hexadecimal result which is greater than F. All carries, in any number system, are handled in the same manner. The base of the number system is subtracted from the result that produced the carry. The result of this subtraction is the new answer for the column that produced the carry and a one is added to the next significant column.

It is difficult to add directly in hexadecimal. When performing an addition, the hexadecimal symbols A through F are converted into their decimal equivalents. Each column is then

Convert 45,346 to hexadecimal.

	Intermediate value	*Hexadecimal sum*
Step (1)	45,346	
	−32,768	8000H
	12,578	
	− 8,192	+2000H
	4,386	A000H
	− 4,096	+1000H
	290	B000H
	− 256	+0100H
	34	B100H
	− 32	0020H
	2	B120H
	− 2	+0002H
	0	B122H

Check: Reconvert B122H to decimal.

11 × 4096 = 45056
1 × 256 = 256
2 × 16 = 32
2 × 1 = 2

$45{,}346_{10}$

Figure 11.12 Decimal to hexadecimal conversion and check.

added. If any column results in an answer greater than 15, 16 (the base value of the hexadecimal number system) is subtracted from that answer. The result of the subtraction is the answer for that particular column and a carry is added to the next significant column. If any columns have a result of 10 through 15, that number is converted into the proper hexadecimal symbol. Refer to the following example.

Hexadecimal Addition with Carry

Problem: 7CH
; 58H

Note: Parentheses indicate decimal numbers.

Step 1 Convert hexadecimal values to decimal.

7 (12)
+ 5 8

Step 2 Add columns.

7 (12)
+ 5 8
(12) (20)

Step 3 Perform carry.

```
        (12)     (20)
carry  + 1     - (16)
        (13)       4
```

Step 4 Reconvert all numbers to hexadecimal.

Answer = D4H

11.5.6 Subtracting Hexadecimal Numbers

The most important part of hexadecimal subtraction is the borrow. When a borrow occurs in decimal subtraction, the value of 1 is subtracted from the next significant column and the value of 10 is added to the column that required the borrow. In a subtraction of any base, the value that is added to the column that requires the borrow is always equal to the base of the number system. Therefore, a hexadecimal borrow will result in a value of 1 being subtracted from the next significant column and the value of 10H (16 decimal) being added to the column that required the borrow.

To make the numbers easier to handle, when we added hexadecimal numbers, hexadecimal symbols were converted to their decimal equivalents. We will perform hexadecimal subtraction in the same manner. Refer to the following example.

Hexadecimal Subtraction with Borrow

Problem:
```
   ABH
 - 5EH
```

Step 1 Convert hexadecimal values to decimal.

```
   (10) (11)
 -   5  (14)
```

Step 2 Perform borrow.

```
(10) - 1 = 9 → 9 (27)
             - 5 (14)
```

Step 3 Subtract.

```
   9 (27)
 - 5 (14)
   4 (13)
```

Step 4 Convert decimal numbers to hexadecimal.

Answer = 4DH

Check:
```
   5E       5 (14)
 + 4D  =  + 4 (13)  =  9 (27)  =  ABH
```

You must practice until you feel confident with the hexadecimal number system. In the microprocessor world, hexadecimal is the most widely used number system. Data bytes and memory locations will be displayed in hexadecimal. When a group of memory locations are "dumped" onto a video display they will be shown in hexadecimal form. Experienced microprocessor technicians can convert back and forth between binary and hexadecimal in their heads. Hexadecimal to K's conversion is also extremely common, as is hexadecimal addition and subtraction. The least used conversion skill is converting between hexadecimal and decimal. Remember that binary and hexadecimal are closely related, whereas in the digital world, decimal is an unnatural number base.

QUESTIONS AND PROBLEMS

Refer to Figure 11.1 for Questions 11.1 to 11.5.

11.1. A scope probe is placed on input A of U6. The oscilloscope displays the waveform shown. Discuss each voltage level as it pertains to the circuitry.

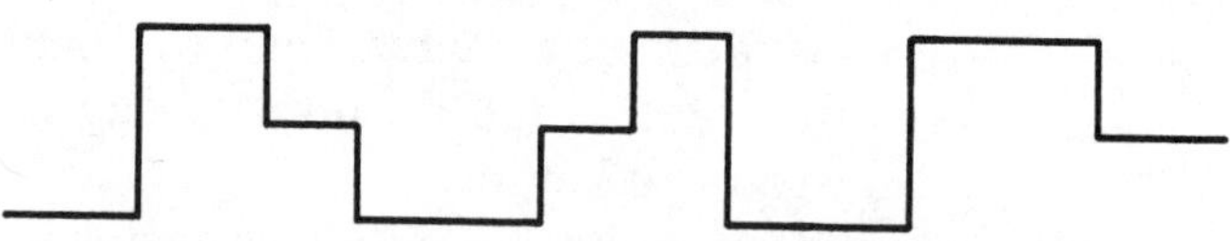

11.2. A scope probe is placed on input D of U6. The oscilloscope displays the following waveform. Discuss each of the four voltage levels. Explain the circumstance in which each voltage will occur.

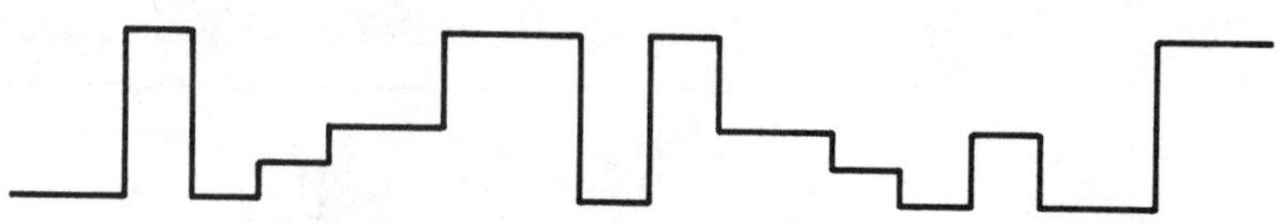

11.3. The circuit in Figure 11.1 works correctly for select codes 00 and 01. When select code 10 is applied to the decoder, the first BCD digit is displayed instead of the third BCD digit. When select code 11 is applied, the second digit is displayed instead of the fourth digit. What is the most likely cause of this circuit malfunction?

11.4. None of the segments of the display will illuminate. What is the most likely cause of such a malfunction?

11.5. Segment "a" of the seven-segment display burns out each time it is replaced. What is the most likely problem?

11.6. Define the term "buffer with three-state output."

11.7. Create a truth table for each of the following circuits. Explain why the common node of an open-collector circuit is called a "wired-OR" or "wired-AND" connection.

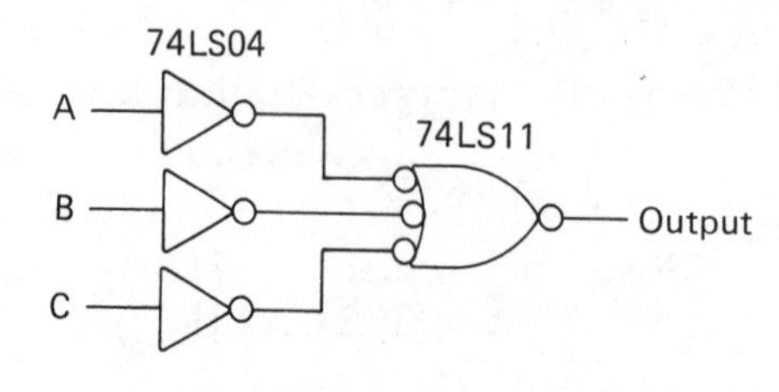

(a)

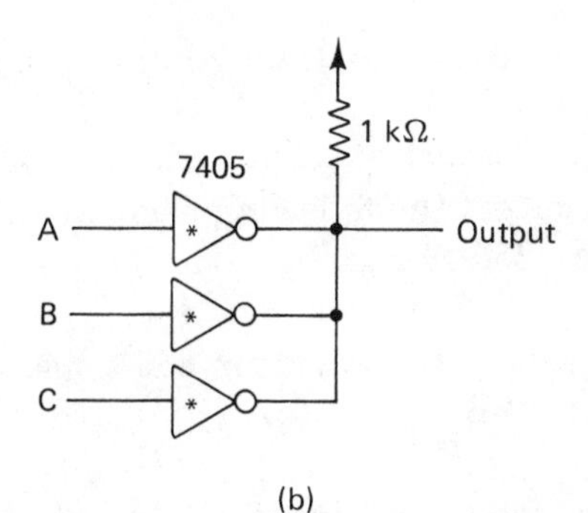

(b)

11.8. In what applications do you expect to encounter each of the following output structures?
(a) Totem poles
(b) Open collectors
(c) Three-state outputs

11.9. What could be a possible reason that the Q outputs of a 4502B are always logic 0s or floating?

11.10. Is there any danger in using three-state devices in CMOS circuitry?

11.11. Create a truth table that describes the circuit shown. The two control inputs should be renamed. Their new names should indicate the functions that they provide.

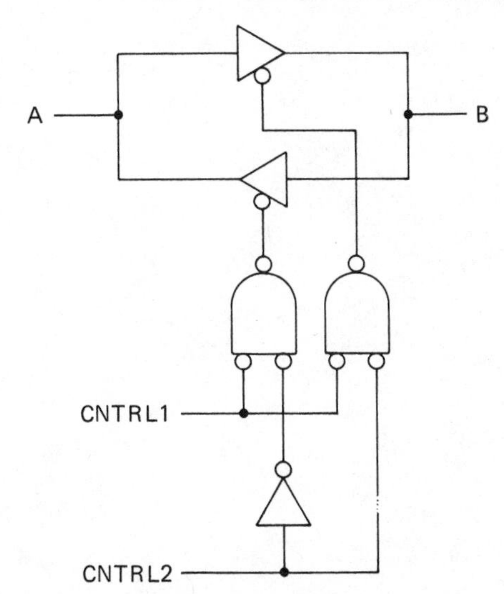

11.12. What is a simple definition of "microprocessor"?

11.13. Draw the block diagram of a computer.

11.14. What is the definition of "bus"?

11.15. Name and describe each bus in a microprocessor-based system.

11.16. Describe the manner in which a microprocessor
(a) Fetches a byte of data from memory
(b) Writes a byte of data into memory

11.17. What advantage do hexadecimal numbers have over decimal and binary numbers?

11.18. Convert the following values.

(a) 40K = ________H (b) 56K = ________H
(c) 64K = ________H (d) 32K = ________H
(e) FFFFH = ________K (f) C000H = ________K
(g) 0C00H = ________K (h) A800H = ________K

11.19. Convert the following hexadecimal values to decimal.

(a) 00FFH = ________ (b) F0A4H = ________

(c) 01FFH = ________ (d) 07FFH = ________

11.20. Convert the following decimal values to hexadecimal.

(a) 255 = ________H (b) 510 = ________H

(c) 1044 = ________H (d) 2560 = ________H

11.21. Perform the following additions.

(a)	(b)	(c)	(d)
10BH	9F8H	998H	F0FDH
+ 88H	+ 8H	+ 42H	+ 704H

11.22. Perform the following subtractions.

(a)	(b)	(c)	(d)
FA0H	770H	4000H	F000H
− 3FH	− F7H	− 7FFH	− 800H

12

IC MEMORY SYSTEMS

In Chapter 11, one block in the block diagram of the computer was labeled "memory." We stated that memory was used to store programs and data. Our first exposure to the concept of memory was in Chapter 8. We discovered that sequential devices have the capacity to "remember" a digital bit of information. Since that chapter, we have used latches and D flip-flops as memory devices in many different applications.

Computers employ many different types of memory devices. Some memory devices are designed to hold large quantities of data. These devices are mechanically driven and store the digital information on some type of magnetic media. IC memories are constructed from D flip-flops or MOS capacitors.

12.1 A 16-BIT READ/WRITE MEMORY

Let's consider each of the seven input and output lines that interface with the circuit in Figure 12.1. The data line is bidirectional. During the write operation, it carries a bit of information from the microprocessor to be stored in the memory; during the read operation it carries a bit of information retrieved from the memory to the microprocessor. The D inputs of each flip-flop and the output of three-state buffer G2 are connected to the data line.

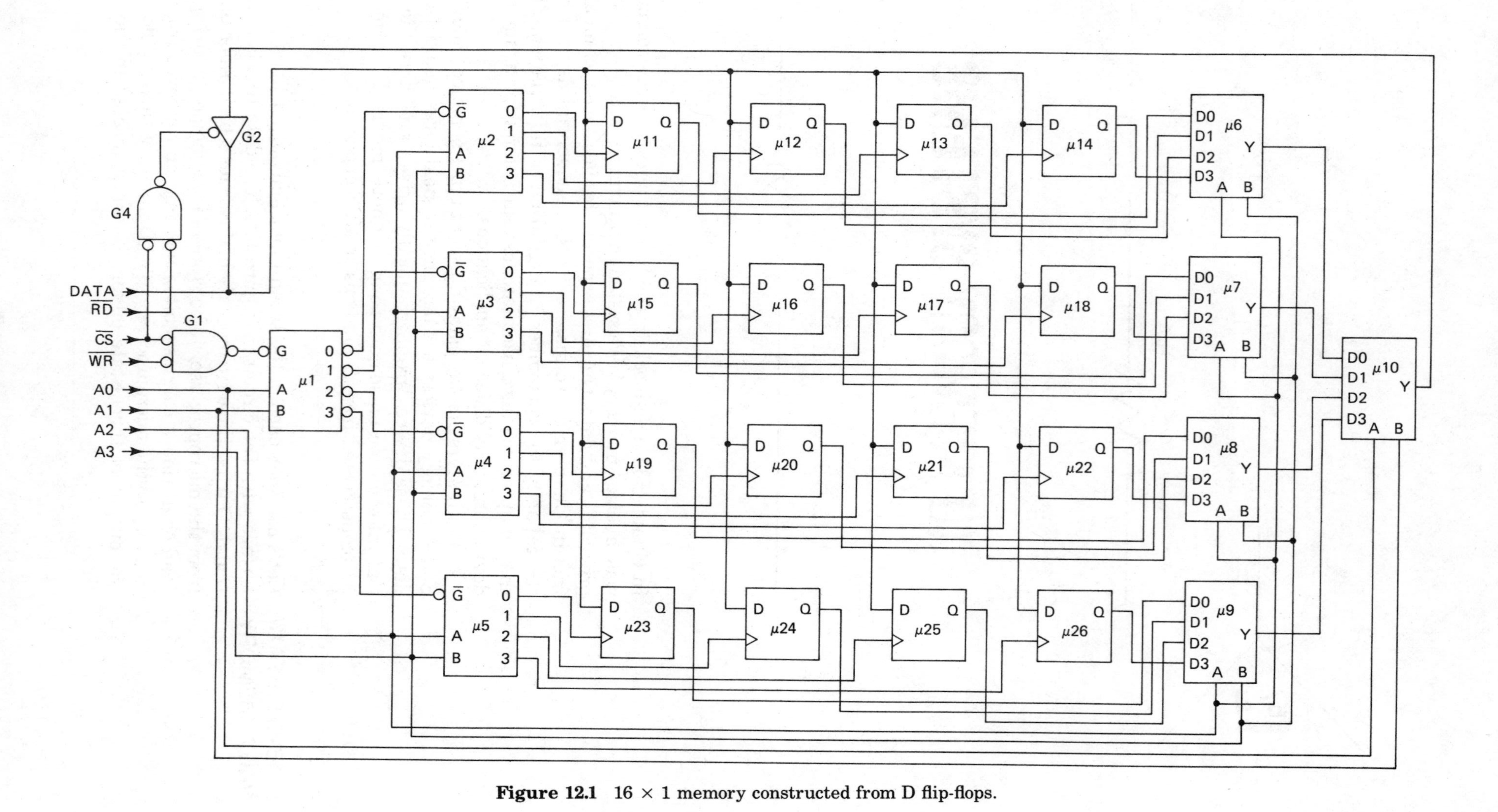

Figure 12.1 16 × 1 memory constructed from D flip-flops.

The $\overline{RD}$ and $\overline{WR}$ lines come from the control bus of the microprocessor. When $\overline{RD}$ goes low, a memory-read operation is designated; when $\overline{WR}$ goes low, a memory-write operation is designated. The $\overline{RD}$ line is connected to an input of G4. If the $\overline{RD}$ line is at a logic 1 level, the output of G2 will be high-Z. $\overline{WR}$ is connected to an input of G1. If $\overline{WR}$ is at a logic 1 level, U1 will be disabled.

The line designated $\overline{CS}$ stands for "chip select." This active-low signal functions like an enable input. If chip select is at an inactive-high level, the memory is disabled; it cannot be accessed by the microprocessor for either a read or a write operation. If the chip select goes active low, the memory is selected and can be accessed by the microprocessor. The $\overline{CS}$ is applied to gates G1 and G4. If the $\overline{CS}$ is high, G2 will be at high-Z and U1 will be disabled.

In Chapter 11 you learned that a typical microprocessor has 16 address lines. In this example, only the first four are used. Four address lines can uniquely address 16 memory locations. A0 and A1 are applied to 2-line-to-4-line decoder U1 and 1-of-4 data selector U10. A2 and A3 are connected to the 2-line-to-4-line decoders U2 through U5 (with active-high outputs) and 1-of-4 data selectors U6 through U9.

This circuit has 16 memory locations; each D flip-flop (U11 through U26) constitutes one memory cell. Notice that the 16 flip-flops are arranged in a square matrix of four rows by four columns. The rows are labeled 0 through 3. Flip-flops U11 through U14 constitute row 0, U15 through U18 row 1, and so on. The columns are also labeled 0 through 3. Flip-flops U11, U15, U19, and U23 constitute column 0; U12, U16, U20, and U24 column 1; and so on.

Each of the 16 flip-flops can be uniquely identified by a row-and-column designation.

Column	*Row*	*Flip flop*	*A3 A2 Column*	*A1 A0 Row*
0	0	U11	00	00
0	1	U15	00	01
0	2	U19	00	10
0	3	U23	00	11
1	0	U12	01	00
1	1	U16	01	01
1	2	U20	01	10
1	3	U24	01	11
2	0	U13	10	00
2	1	U17	10	01
2	2	U21	10	10
2	3	U25	10	11
3	0	U14	11	00
3	1	U18	11	01
3	2	U22	11	10
3	3	U26	11	11

The last two columns, labeled "column" and "row," designate the actual binary address of each memory cell. Notice that the lower 2 bits are the row address and the upper 2 bits are the column address.

The function table of a D flip-flop indicates that if the set and preset inputs are held at inactive levels, the Q output can change only on the rising edge of the clock input. During a write operation, the data that will be stored in memory is placed on the input of every D flip-flop in the memory matrix. The row and column decoders will steer a rising edge to the clock input of the addressed memory cell. In this manner, only the addressed memory cell will be affected.

During a write operation, the output of three-state buffer G2 must be held at high-*z*. This is because during the memory-write operation, the memory device is a bus "listener" and must not output any logic levels onto the data bus. The only time a memory device will become a bus "talker" is during a read operation. The $\overline{\text{RD}}$ input must be held at an inactive high level during the write operation.

Let's examine a typical write operation (Figure 12.2). Let's analyze each event in the write-cycle timing diagram as it is carried out in the circuitry of Figure 12.1. Assume that the $\overline{\text{RD}}$ input is held high during the write operation.

Event 1 The 4-bit address designating a unique memory location is placed on the address bus (A0 through A3). A0 and A1 designate the row address, and A2 and A3 designate the column address.

Event 2 The $\overline{\text{CS}}$ input goes active low. This indicates that the memory is now selected.

Event 3 The microprocessor places the data bit to be stored into memory on the data input.

Event 4 The $\overline{\text{WR}}$ input goes active low. The active-low CS and WR inputs cause the output of G1 to go active low; U1 is now enabled. U1 is called the *row decoder*. Each of the four decoded

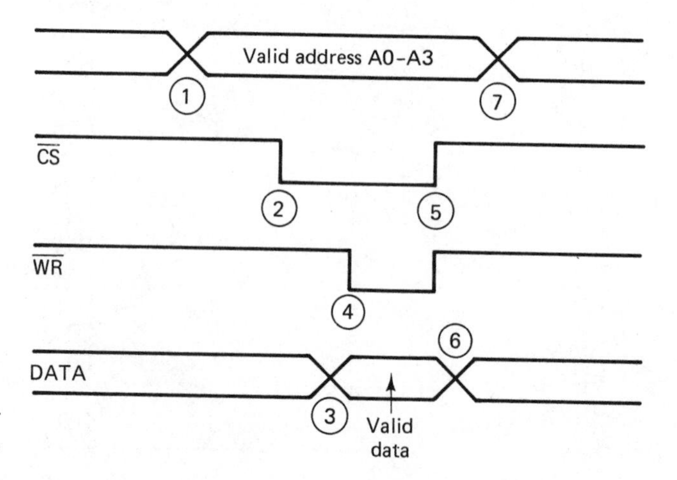

Figure 12.2 Typical write-cycle timing diagram.

outputs of U1 are associated with a particular row of memory cells. The row address of A0 and A1 will force a decoded output of U1 to an active-low level.

U2 through U5 are the column decoders. Now that a row has been selected by the output of U1, a column within that row must be selected. The column address is applied to column decoders. Notice that these 2-line-to-4-line decoders have active-high outputs. When the selected column output goes to an active-high level, the rising edge on the clock input of the addressed memory cell will transfer the bit on the D input onto the Q output. The data bit has now been stored in memory.

Events 5–7 These events show the address, data, and control lines returning to inactive states.

Notice that the data must be given a short period to stabilize before the $\overline{WR}$ input goes active low. If this period is too short, the wrong logic level may be stored. The data input also has to be held for a short time after the falling edge of $\overline{WR}$. This is due to the setup and hold times of flip-flops that were examined in Chapter 8.

After the falling edge of $\overline{WR}$, the propagation times of the OR gate (G1), row decoder, column decoder, and a flip-flop must occur before the data is actually written. Remember these delays when we encounter actual IC memory devices.

Let's consider a typical memory-read operation. Assume that the $\overline{WR}$ input is held at an inactive high level during the read cycle (Figure 12.3).

Event 1 The address of the memory cell to be read is placed on A0 through A3 by the microprocessor.

Event 2 The chip select goes to an active-low level. U1 will be disabled throughout the read operation. Flip-flops need not be clocked to be read.

Event 3 The $\overline{RD}$ input goes active low. The active-low $\overline{CS}$ and

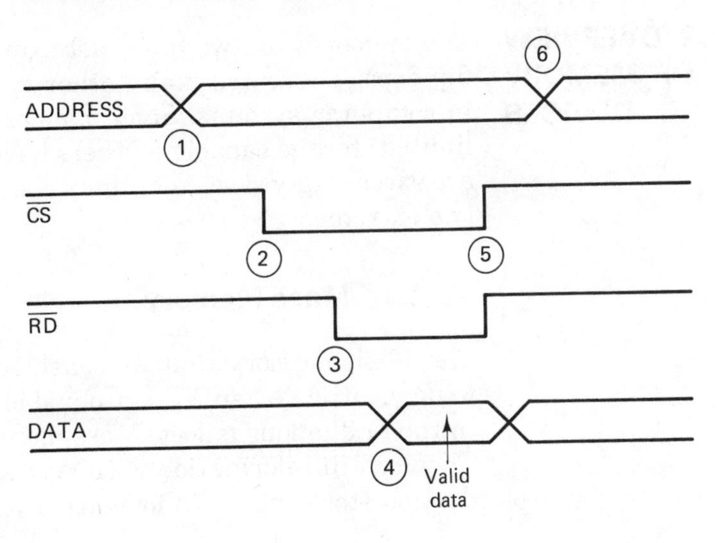

Figure 12.3 Typical read-cycle timing diagram.

$\overline{RD}$ inputs will cause the output of G4 to go active low. G2 is now enabled to pass data.

G2 is driven by the output of data selector U10. U10 is the memory-read counterpart of U1. U1 is the row-write decoder, and U10 is the row-read data selector. The select inputs of both U1 and U10 are driven by the row address A0 and A1.

U6 through U9 are the counterparts of U2 through U5. U6 through U9 are the column read data selectors. The selects of U2 through U5 and U6 through U9 are driven by the column address bits A2 and A3.

The Q output of the selected memory cell will first be transferred through a column-read data selector. Then it will be transferred through the row-read data selector and placed on the input of G2. This read process will be subject to the propagation delay of a column-read data selector, the row-read data selector, and G2.

Event 4 After the sum of all the propagation delays, the valid data appear on the output of G2.

Event 5 The $\overline{CS}$ and $\overline{RD}$ inputs are taken inactive high. After a slight delay the output of G2 will return to high-*z*.

Event 6 The microprocessor changes the address on A0 through A3.

Constructing memories from discrete flip-flops is impractical. This 16-bit memory has been used to illustrate some important points concerning IC memories. Memories are constructed in square matrices where the lower half of the address bits represent a row address and the upper half of the address bit represent a column address. You can now appreciate the complex timing parameters that describe the operation of IC memory devices.

12.2 OVERVIEW OF MEMORY DEVICES

The memory that we have just examined is constructed from D flip-flops. There are many other types of memory devices used in computer systems. Some memory devices are fast but have limited storage capacity. Others have vast storage capacities but are exceedingly slow. Yet others are a compromise between these two extremes.

12.2.1 Mass Memory

The 16-bit memory that we constructed with D flip-flops is said to be volatile. When V_{cc} is turned off, all the information stored in the D flip-flops is lost. Mass-memory devices are nonvolatile. They retain information without the need of V_{cc}. Mass-memory devices store digital information on magtnetic media. As the name implies, mass-memory devices are used to store large quantities of data. Because mass-memory devices are mechanical in nature, they operate too slowly to be accessed directly by microprocessors. The digital information stored in mass-memory devices must first be loaded into semiconductor memory before it can be used by the microprocessor.

Many inexpensive home computers use audiocassette recorders as mass-memory devices. They store logic 0's and logic 1's as different tones. Cassette storage is inexpensive, but it is also extremely slow. Many manufacturers build high-speed digital cassette drives designed specifically for computer applications, but they are much more expensive than standard audiocassette drives.

Cassettes are an example of serial memories. Data are stored in a long line of sequential bits. One cassette may have 10 programs stored on it. If you desire to place the eighth program into the computer's memory, you must first read sequentially through the first seven programs. This can be a slow and frustrating process.

The 16-bit memory that we constructed from D flip-flops is called a *random access memory*. If we want to read the memory cell in the second row, third column, we can access it directly without having to read through the other memory cells. Every location in a random access memory takes the same length of time to access. The period required to access information in a serial memory device depends on the location of the information.

Large computers use magnetic-tape mass memories. You have surely seen pictures of large reel-to-reel magnetic tapes in computer rooms. Magnetic tape is an old, but still useful form of mass memory. Its use is limited to large computers in the data processing industry.

The *hard disk* is a high-speed mass memory. A metal disk is coated with a ferrite material. Several of these metal disks are mounted on a central spindle. Each disk is called a *platter*. The platters spin at an extremely high speed. Digital information is stored on the platters by magnetic write heads. The information is read with magnetic read heads. The read/write heads never touch the surface of the platters. They ride on a few microns of air. Hard disks are available for both large computers and microcomputers.

The most popular form of mass storage for microcomputers is the *floppy disk*. Instead of being constructed from a metal plate, a floppy disk is a circular piece of plastic coated with a ferrite material. After a hole is punched out of the center of the floppy disk, it is inserted into a protective cover. This cover has a slot where the read/write head can access the surface of the floppy disk. Unlike hard disks, the read/write head actually contacts the surface of the floppy disk. Standard sizes of floppy disks are 8, 5.25, and 3.5 inches.

Hard and floppy disks are pseudo-random-access memories.

The disk is formatted in a series of circular tracks. Each track is further broken into blocks called *sectors*. A particular track can be accessed without reading all the previous data; that is the random access portion. To reach a particular sector within the track requires that the entire track be read. This is the serial access portion. The speed at which disks operate is limited by their mechanical drives. IC memories speeds are limited only to the propagation delay of their circuitry.

Like cassettes, floppy disks are a removable medium. A typical microcomputer user will have many floppy disks. Each one can contain several programs and files of data. To be accessed by a computer, the floppy disk must be inserted into a floppy disk drive.

12.2.2 Main Memory

Microprocessors operate at high speeds. The memories that the microprocessor accesses must be capable of operating at the speed of the microprocessor. They contain the current programs and data that are being processed. These memories are called *main memory*.

Each type of mass memory was driven by a mechanical means that limited their response time. Main memory devices are constructed from high-speed semiconductors. In the early days of computers main memories were constructed from small iron doughnuts called *magnetic cores*. The modern use of magnetic core memory is limited to extremely specialized applications. To this day the main memory in computers is still often called "core memory" even though it has long since been replaced by IC memories.

We have already examined the most common form of main memory: the random access read/write memory. Read/write main memory devices are called RAMs. As we have previously stated, RAMs are volatile memory devices. When V_{cc} is turned off, the contents of the RAM are lost. RAMs are used as temporary storage for programs and data loaded into the computer from mass memories.

There are two types of RAMs: static RAMs and dynamic RAMs. RAMs constructed from flip-flops are called static RAMs. Once a bit of information is written into a static RAM, no further action (other than maintaining V_{cc}) is required to assure that it is not lost. Dynamic RAMs are constructed from MOS capacitors. When a logic 1 is stored in a dynamic RAM cell, a MOS capacitor is charged to a logic 1 level. When a logic 0 is stored in a dynamic RAM cell, a MOS capacitor is discharged to 0 V. All capacitors suffer from leakage current. In a matter of a few milliseconds the logic 1 stored on the MOS capacitor may discharge to a logic 0 level. Because of these leakage currents, dynamic memories must be "refreshed" at least once every few milliseconds. The refresh process will read the logic level stored on a MOS capacitor memory cell, and then rewrite it. This will prevent logic 1-level charges from leaking down to logic 0 levels. Although refreshing dynamic RAMs may sound inconvenient, you will discover that they have many advantages over static RAMs.

Another type of main memory device is the read-only memory. Read-only memories are called ROMs. A ROM cannot be written into by the microprocessor. ROMs are nonvolatile; they are used to store programs, data, and tables of information that must not be lost when V_{cc} is turned off.

There are many different types of ROMs. Some must be ordered from the factory, while others can be programmed by the user. (Here the term "program" describes the process of burning information into the memory cells of the ROM.) Some ROMs can be programmed only once; others can be erased and reprogrammed many times.

12.3 THE STATIC RAM

Because static RAMs do not require refreshing, they are often used in simple microprocessor applications. Static RAMs have two major disadvantages:

1. They are limited to small storage capacities. D flip-flops require much space on the surface of an IC. This greatly limits the number of memory cells that can be integrated onto a single chip. One or two circuit boards of static RAMs must be used to achieve the same storage capacity of eight dynamic RAM ICs.
2. Because more ICs are required to gain the same storage capacity, static RAMs tend to consume great amounts of power. As power supplies grow in capacity, they also increase in price, size, and weight.

Even with these limitations, static RAMs are ideal for small memory systems.

12.3.1 The 6116 2K × 8 High-Speed CMOS RAM

The 6116 uses 16K D flip-flops for its storage cells. These flip-flops are organized in 2K groups of 8. That is exactly what the description 2K × 8 states.

In general, the description of the storage capacity of memory ICs will take the following form:

unique locations × memory cells at each location

The product of the number of unique locations and the number of memory cells at each location will yield the total number storage devices in the memory IC.

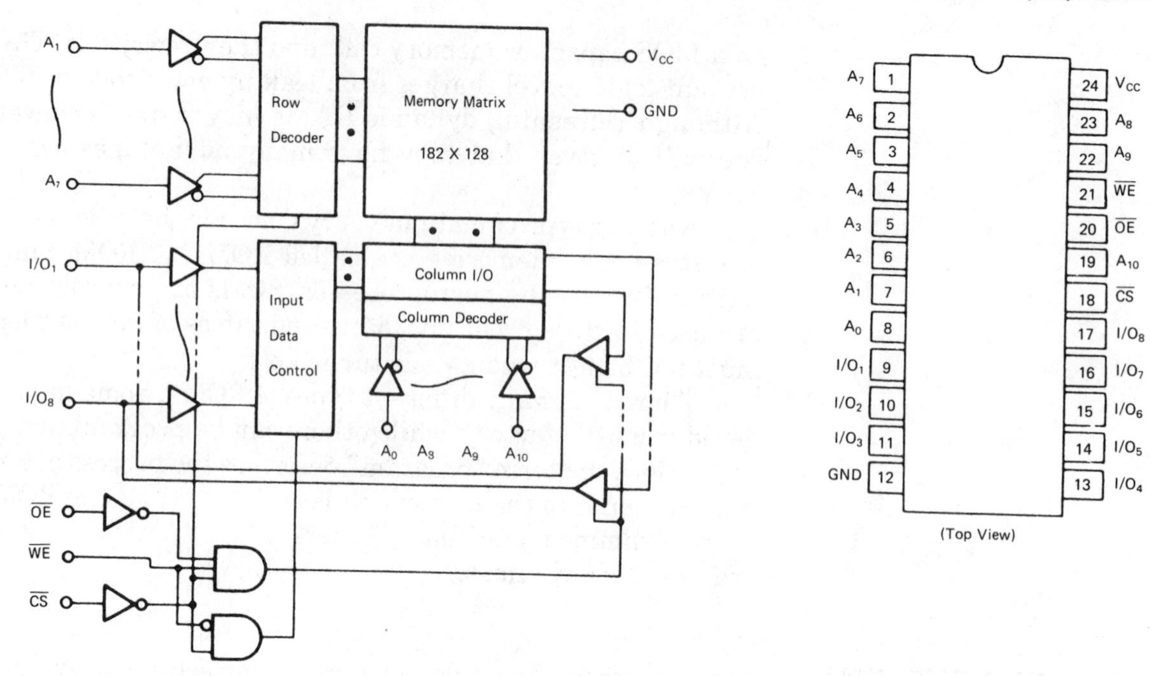

Figure 12.4 Functional block diagram and pinout of a 6116 static RAM.

Refer to Figure 12.4. By observing the number of address inputs you should be able to calculate the total number of unique storage locations within a memory device. The pinout in Figure 12.4 indicates that the 6116 has 11 address pins (A0 through A10). 2^{11} = 2K unique memory locations. This agrees with the description of the 6116.

Counting the number of data inputs or outputs will tell us the number of memory cells at each unique location. You will not find any pins on the 6116 labeled as data in or data out. Instead, there are eight pins labeled "I/O." This is the method used to indicate data pins on memory devices that are bidirectional—that function as both inputs and outputs. Because the 6116 has eight I/O pins (I/O1 through I/O8), there must be eight storage cells at each unique location. Once again this agrees with the 2K $\times$ 8 description of the 6116.

Refer to the functional block diagram in Figure 12.4. Notice the odd inverter-like devices that the 11 address lines are driving. This is a way of illustrating that the address lines are buffered and require only minimal drive currents. The inverter-like drivers imply that both the true and inverted values of the address lines are applied to the row and column decode circuitry. (Remember the discrete 2-line-to-4-line decoder that we built in Chapter 4? Both the true and complemented forms of all the inputs were required to drive the internal circuitry. You should refer to Figure 4.12 to jog your memory.) The memory matrix (D flip-flop storage cells) is described as 128 $\times$ 128. This calculates to a total number of 16K flip-flops. This agrees with the product of 2K $\times$ 8.

The data I/O lines are buffered on the input and driven by three-state drivers on the output. This tells us that the I/O lines are capable of going to high-*z*. You know that any talker on a microprocessor bus must be capable of going to high-*z*.

The only power requirements of the 6116 is +5 V (V_{cc}) and ground. This makes it compatible with TTL and microprocessor circuitry. Although the 6116 is a CMOS device, this fact will usually be transparent to the design engineer and the technician.

The last part of the functional diagram that must be analyzed consists of the control bus inputs: $\overline{OE}$ (output enable), $\overline{WE}$ (write enable), and $\overline{CS}$ (chip select). Notice that these three signals are barred; they are active low. The output of the top AND gate appears to control the enable lines on the three-state data output buffers. If the output of the top AND gate goes active high, the three-state data output buffers are enabled. This AND gate must control the memory-read operation.

The active-high output of the bottom AND gate is connected to the "input data control" block. It must control the memory-write operation. The control circuitry is redrawn in Figure 12.5. Notice that the only signal common to both NOR gates is the chip select input. The chip select must be at an active-low level before the RAM will respond to either read or write commands.

The read operation will be true when chip select and output enable are at active-low levels and the write enable input is at an inactive-high level. The write operation will occur when both the chip select and write enable inputs are active low. The only signal that is not applied to both AND gates (in true or complemented form) is output enable. In many applications the output enable pin is hardwired to ground. When this is true, the 6116 will either be reading or writing whenever Chip Select goes active (depending on the status of the write enable input). In our application of the 6116, the output enable input will be driven by a memory-read signal.

12.3.2 Timing Parameters of the 6116

The timing parameters describing memory ICs are complex. A typical read or write cycle has at least 10 specifications. These specifications describe minimum pulse widths and setup and

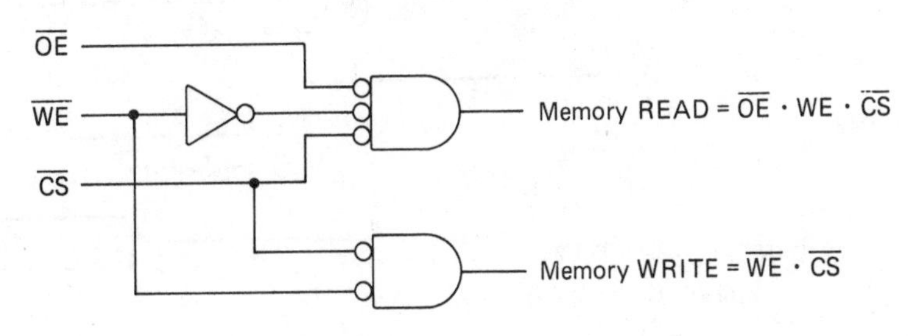

Figure 12.5 Internal decode circuitry of the 6116.

hold times for the address, data, and control signals. Design engineers must consider each specification carefully when they are designing memory systems. Technicians only need to know the most basic of these specifications. The words and phrases used to describe these timing parameters are not standardized. They differ greatly between manufacturers. After you study and understand the timing parameters of the 6116, you should be able to read and moderately understand the timing parameters for any static RAM (see Figure 12.6). The microprocessor performing the read and write operations will have an 8-bit bidirectional data bus, 16-bit address bus, and a control bus with these active low signals: $\overline{\text{MEMRQ}}$ (memory request), $\overline{\text{WR}}$ (write), and $\overline{\text{RD}}$ (read).

Figure 12.6 describes a typical read cycle of the 6116. Assume that the $\overline{\text{WE}}$ signal is held at the inactive high level throughout the read cycle.

Event 1 The microprocessor will provide a 16-bit address denoting the memory location to be read. Address bits A0 through A10 are directly connected to the 6116. Address bits A11 through A15 and $\overline{\text{MEMRQ}}$ will be used to create the chip select pulse. A complete schematic and explanation of the chip select derivation will be presented later in this chapter.

Event 2 The $\overline{\text{OE}}$ signal goes active low. The output enable of the 6116 is driven by the $\overline{\text{RD}}$ output of the microprocessor.

Event 3 The chip select goes active low. Our discussion of the internal 6116 control circuitry indicated that this combination of signals ($\overline{\text{OE}}$ and $\overline{\text{CS}}$ active—$\overline{\text{WE}}$ inactive) enabled the three-state output buffers for a memory-read operation.

Event 4 After a short delay, the data output is available on pins I/O1 through I/O8. Notice that the line before and after the data valid indication is neither high nor low but appears at a floating level. This indicates that before and after the data valid time, the I/O pins are held at high-Z. Unless a bus talker is selected, its outputs will remain at high-*z*. The I/O lines are the

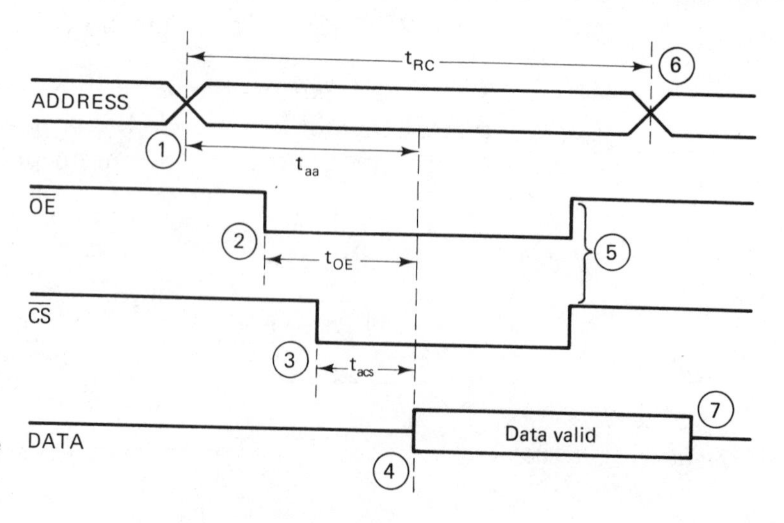

Figure 12.6 Read cycle of 6116.

only outputs of the 6116; the other signals—address and control—are inputs; only output lines need to be taken to high-*z* when a device is deselected.

Event 5 The $\overline{\text{OE}}$ and $\overline{\text{CS}}$ lines are taken to inactive levels.

Event 6 The address bus is taken to another address. The read operation is now complete.

Event 7 There will be a delay after the address bus changes and the I/O outputs return to the high-*z* levels.

Let's consider the major timing parameters indicated in the timing diagram.

Read-cycle time, t_{RC} This is the most important specification of a memory IC. It specifies the minimum time in which a read cycle can be completed. When an engineer or technician references a particular RAM, two numbers must be indicated: the part number and the read access time. For example, 6116s are available with 120-, 150-, and 200-ns read access times. When you say that you need a 6116, the read access time must also be specified: for example, 150-ns 6116. Faster memory ICs are proportionally more expensive than slower memory ICs.

Address access time, t_{AA} This is the maximum time required between a valid address and a valid data output.

Output enable to output valid, t_{OE} This is the output enable equivalent to the address access time; t_{OE} is the maximum time required between an active output enable and valid data.

Chip select access time, t_{ACS} Like t_{AA} and t_{OE}, chip select access time is the maximum time required between an active chip select and valid data.

All three of the previous specifications must be met before the data output is valid.

Figure 12.7 illustrates a typical 6116 memory-write cycle. There are two possible ways to perform a 6116 write. We are

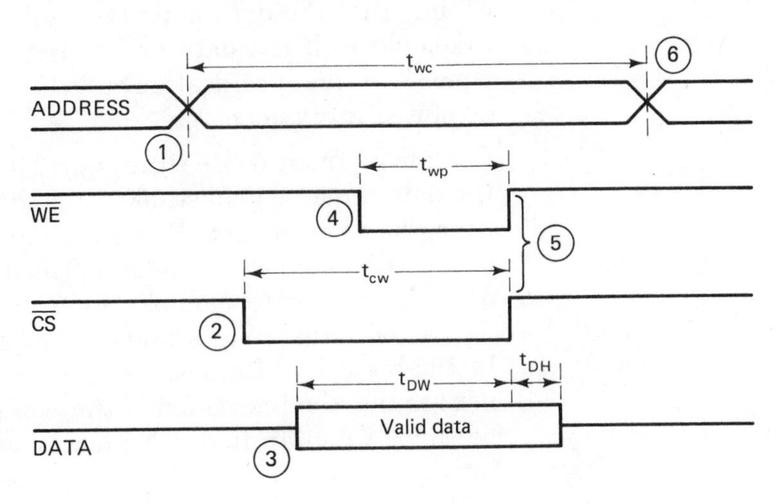

Figure 12.7 Write cycle of 6116.

going to assume that the $\overline{OE}$ is held at an inactive logic 1 level. This will assure that the three-state output buffers in the 6116 are never enabled when the microprocessor is outputting data onto the data bus. Another method of performing a write operation is to hardwire the $\overline{OE}$ to ground. When the chip select goes active, the three-state buffers will momentarily be enabled until the write enable goes active. An overlap of active chip select and write enable inputs designates a write operation.

Event 1 The microprocessor outputs the address of the memory location that will be written into.

Event 2 The chip select is taken to an active-low level.

Event 3 The microprocessor places the byte of information to be stored onto the data bus.

Event 4 The write enable input on the 6116 is driven by the $\overline{WR}$ output of the microprocessor. When chip select and write enable are both low, the input buffers on the 6116 are enabled to receive data. Notice that the microprocessor placed valid data on the data bus before write enable was taken active.

Event 5 $\overline{WE}$ and $\overline{CS}$ are taken to inactive levels.

Event 6 The microprocessor removes the address and data from the buses. This completes the write operation.

Let's consider the major specifications describing the memory-write operation.

Write-cycle time, t_{WC} This is the write equivalent of the read-cycle time. Read- and write-cycle times are equal. A 200-ns memory IC has read- and write-cycle times of 200 ns.

Write pulse width, t_{WP} This specifies the minimum active pulse width of $\overline{WE}$.

Chip select width, t_{CW} This is the minimum active pulse width of the chip select signal.

Data to write time overlap, t_{DW} This is the minimum length of time that the data must be stable before the write operation takes place. If the data are on the bus for a shorter length of time than this specification calls for, an incorrect data byte may be stored into memory.

Data hold from write time, t_{DH} This is the length of time that the data must be held stable after the specified minimum write times have been met. Remember that we examined the hold times associated with flip-flops. The data hold time in a memory device is similar to the hold times in flip-flops.

This timing diagram and specifications may seem complex. In the majority of cases, troubleshooting memory boards does not require the inspection of these specifications, but there will be times when an in-depth knowledge of these specifications is invaluable.

12.3.3 Constructing Memory Systems with 6116 RAMs

Several memory ICs can be connected together to form large blocks of memory (Figure 12.8). Because so many connections are involved, schematics of memory boards can get complex to the point of being unreadable. To avoid this confusion the signals that are common to all the RAMs are only shown on this first RAM. The four memory ICs in Figure 12.8 share these common lines: data bus (D0 through D7), first 11 bits of the address bus (A0 through A10), and the microprocessor's read and write control lines.

The only microprocessor signals that are not connected directly to the RAMs are $\overline{\text{MEMRQ}}$ and address bits A11 through A15. These are the signals from which the independent chip selects are derived. Notice that address lines A11 and A12 and memory request are connected to the 74LS138. With the C select input grounded, the 74LS138 performs the function of a 2-line-to-4-line decoder. Each of the decoded outputs drives a 6116's chip select. From your knowledge of decoders, you know that it is possible for only one decoded output to be active low at any time. This assures that it is impossible for more than one RAM to be selected simultaneously.

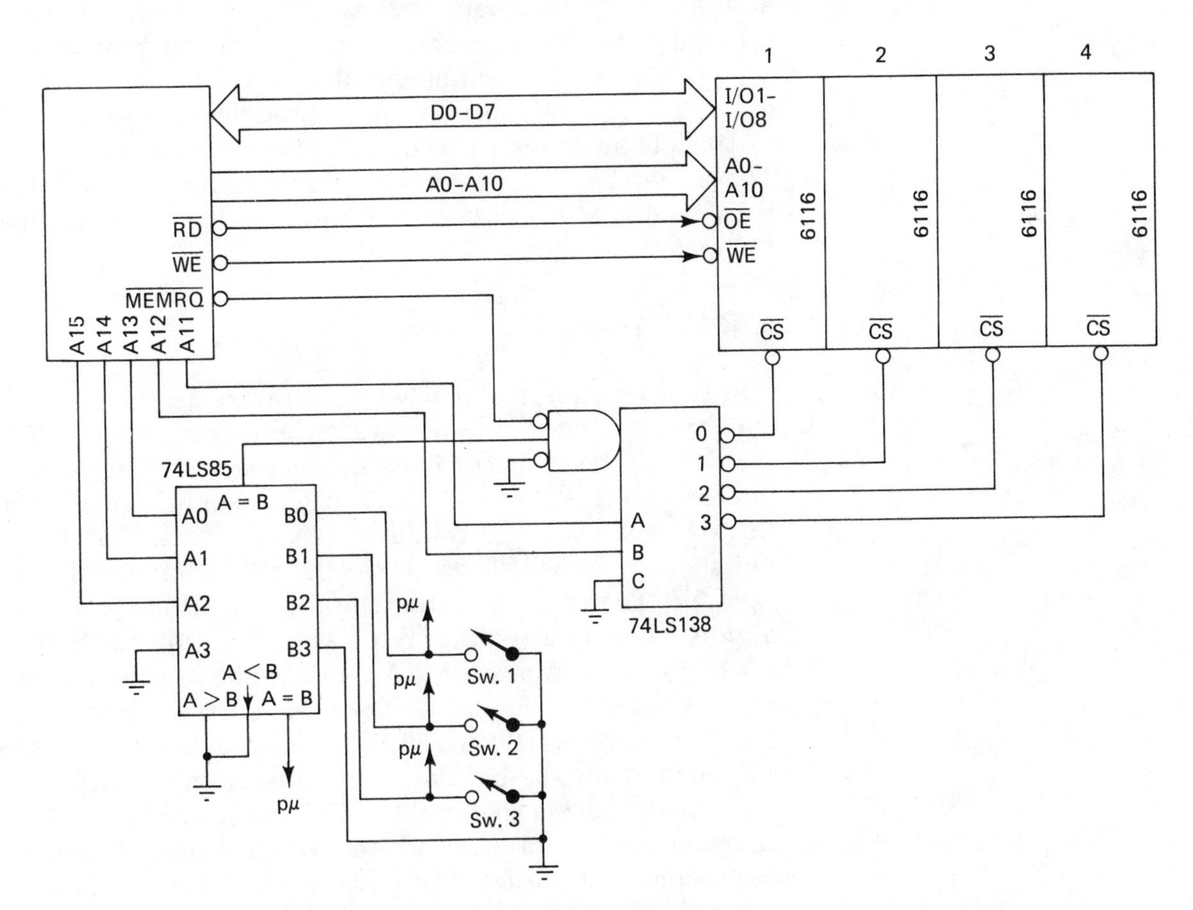

Figure 12.8 8K × 8 RAM system constructed from 6116's.

The most significant address bits (A13 through A15) are connected to the A inputs of the 74LS85 4-bit magnitude comparator. Because inputs A3 and B3 are both grounded, the 74LS85 functions as a 3-bit magnitude comparator.

Now that we have separately examined each device in Figure 12.8, we must consider how they interact to form an 8K memory system. Another important question concerns the actual address of each RAM. With 16 address bits, the microprocessor can address 64K of memory. Where in this 64K space will the 8K of 6116 RAM appeaer? To solve this problem, you must create a *memory map*. Memory maps are diagrams that illustrate the actual address of each memory IC within the 64K memory space.

To understand how the system functions we must work backwards from the chip selects. An individual RAM will perform a read or write function only when it is chip selected. The CS inputs are driven by the output of the decoder. The decoded outputs can become active only if the decoder is enabled. One of the decoder's active-low enable inputs is tied to ground. Another of the active-low enable inputs is driven by the $\overline{\text{MEMRQ}}$ line from the microprocessor. This means that the decoder will be enabled only when the microprocessor is requesting a memory access. This makes perfect sense. The RAMs should never be selected unless the microprocessor is requesting memory.

The active-high enable input is driven by the A = B output of the 74LS85. Address bits A13 through A15 must match the settings of switches 1 through 3 before the decoder is enabled. These switches will give us the ability to move our 8K block of RAM to different locations within the 64K addressing space of the microprocessor.

12.3.4 The Memory Map

Now that we understand how the hardware functions, a memory map must be created to illustrate the location of each RAM in the 64K memory space of the microprocessor. The easiest way to approach this is to make an initial assumption about the settings of switches 1 through 3. Let's assume that the three switches are closed. That makes the four B inputs all equal to logic 0's. The only time that the A = B output will go active high is when address lines A13, A14, and A15 are all at logic 0 levels. Refer to Figure 12.9. Assume that switches 1 through 3 are closed. What happens when the microprocessor requests memory at address 0000H? The A = B output of the comparator will go active high, and the memory request line will go active low. The decoder is now enabled. The select inputs are driven by addresses A11 and A12, which are currently logic 0's. This causes decoded output 0 to go active low. RAM 1 is now chip selected. It will be chip selected starting at memory location 0000H. The 6116 is a 2K RAM. 2K is equal to 800H. Therefore,

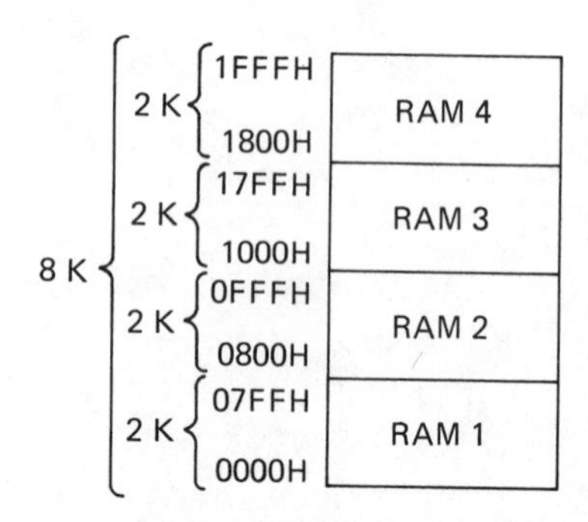

Figure 12.9 Memory map of 8K system in Figure 12.8.

RAM 1 will be chip selected for addresses 0000H through 07FFH, a total of 2K (800H) locations.

Address line A11 will go high at address 800H. This will cause decoded output 1 to go active low, chip-selecting RAM 2. RAM 2 will be selected from memory location 800H through 0FFFH: a range of 800H. At address 1000H, A11 will return low and A12 will go high. This will cause decoded output 2 to go active low, chip-selecting RAM 3. RAM 3 will be active from address 1000H through 17FFH: again a range of 800H. At address 1800H, both A11 and A12 are high. This will cause decoded output 3 to go active low, chip-selecting RAM 4. RAM 4 will be selected from address 1800H through 1FFFH, a range of 800H.

At address 2000H, A13 will go high. This will cause the A = B output to go inactive low. The 74LS138 will be disabled for addresses 2000H through FFFFH. Notice that although the data, addresses, and read and write signals are applied to the 6116's for the full range of 64K, they will not be visible except for the 8K space (0000H through 1FFFH) where they are chip-selected.

What happens if we open switch 1? Instead of being enabled at addresses 0000H through 1FFFH, the decoder will now be enabled from the 2000H through 3FFFH; The 8K of RAM will then appear to occupy that memory space. (Refer to Figure 12.10.)

Let's draw some general conclusions concerning the 16 address lines and their effect on the actual address of the 8K bank of RAM.

Address lines A0–A10 These address lines are applied to each memory device. They have no influence on the status of the decoder. 2^{11} = 2K: The function of these address lines is to choose the particular memory cell within the presently chip selected RAM.

Address lines A11 and A12 (2^2 = 4) These address lines select one of four RAMs to chip select. 2^{13} = 8K: This indicates that the 13 address lines (A0 through A12) can control 2000H of memory space. That is the size of our memory system.

Address lines A13–A15 (2^3 = 8) These three address lines select the base address of the 8K of memory. The base address refers to the first physical address where RAM 1 is chip selected. As Figure 12.10 illustrates, we have the choice of eight different base addresses in which to place our 8K of memory.

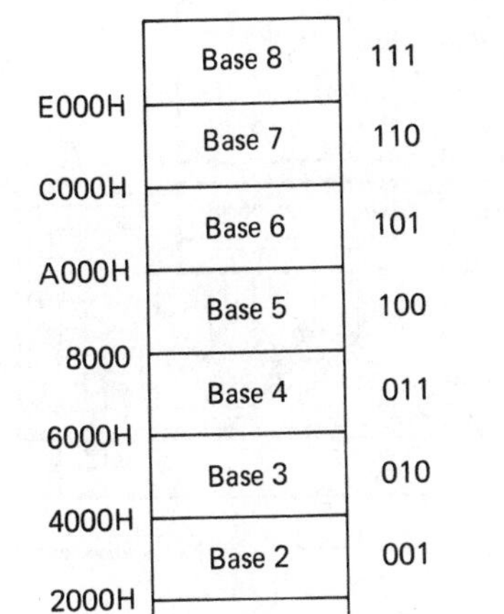

Figure 12.10 Memory map indicating base addresses.

These switches provide our system with an extremely important flexibility. Assume that you have a microprocessor system that already has 24K of memory and you would like to add another 8K of memory, for a total of 32K. The switches allow you to install the new memory board on any 8K boundary in your system. This is a great advantage over memory boards with fixed addresses.

12.4 OTHER STATIC RAMS

The 6116 is a "byte-oriented" memory IC. This means that at each unique memory location resides a group of eight memory cells. There are many RAMs that are not byte oriented. A 4-bit quantity (half a byte) is called a *nibble*. Many popular RAMs are nibble oriented, having four memory cells at each unique address. Other RAMs are bit oriented. The RAM in Figure 12.1 is an example of a bit-oriented memory. These memories have only one memory cell at each unique location. Let's examine a nibble-oriented memory device and discover how it must be connected to interface with a byte-oriented data bus.

12.4.1 The 2114 1K × 4 Static RAM

The 2114 has 1K unique memory locations. At each unique address, there are four memory cells. Because the 2114 is a static RAM, the memory cells are flip-flops (see Figure 12.11). A 1K × 4 RAM should have 10 address lines ($2^{10} = 1K$) and four data input/output lines. The block diagram in Figure 12.11 supports this conclusion. The block diagrams of the 6116 and 2114 are extremely similar, but the 2114 does not have an output enable pin.

Refer to the two gates at the bottom of the block diagram. When $\overline{CS}$ and $\overline{WE}$ are both active low, the input data buffers are enabled and a write operation occurs. When $\overline{CS}$ is active low and $\overline{WE}$ is inactive high, the output buffers are enabled. This designates a read operation. This is the type of read operation that the 6116 would use if the $\overline{OE}$ pin were hardwired to ground. When $\overline{CS}$ is inactive, the output of both control gates will be inactive logic 0's.

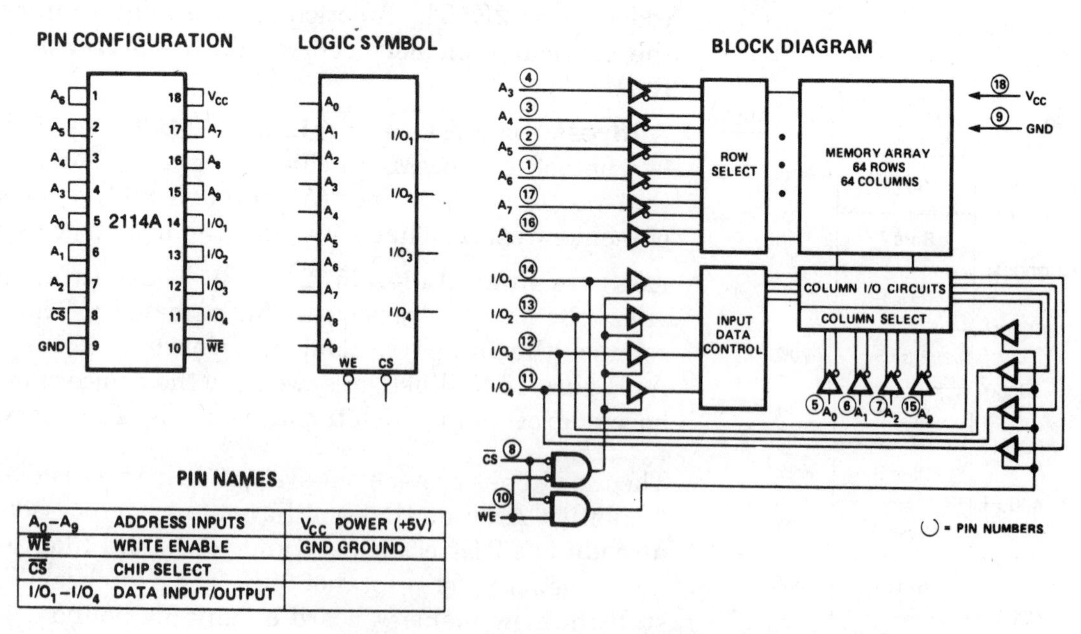

Figure 12.11 Block diagram, logic symbol, and pinout of 2114.

The read and write cycles are similar to those of the 6116. Other than different specifications, all static RAMs will have similar timing diagrams.

Our major concern is: How do we interface a nibble-oriented memory IC with a byte-oriented microprocessor data bus? The answer is simple: we must use two 2114's to create 1K bytes of storage. The first 2114 will be connected to the least significant nibble (D0 through D3) and the second 2114 will be connected to the most significant nibble (D4 through D7). We will want both memory ICs to be active at the same time. Therefore, the chip selects for each pair of 2114s must be connected together. See Figure 12.12, where the major part of the circuit is constructed from 2114 1K × 4 bit static RAMs. Together they form an 8K × 8 block of memory. Notice that we accomplished the same storage capacity using 6116 2K × 8 RAMs with a total of four instead of 16 memory ICs.

The top row of RAMs will handle the least significant nibble of the data byte. The second row of RAMs will handle the most significant nibble of the data bus. Address lines A0 through A9 and $\overline{\text{WE}}$ are applied to all 16 memory ICs. The 2114s in each column have their chip select inputs tied together. You can think of each pair of 2114s as one 1K × 8 memory IC.

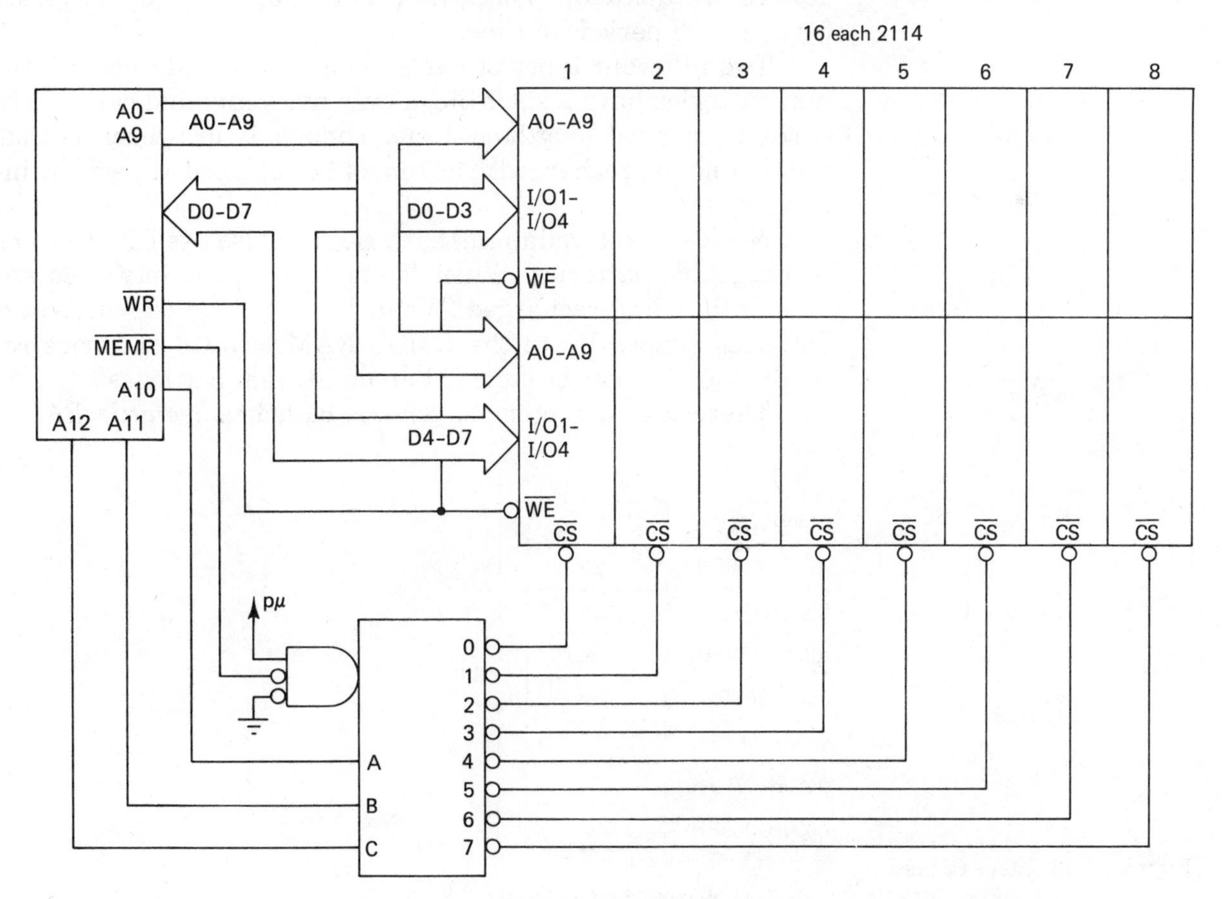

Figure 12.12 8K memory system constructed from 1K × 4 memory ICs.

The chip selects are driven by a 74LS138. Because each chip select represents 1K of memory (instead of 2K as in Figure 12.8), all eight decoded outputs must be used to access the 8K of memory. Address lines A10 through A12 are used to drive the select inputs of the decoder. We have chosen not to use address bits A13 through A15. In this application they can be considered as don't-care values. A magnitude comparator could easily be added to the circuit for greater addressing flexibility.

The decoder will be enabled whenever memory request is active. Refer to Figure 12.13. Because address bits A13, A14, and A15 are not used, this system is said to be "partially decoded." No expansion memory can be added to the system until these upper-order address bits are decoded in some manner.

12.4.2 Special-Purpose RAMs

There are many types of special-purpose RAMs in common usage. Many systems employ battery backed-up CMOS RAMs. When V_{cc} is turned off, the information in static RAM is lost. There are many applications when information vital to the system's operation must not be lost. In those situations CMOS RAMs and some types of battery backup is used. Because CMOS devices are ultra-low power, they can be operated by batteries for extended periods of time.

Two different types of batteries are commonly used. Lithium batteries have a shelf life of over five years and can supply current for great lengths of time. Their only drawback is that they cannot be recharged. They must be replaced at periodic intervals.

Ni-Cd (nickel–cadmium) batteries are also used. Ni-Cds are rechargeable batteries. When V_{cc} is applied, the batteries are constantly being recharged. When V_{cc} is turned off, the Ni-Cd batteries supply V_{cc} to the CMOS RAMs. Ni-Cd batteries require more support circuitry than do lithium batteries.

There are also memory devices called *nonvolatile RAMs.*

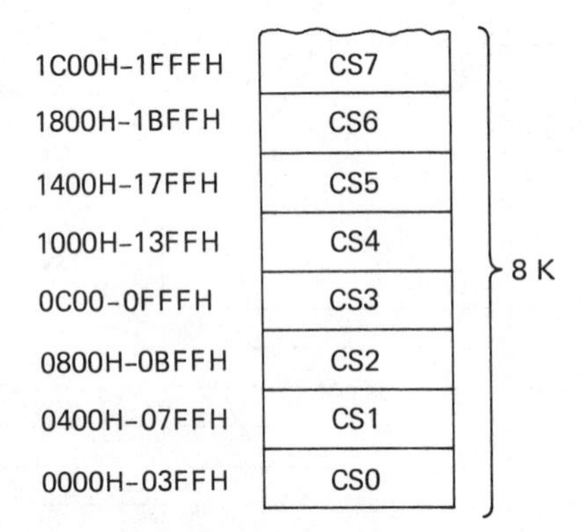

Figure 12.13 Memory map of Figure 12.12.

These CMOS RAMs have small batteries built into them. When the battery wears out, the whole package must be replaced.

12.5 DYNAMIC RAMS

Instead of using a flip-flop as the memory cell, dynamic RAMs use a MOS capacitor. Dynamic RAMs can pack much more storage capacity in a smaller space using less power than a standard static RAM. Dynamic RAMs are extremely popular.

12.5.1 A Dynamic RAM Cell

Figure 12.14 illustrates a possible way to implement a dynamic memory cell. Notice that the data-in and data-out lines are separate. The static RAMs that we examined before had common data input/output lines. The input buffer is enabled when the write input goes active low. If the input data is logic 1, the capacitor will charge to a positive voltage; if the input data is a logic 0, the capacitor will discharge to 0 V.

The sense amplifier is a voltage comparator with an extremely high input impedance. (A high input impedance will ensure that the sense amplifier does not load down the storage capacitor.) If the voltage on the storage capacitor is greater than the reference voltage, the output of the sense amplifier will be a logic 1; if the voltage on the capacitor is less than the reference voltage, the output of the sense amplifier will be a logic 0.

The output of the sense amplifier drives the output buffer and the refresh amplifier. Three things occur each time the read input is taken to its active level: (1) the output buffer is enabled, (2) the output of the sense amplifier passes onto the data output pin, and (3) the refresh amplifier rewrites the output of the sense amplifier onto the capacitor. All capacitors suffer from leakage

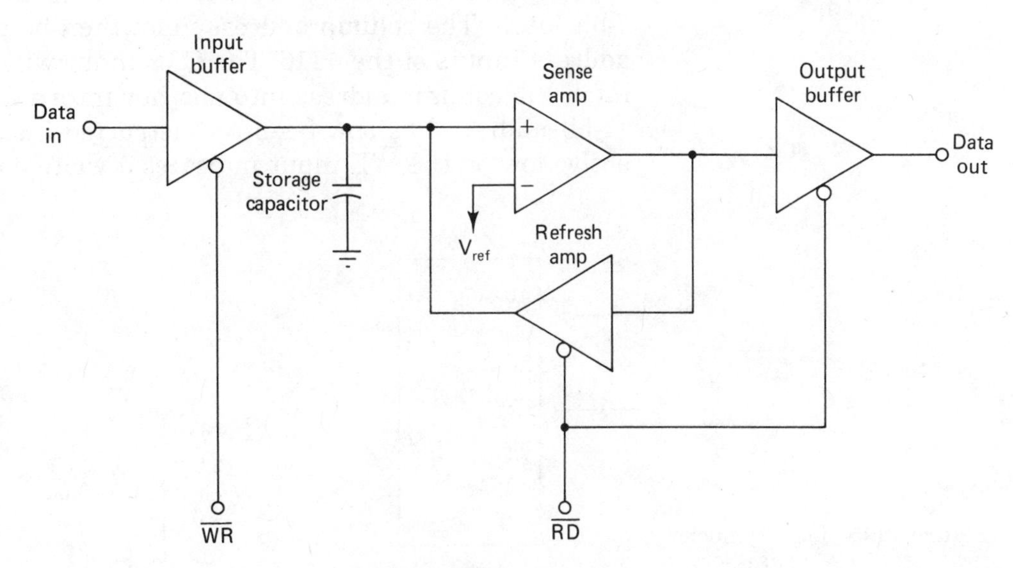

Figure 12.14 Dynamic memory cell.

current. After a few milliseconds, a logic 1 on the storage capacitor could leak down to a logic 0 level. By reading the memory cell, the voltage on the storage capacitor is refreshed.

12.5.2 The 4116 16K × 1 Dynamic RAM

Figure 12.15 depicts the logic symbol of a 16K dynamic RAM. Dynamic RAMs are bit-oriented memory ICs. At each unique address is only one memory cell. We have discovered that two 1K × 4 RAMs must be used to create 1K bytes of memory. In the same manner, eight 16K × 1 RAMs must be used to construct 16K bytes of memory. Each RAM in the eight-IC array will handle 1 bit of the data bus. Notice that the 4116 has separate data-in and data-out pins. In some applications they can be connected to form a common I/O line, as we saw in static RAMs. In other applications, an 8-bit bidirectional bus driver will be required to interface the 4116s with the bidirectional data bus.

2^{14} = 16K: We would expect to find 14 address lines on the 6116; instead, only seven are indicated. The 14-bit address must be multiplexed onto the seven address bits of the 4116. This will save seven pins, which will allow the 4116 to be manufactured in a standard 16-pin DIP package. You will remember that memory cells are constructed in arrays. Each memory cell is uniquely identified by a row and a column address.

Two control signals are used to multiplex the row-and-column addresses: $\overline{RAS}$ (row address strobe) and $\overline{CAS}$ (column address strobe). The row address is the least significant 7 bits of address (A0 through A6); the column address is the most significant 7 bits of address (A7 through A13). The row address will first be presented to the address inputs of the 4116. The $\overline{RAS}$ input will be taken low to latch the row address into an internal 7-bit latch. The column address must then be presented to the address inputs of the 4116. The $\overline{CAS}$ input will be taken low to latch the column address into another internal 7-bit latch. The 14-bit address has now been reconstructed inside the 4116. An active low on the $\overline{WE}$ input indicates a write operation.

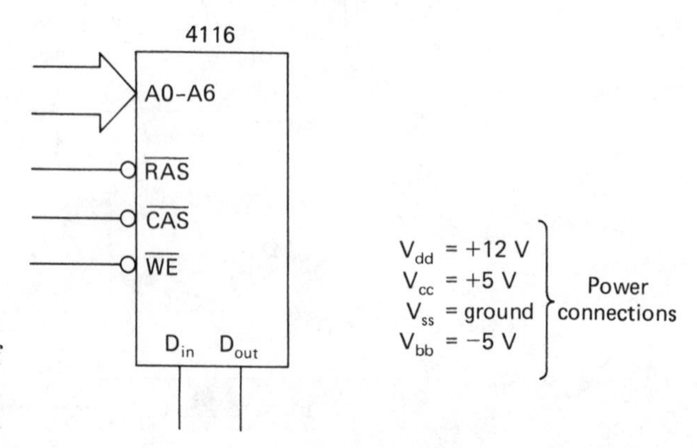

Figure 12.15 Logic symbol of the 4116 16K × 1 dynamic RAM.

The 4116 requires three voltages: +12, +5 and −5. At a higher price, improved versions of the 4116 (like the Intel 2118) are available that require only a single +5-V power supply.

12.5.3 A Typical Dynamic RAM Read Cycle

Figure 12.16 illustrates a typical read-cycle timing diagram. Assume that the $\overline{WE}$ input is held at a high level throughout the read cycle. The signal labeled MUX is the multiplexer select input. When MUX is low, the row address will appear on A0 through A6 of the 4116; when MUX is high, the column address will appear on A0 through A6. The MUX signal and multiplexing of the address will be generated by additional MSI devices. The multiplexing circuitry will be examined in a few pages. Let's examine each event in Figure 12.16.

Event 1 The MUX signal is low, so the row address is on A0 through A6 of the 4116. The falling edge of $\overline{RAS}$ will latch the row address into the 4116.

Event 2 The multiplex signal goes high. The column address will now appear on A0 through A7 of the 4116.

Event 3 The falling edge of $\overline{CAS}$ will latch the column address into the 4116. The 4116 now has all the information required to access the addressed memory location.

Event 4 The valid data is now on the data-out pin.

Events 5–7 The read operation will end by all control signals returning to their inactive levels.

Dynamic RAMs function under many critical timing constraints. A 4116 has approximately 15 timing specifications that must be met to accomplish a read or write cycle. At this point in your digital education, these complex timing parameters would probably cause more confusion than enlightenment.

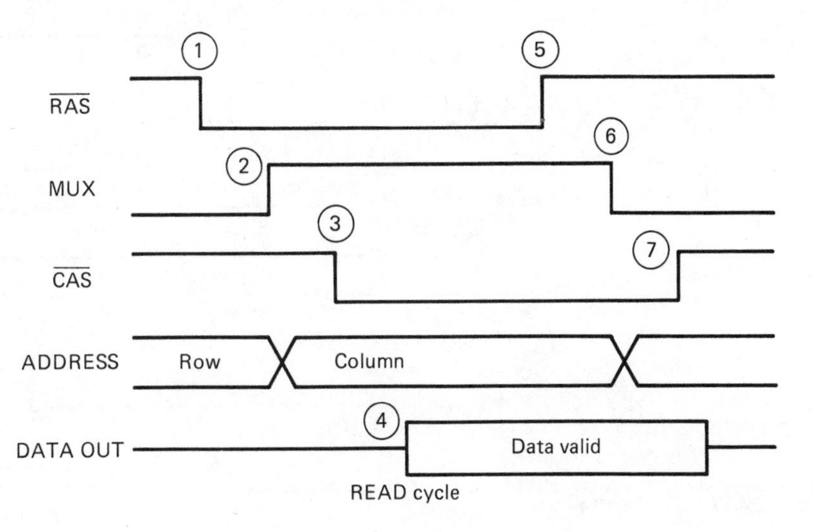

Figure 12.16 Read-cycle timing diagram.

12.5.4 A Typical Dynamic RAM Write Cycle

There are many ways to write information into a dynamic RAM. Figure 12.17 illustrates a *late write cycle.* A late write cycle uses the $\overline{\text{WR}}$ control signal from the microprocessor. An *early write cycle* uses the inverted $\overline{\text{RD}}$ signal to drive the $\overline{\text{WE}}$ pin on the dynamic RAM. The late write cycle is actually a read and write access. The output data of the addressed location will also be available on the data-out pin (although it is not normally used). To avoid indeterminate levels on the data bus, a microprocessor using a late write cycle must employ a bidirectional bus driver to isolate the data input and data output pins. The circuitry required for such a system will be illustrated in a few pages.

Let's examine each event in Figure 12.17.

Event 1 The row address is latched into the 4116 on the falling edge of $\overline{\text{RAS}}$.

Event 2 The multiplexer select goes high and the column address is steered onto the A0 through A6 inputs of the 4116.

Event 3 The falling edge of $\overline{\text{CAS}}$ latches the column address into the 4116.
The first three events of the read and write cycles are exactly the same.

Event 4 The data to be written into the memory must be given a short period to stabilize before the WE goes active.

Event 5 Although we are not going to use it, the data that happen to be residing in the addressed memory location (before the write occurs) will appear on the data-out pin.

Event 6 The write enable goes active. The data bit on the data-in pin will be written into the addressed memory location.

Event 7 The $\overline{\text{WE}}$ pin is returned to its inactive level.

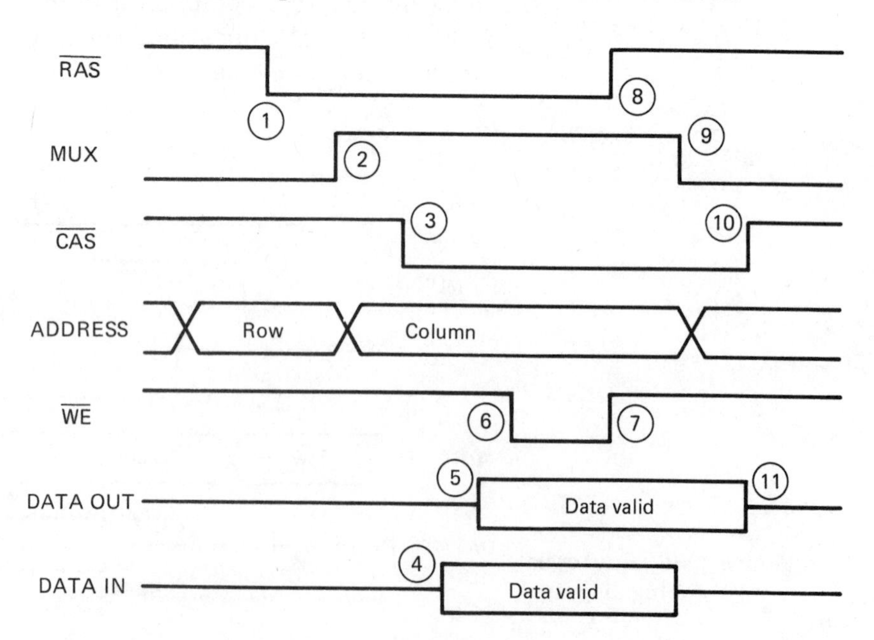

Figure 12.17 Late write cycle.

Events 8–11 The control signals and data in/out return to their inactive levels.

It is important to realize that the data in and data out are valid at the same time. Any system that uses the late write cycle must isolate the data-in and data-out pins.

12.5.5 Refresh Operation

Each time a dynamic RAM is read, all the memory cells in the addressed row are automatically refreshed. If we were sure that a memory location in each row would be read at least once every millisecond, a special refresh operation would not be required. A memory that is used to refresh a high-speed computer display is an example of a dynamic RAM system that does not require refresh.

Most systems have to employ some form of dynamic RAM refresh. The memory cells in a 4116 are arranged in 128 rows of 128 columns ($128 \times 128 = 16K$). There are many special ICs that are used to supply refresh operations for dynamic RAMs. At periodic intervals they execute the following steps:

1. The dynamic RAM controller must first indicate to the microprocessor that a refresh operation is in progress. During the refresh operation the microprocessor cannot access the memory.
2. The dynamic RAM controller will connect the outputs of a 7-bit counter onto the address bus. The counter will start at the count of 0. This would correspond to row 0 of the 4116. When the $\overline{RAS}$ line is pulsed low, the 128 memory cells in row 0 will be refreshed.
3. Step 2 is repeated for each row (0 through 127) in the 4116.
4. After all 128 rows have been refreshed, the dynamic RAM controller indicates to the microprocessor that it may once again access the RAM.

Figure 12.18 illustrates the simple refresh-only cycle. While the MUX line is held low, the row address is output from a 7-bit counter in a dynamic RAM controller. The $\overline{RAS}$ line is pulsed low. The entire 128 memory cells in the addressed row have been refreshed. Approximately 1 to 3% of the systems time is spent refreshing the dynamic RAM. This is a small price to pay for dense low-cost, low-power memories.

The Z80, manufactured by Zilog Corporation, is an ex-

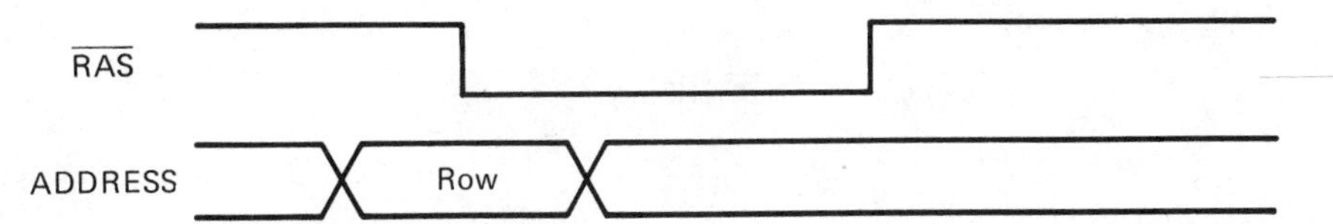

Figure 12.18 Refresh-only cycle.

tremely popular microprocessor. The capacity to refresh dynamic RAM is built into the Z80. When the microprocessor is involved with internal operations, it uses the few spare microseconds to refresh a row of dynamic RAM. The refresh process is transparent to the operation of the Z80, and no extra hardware is required.

12.5.6 A Multiplexing Scheme for Dynamic RAMs

We require a circuit that receives 14 address bits from the microprocessor and outputs the 7-bit row and column addresses at the correct moment, as indicated by the read and write cycle timing diagrams (Figure 12.19). This circuit is constructed around the 74LS157 quad 2-line-to-1-line multiplexer. Examine the function table for the 74LS157 in Figure 12.19. Each 74LS157 contains four 2-line-to-1-line multiplexers with common enable

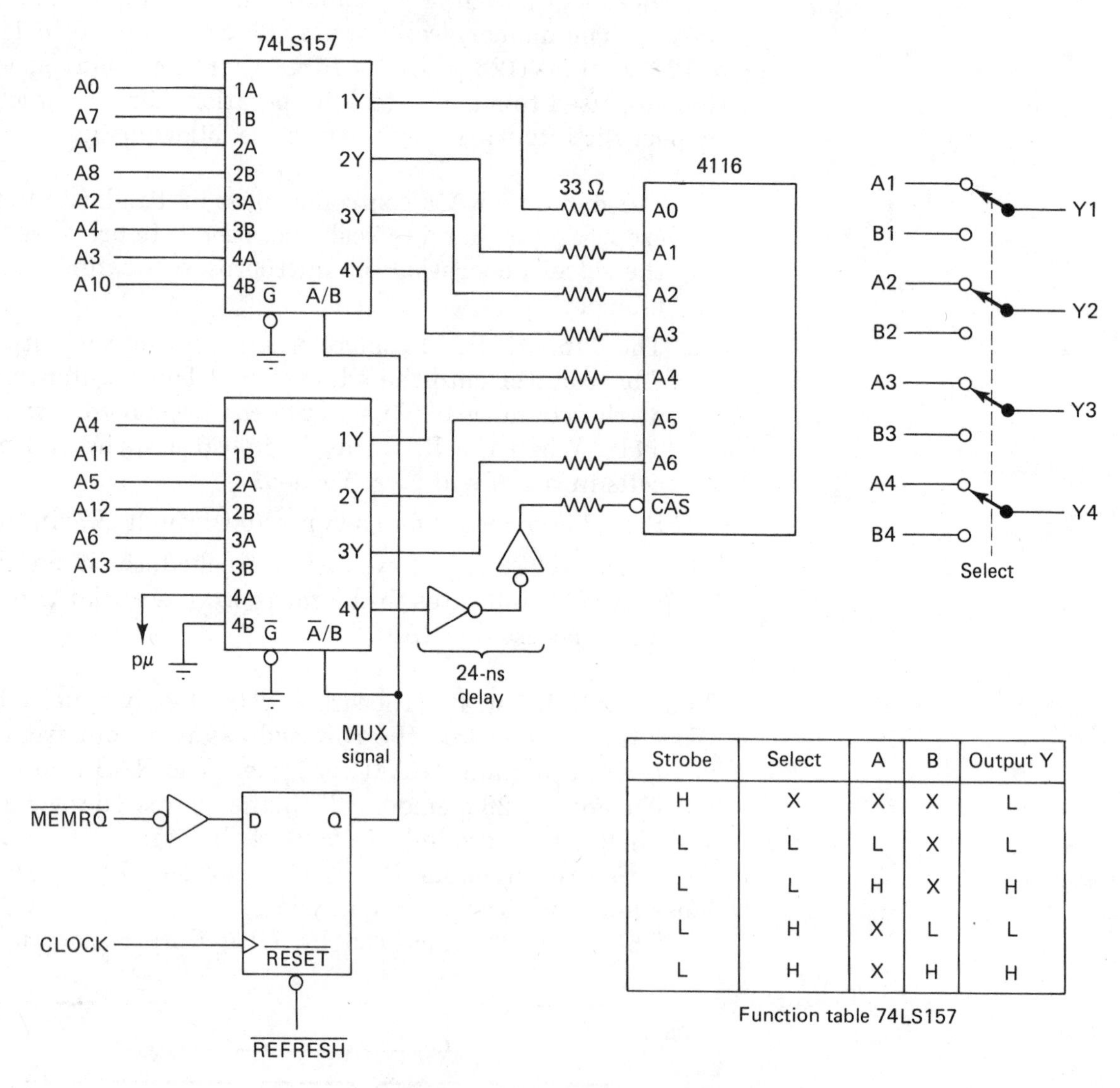

Strobe	Select	A	B	Output Y
H	X	X	X	L
L	L	L	X	L
L	L	H	X	H
L	H	X	L	L
L	H	X	H	H

Function table 74LS157

Figure 12.19 Row/column multiplexing circuit for the 4116 dynamic RAM.

and select. The data inputs are divided into two groups of four: A1 through A4 and B1 through B4. When the select input is low, the A inputs are passed onto the Y outputs; when the select input is high, the B inputs are passed onto the Y outputs. This is indicated on the schematic by labeling the select input as $\overline{\text{A}}$/B. The 74LS157 operates as the digital equivalent of a quad SPDT switch. The function table describes the operation of each multiplexer.

Line 1 If the strobe input (G) is at an inactive-high level, the select and data inputs become don't cares and the Y output is forced to a logic 0. The strobe is held at an active level for lines 2 through 5.

Lines 2 and 3 The select input is low, indicating that the data on the A input will be passed onto the Y output. When the select is low, the level on the B input is in a don't-care condition.

Lines 3 and 4 The select input is high, indicating that the data on the B input will be passed onto the Y output. When the select is high, the level on the A input is in a don't-care condition.

It will require two 74LS157s to handle the 14 address lines and the $\overline{\text{CAS}}$ signal. The row address (A0 through A6) will be connected to the A inputs and the column address (A7 through A13) will be connected to the B inputs. The next question is: How is the MUX signal (which drives the select inputs of the 74LS157s) generated?

Remember that the MUX signal stays low until after $\overline{\text{RAS}}$ has a falling edge. The MUX signal will then go high and pass the column address (A7 through A13) onto the address inputs of the 4116s. After a short delay the CAS input will be taken low.

Every microprocessor has a clock input. The microprocessor uses this clock input as a timing reference. A memory read or write cycle may take three microprocessor clock pulses. Consider the D flip-flop shown in Figure 12.19. The D input is driven by the inverted memory request signal. Normally, the $\overline{\text{MEMRQ}}$ signal is high; this will place a low on the D input of the flip-flop. Therefore, the MUX signal is normally low. That fact agrees with the read and write cycle timing diagram. The microprocessor will initiate a memory access on a rising edge of the clock by placing the required address onto the bus and taking $\overline{\text{MEMRQ}}$ low. For the first clock cycle the MUX signal will be low. On the next rising edge of the clock, the inverted $\overline{\text{MEMRQ}}$ signal will cause the MUX signal to go high. This will place the column address onto A0 through A6 of the 4116.

The CAS signal is generated from the second 74LS157. Input 4A is pulled up to V_{cc} and input 4B is tied to ground. When the MUX signal is low, the $\overline{\text{CAS}}$ signal (4Y) is high. This agrees with our timing diagrams. When the MUX signal goes high, output 4Y goes low. But the timing diagram indicates that there must be a short delay before the $\overline{\text{CAS}}$ signal should have a falling edge. This delay is provided by the two inverters. This 24-ns delay gives the column address a chance to settle down before it is latched into the 4116.

When the memory-read or memory-write operation is completed, the microprocessor will return the $\overline{\text{MEMRQ}}$ signal to its inactive-high level. On the next rising edge of the clock, the MUX signal will return to a logic 0 level. It appears that this circuit will satisfy all the demands of the read and write timing diagrams. Notice that an active-low signal called "refresh" will asynchronously reset the MUX line. This will provide us with the proper "refresh only" timing parameters. We will soon see how $\overline{\text{RAS}}$ is generated.

The 33-Ω resistors are commonly used to help impedance match the outputs of the multiplexers with the inputs of the dynamic RAM. This will ensure that the address lines settle down to valid levels in a minimal length of time.

12.5.7 A 16K DRAM System

The term *DRAM* is commonly used in place of "dynamic RAM." We now need to create a 16K byte memory system using eight 4116s and a bidirectional bus driver. Figure 12.20 illustrates a 16K × 8 DRAM memory system. Each 4116 handles 1 bit of the 8-bit data byte. A0 through A7, $\overline{\text{RAS}}$, $\overline{\text{CAS}}$, and $\overline{\text{WE}}$ are common to all eight 4116s in the 16K bank. Remember that during the late write operation, both data-in and data-out pins are active. A method of interfacing the data-in and data-out pins with the microprocessor's bidirectional data bus is required.

The 8216 4-bit bidirectional bus driver was designed specifically for that purpose. Refer to the block diagram of the 8216 in Figure 12.20. On the right side of the block diagram, the input and output of each pair of drivers/receivers are connected. This side of the 8216 is interfaced with the bidirectional data bus. The left side of the 8216 is used to interface with memory ICs that have independent data-in and data-out pins. The pins labeled Out0 through Out3 will be connected to the data output pins of four 4116s. The pins labeled In0 through In3 are connected to the data input pins of four 4116s. Two 4116s are required to interface to an 8-bit data bus.

Consider the control gates at the bottom of the block diagram. If $\overline{\text{CS}}$ (chip select) is at an inactive-high level, all outputs of the 8216 will be at high-*z*. The control pin labeled "$\overline{\text{DIEN}}$" is the data input enable control. When $\overline{\text{CS}}$ and $\overline{\text{DIEN}}$ are both low, the logic levels on the data-out pins of the 4116s are routed onto the data bus; this would happen during a memory-read operation. When $\overline{\text{CS}}$ is low and $\overline{\text{DIEN}}$ is high, the logic levels on the data bus are placed onto the data input pins of the 4116s; this would occur during a memory-write operation.

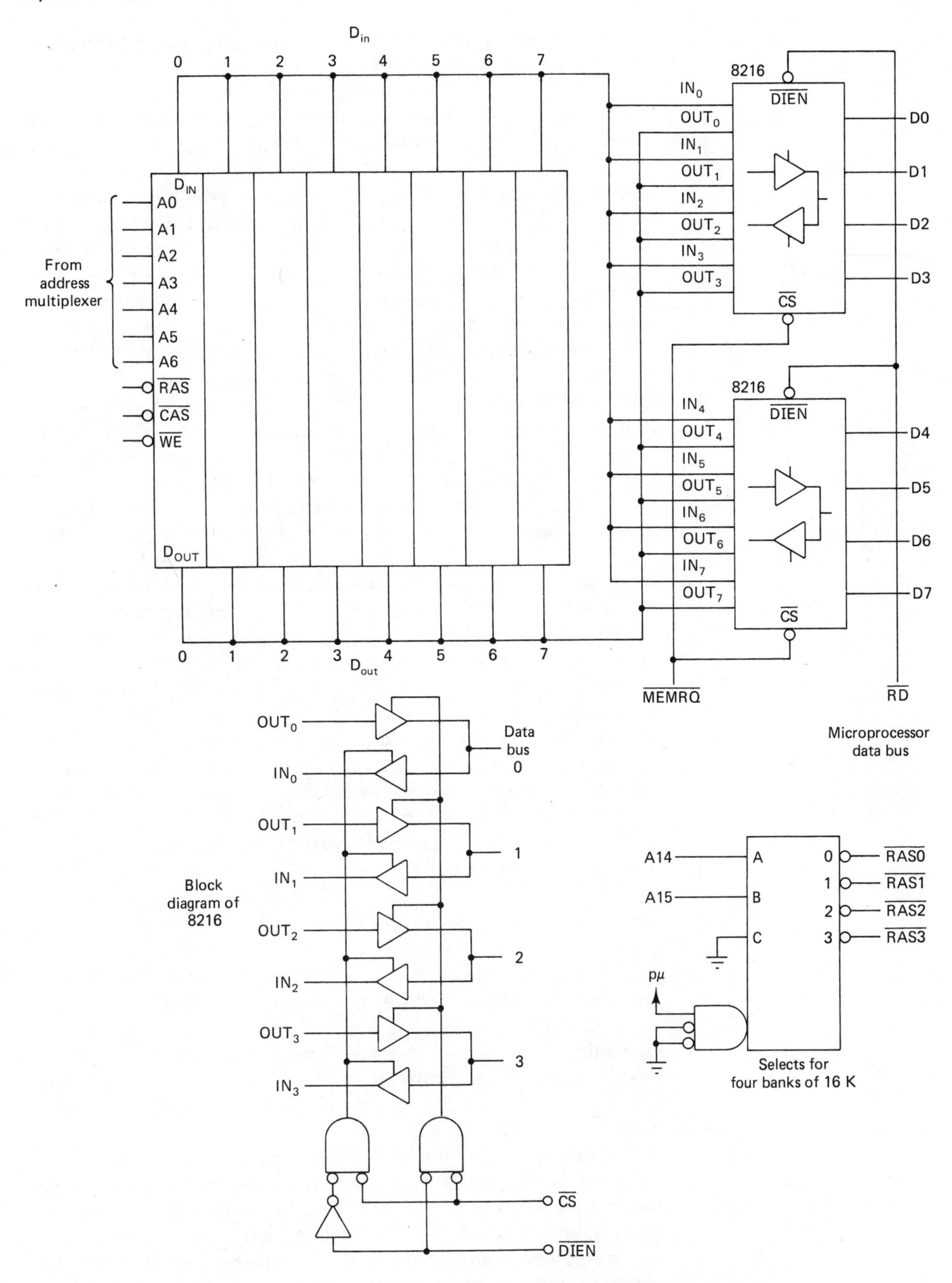

Figure 12.20 16K DRAM memory system.

Once again, in an attempt to avoid confusion, all the data-in and data-out lines of the 4116s are not illustrated. Notice that the D_{in} lines are drawn together onto line line. This is also true for the D_{out} lines. This does not mean that the eight lines are connected electrically. It is just a convenient way of drawing the D_{in} and D_{out} buses of the 4116s without complicating the schematic with 16 separate lines. Each D_{in} of the 4116s is connected to an input pin on the 8216; each D_{out} pin of the 4116s is connected to an output pin on the 8216. In this manner the D_{in} and D_{out} pins of the 4116s are isolated; there will be no conflict when D_{in} is being driven by the microprocessor and the 4116 is placing data onto the D_{out} pin, as occurs during the late write operation.

The $\overline{CS}$ inputs of the two 8216s are driven by $\overline{MEMRQ}$; unless the microprocessor is requesting memory, the buffers of the 8216 will be held at high-*z* outputs. The data-in enable ($\overline{DIEN}$) is driven by the $\overline{RD}$ output of the microprocessor. When the microprocessor is performing a read operation, the 8216s will pass the data-out values of the 4116s onto the data bus.

Now that we have created one bank of 16K bytes of memory, how can we expand the capacity to all 64K of the microprocessor's address space? We have seen that static RAM ICs are chip selected by an active-low output of a decoder. DRAMs do not have $\overline{CS}$ inputs, but $\overline{RAS}$ provides almost the same function. To construct 64K bytes of memory using four 16K banks of 4116s, A0 through A6 (from the address multiplexer), $\overline{CAS}$, and $\overline{WE}$ will be connected to each 4116. We will use the 74LS138 illustrated in Figure 12.20 to generate four $\overline{RAS}$ signals: $\overline{RAS0}$ through $\overline{RAS3}$. $\overline{RAS0}$ will be applied to the eight 4116s in bank 0; $\overline{RAS1}$ will be connected to the eight 4116s in bank 1, with similar connections for banks 2 and 3. The $\overline{RAS}$ actually functions as a modified chip select. If a 4116 does not receive the falling edge of a $\overline{RAS}$ signal, it will ignore all activity on the A0 through A6, $\overline{CAS}$, and $\overline{WE}$ inputs.

12.5.8 Address and Data Buffers

The DRAM memory system in Figure 12.20 used the 8216 to interface the separate data inputs and outputs of the DRAMs with the bidirectional data bus. Address and data bus buffers are employed for many reasons. They are designed to drive the many inputs that occur in bit-oriented DRAM memory systems. This prevents the microprocessor's address and data bus from being excessively loaded.

Another important reason for using these buffers is to isolate banks of memory from the main buses in the microprocessor system. This aids the technician in troubleshooting memories. If banks of RAM are not buffered, one short can drag down the whole bus. Finding this problem would be extremely difficult because of the many inputs sharing each trace on the bus. Data and address buffers allow the technician easily to isolate the memory problem to a specific bank of RAM. Address bus buffers are normal drivers or three-state drivers, depending on the application. Data bus buffers will always be bidirectional, allowing both memory-read and memory-write operations.

12.5.9 Advanced DRAMs

DRAMs are now available in 64K × 1 and 256K × 1 capacities. The 4164 (64K DRAM) is pin-for-pin compatible with the 4116. One address line is added to the 4164. Because the addresses are multiplexed, the addition of one address line increases the addressing capacity by a factor of 4, instead of a factor of 2 as would occur in a nonmultiplexed RAM. An important improvement is that it only requires a +5-V V_{cc}. This creates an interesting problem; to keep compatibility with the 4116, V_{cc} is pin 8 and ground is pin 15. This is backwards from normal digital ICs. Beware of this strange standard pinout when you are troubleshooting memory systems.

4164s have displaced the 4116 in all new designs. Soon the 41256 will replace the 4164s in new designs. The newer-generation microprocessors have address buses up to 24 bits wide. This would enable the microprocessor to address directly 16MB (megabytes) of memory. The demand for larger memory systems will be answered by even-higher-density RAMs.

12.6 THE READ-ONLY MEMORY

All the memory devices that we encountered have been read/write devices. Another type of memory IC that is extremely important is the ROM (read-only memory). ROMs are nonvolatile memory ICs that store permanent digital information. This information may be programs or other forms of data. How is the information initially stored in a ROM? The process of writing information into a ROM is called *programming* the ROM. Certain ROMs are programmed by the manufacturer; others can be programmed by the user.

12.6.1 Applications of ROMs

ROMs are used for many applications. Every microcomputer must have some basic program stored in ROM. When the power is first turned on, the microprocessor will access this "startup or boot" ROM. The boot ROM will contain simple programs that enable the computer system to read further instructions from a mass-memory device (typically a floppy disk) into standard RAM. Some inexpensive microcomputers actually contain all the required system's programs entirely in ROM. Video games are an example of such a ROM-based system. The game cartridge that is inserted into the video game is nothing more than a small circuit board containing a game program in ROM.

Another application of ROMs is in creating custom decoders. Consider the 74LS47. A BCD-to-seven-segment decoder can be implemented as a ROM. The BCD input could be considered as a 4-bit address, and the seven-segment output as the data outputs. Each of the 16 memory locations would contain a group of seven read-only memory cells. ROMs are often created to provide unique or custom decoder functions.

ROMs are also used as look-up tables. Instead of having a microprocessor calculate the square root of a number, it could look it up at a certain memory location. The microprocessor would use the look-up ROM in the same way that we use standard tables in textbooks. Having the microprocessor look up the value of a particular function is much faster than calculating it with a program.

Character generators are ROM-based devices that are used to create the alphanumerics and graphic symbols on CRT displays. The ROM holds the information concerning which dots in a matrix must be illuminated to create a particular character or graphic symbol.

12.6.2 A Typical Pinout of a ROM

The pinout of a ROM resembles the pinout of a static RAM, minus the $\overline{WE}$ input. Refer to Figure 12.21. The number of address lines will indicate how many unique memory locations the ROM contains. ROMs are available with capacities of up to 256K bits. The eight data outputs indicate that the ROM is a byte-oriented memory device. Notice that this particular ROM has more than one chip select. It is common for ROMs, and even some RAMs, to have upward of four chip selects, both active low and active high. The chip selects are internally ANDed together. All chip selects must be at their active levels before the device is internally chip selected.

The read-cycle timing diagram of a ROM is almost identical to the read-cycle timing diagram of a static RAM. Each ROM will have an associated read cycle time that describes its speed.

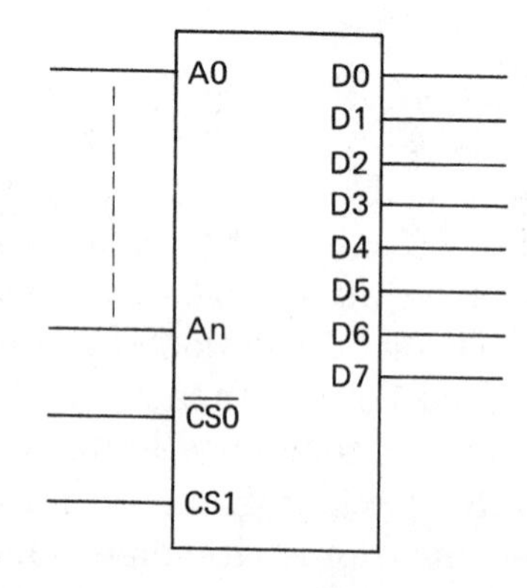

Figure 12.21 Logic symbol of a typical ROM.

12.6.3 The Mask-Programmed ROM

Mask-programed ROMs must be ordered from the manufacturer. A design engineer will create a truth table that describes the required data output for each address. The manufacturer will use this truth table to create the ROM. This is an extremely expensive process. ROMs are used only for super-high-volume products that have been thoroughly debugged.

Technicians must consider these ROMs as black boxes that output a byte of information for each address. ROMs are usually placed in DIP sockets. If a ROM is suspect, it is swapped with a known-good ROM.

12.6.4 The PROM (Programable ROM)

For many applications, the initial cost of manufacturing mask-programmed ROMs is cost prohibitive. An alternative is the PROM. The advantage of PROMs is that they can be programmed by the user. PROMs contain memory cells that have fusible links. When a PROM is programmed, selected fuses are blown out, creating the required memory pattern. Because of this, the process of programming PROMs is often called "burning" PROMs.

PROMs are programmed with a device called a PROM *programmer.* The memory pattern that will be burned into the PROM is downloaded from a computer or read into the PROM programmer by a known-good PROM. A new PROM is inserted into a socket on the PROM programmer. A pushbutton will initiate the programming process. A typical PROM takes 60 to 90 seconds to program.

Each PROM has a *signature.* The signature is a unique four- to six-place hexadecimal number. All PROMs that have been programmed with the same information will have the same signature. If a PROM is suspect, it is placed into the PROM programmer and its signature is checked. If the PROM programmer displays the wrong signature, the PROM is bad and should be discarded.

After the PROM is programmed, a new part number must be attached to it. Because a blank PROM can be programmed an almost infinite number of different ways, a new part number must be associated with each unique programmed PROM.

Blank PROMs are available in many different capacities and speeds. Most PROM programmers have the ability to program all the popular types of blank PROMs.

Although PROMs are more expensive than their ROM counterparts, they have important applications in small-scale production or in areas where the design is continually changing. The bit pattern programmed into the PROM can be easily modified to create new revisions.

12.6.5 The EPROM (Erasable PROM)

PROMs can be programmed only once. If the design of the circuit is modified, all the previously burned PROMs must be thrown away and new PROMs programmed. In an R&D environment, design changes are everyday occurrences. The use of PROMs in such an environment would be extremely expensive. The EPROM is a type of ROM that can be erased and reprogrammed hundreds of times. This is the perfect solution in an environment where design changes happen on a monthly or even weekly basis.

In the center of an EPROM is a clear, round window. When

exposed to UV (ultraviolet) light, the previously programmed information will be erased. To be erased, EPROMs are placed into devices called *PROM erases,* which expose the PROM to concentrated UV. A typical EPROM takes 30 minutes to be erased.

After the EPROM is erased it can be reprogrammed. Standard PROM programers will also program EPROMs. After the EPROM is programmed, it must also have a unique part number attached to it. This part number is written on a label that is placed over the clear window of the EPROM. If the window is not covered, stray UV, from fluorescent lights or sunlight, can erase the contents of the PROM. Direct sunlight will erase the contents of a PROM within a couple of weeks; fluorescent light would take a few years to erase the EPROM.

EPROMs, like their PROM counterparts, have a signature associated with them. If an EPROM is suspect, its signature must be verified with the PROM programmer.

12.6.6 The 2700 Family of EPROMs

The 2700 family of EPROMs are designed to be upward compatible. This means that as memory requirements increase, the users can easily upgrade the system with the next-larger-capacity memory of the 2700 family.

Part number	*Capacity*
2708	1K × 8
2716	2K × 8
2732	4K × 8
2764	8K × 8
27128	16K × 8

Notice that the part number indicates the total number of memory locations. Divide this number by 8 to calculate the number of kilobytes that the device can store. For example, 2716 ⇒ 16/8 = 2K bytes.

Refer to the pinout of the 2716 in Figure 12.22. The 11 address inputs (A0 through A10) prove that the 2716 has 2K unique locations. There are also eight output data lines (O0 through O7). V_{cc} and ground indicate that this device is powered from a standard +5-V power supply.

The pin labeled V_{pp} is the programming voltage pin. When +25 V is applied to this pin, the 2716 can be programmed. This voltage will be required only during the programming process; it will be furnished by the PROM programmer.

There are two control signals: $\overline{OE}$ and $\overline{CE}$. Output enable functions just like the $\overline{OE}$ pins on static RAMs. Chip enable ($\overline{CE}$) is equivalent to chip select. Both $\overline{OE}$ and $\overline{CE}$ must be at logic 0 levels during the 2716 read cycle. The timing diagram describing the 2716 read cycle is similar to the static RAM read-

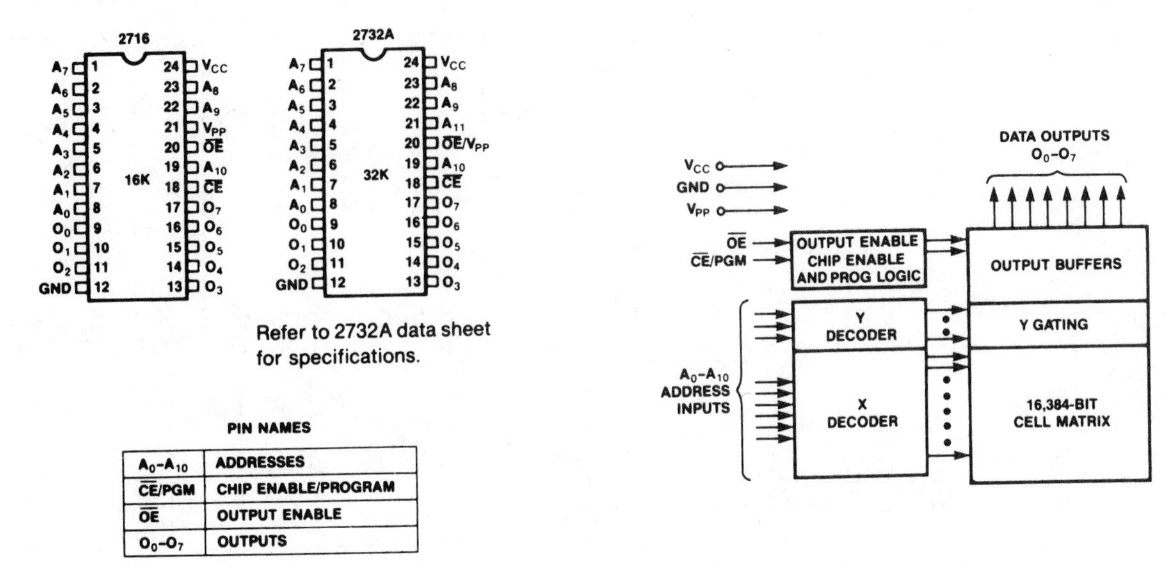

Figure 12.22 Pinout and block diagram of 2716.

cycle timing diagrams. Both $\overline{OE}$ and $\overline{CE}$ are furnished to simplify the expansion of ROM systems. All the output enables are connected to the $\overline{RD}$ output of the microprocessor. The $\overline{CE}$s are generated using the upper address bits and a decoder.

Notice that the 2716 is pin-for-pin compatible with the 2732. The only difference is that the V_{pp} pin of the 2716 will become the extra address line required for the 2732. The $\overline{OE}$ on the 2732 will double as the V_{pp} pin.

The 2716 is also pin-for-pin compatible with the 6116 2K × 8 static RAM. Notice the similarity between the two part numbers: 6116 and 2716. This enables a manufacturer to design a memory system that can be either ROM, RAM, or a combination of both.

12.6.7 A 16K ROM/RAM Memory System

Figure 12.23 illustrates a 16K system that can be populated with either 6116 RAMs or 2716 EPROMs. Notice the two-position jumper labeled J1. If the system is populated with RAM, J1 should be in the A-B position. This will allow the $\overline{WR}$ signal from the microprocessor to reach the WE (pin 21) of the 6116. If the system is stuffed with 2716s, J1 should be in the A-C position. Pin 21 in the 2716 is the V_{pp} pin and should be pulled up to +5 V when the device is not being programmed. To add even more flexibility to the system, eight separate jumpers could be used on pin 21 of each memory device. This would let us mix RAMs and ROMs in the same 16K bank of memory.

Address lines A14 and A15 drive the enable inputs of the 3-line-to-8-line decoder. The memory map for this system will be prepared as an exercise at the end of this chapter.

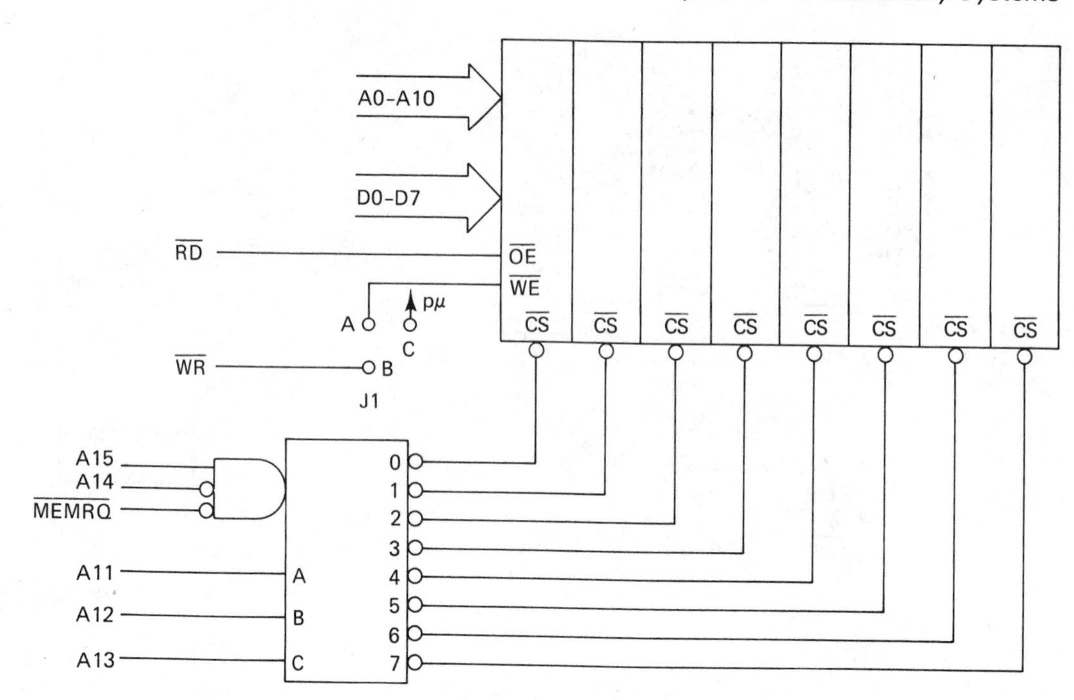

Figure 12.23 16K RAM/ROM system.

12.6.8 The E^2PROM (Electrically Erasable PROM)

The E^2PROM has characteristics of both ROMs and RAMs. It is nonvolatile like a ROM, but it can also be written into like a RAM. The write cycle of the E^2PROM is extremely long. This device is an alternative to CMOS battery-backed-up RAMs; it is not intended to be used as a general-purpose read/write memory device. The E^2PROM is gaining much greater popularity as the technology improves.

12.7 PROGRAMMABLE LOGIC DEVICES

As IC manufacturing technology improves, LSI devices can accomplish more complex functions. We have stated that SSI/MSI devices are the "glue" that bonds the LSI devices. There are many times when a design engineer wants to reduce the IC count in a new circuit design. This will result in space savings, lower manufacturing costs, and improved reliability.

Custom ICs can be manufactured that provide the equivalent of many SSI/MSI devices in a single IC. But custom devices suffer from the same cost and inflexibility faults as those of mask-programmed ROMs. A programmable combinational/sequential equivalent of the PROM is needed.

PALs (programmable array logic) and IFLs (integrated fuse logic) fulfill these needs. Refer to Figure 12.24. The simple PROM illustrated is an 8 × 1 memory device. Think back to Chapter 5, where we introduced the sum-of-products method of creating combinational circuits. The PROM architecture is a standard sum of products with programmable fuses on the OR input lines. The 3-bit address creates eight different combinations. Each

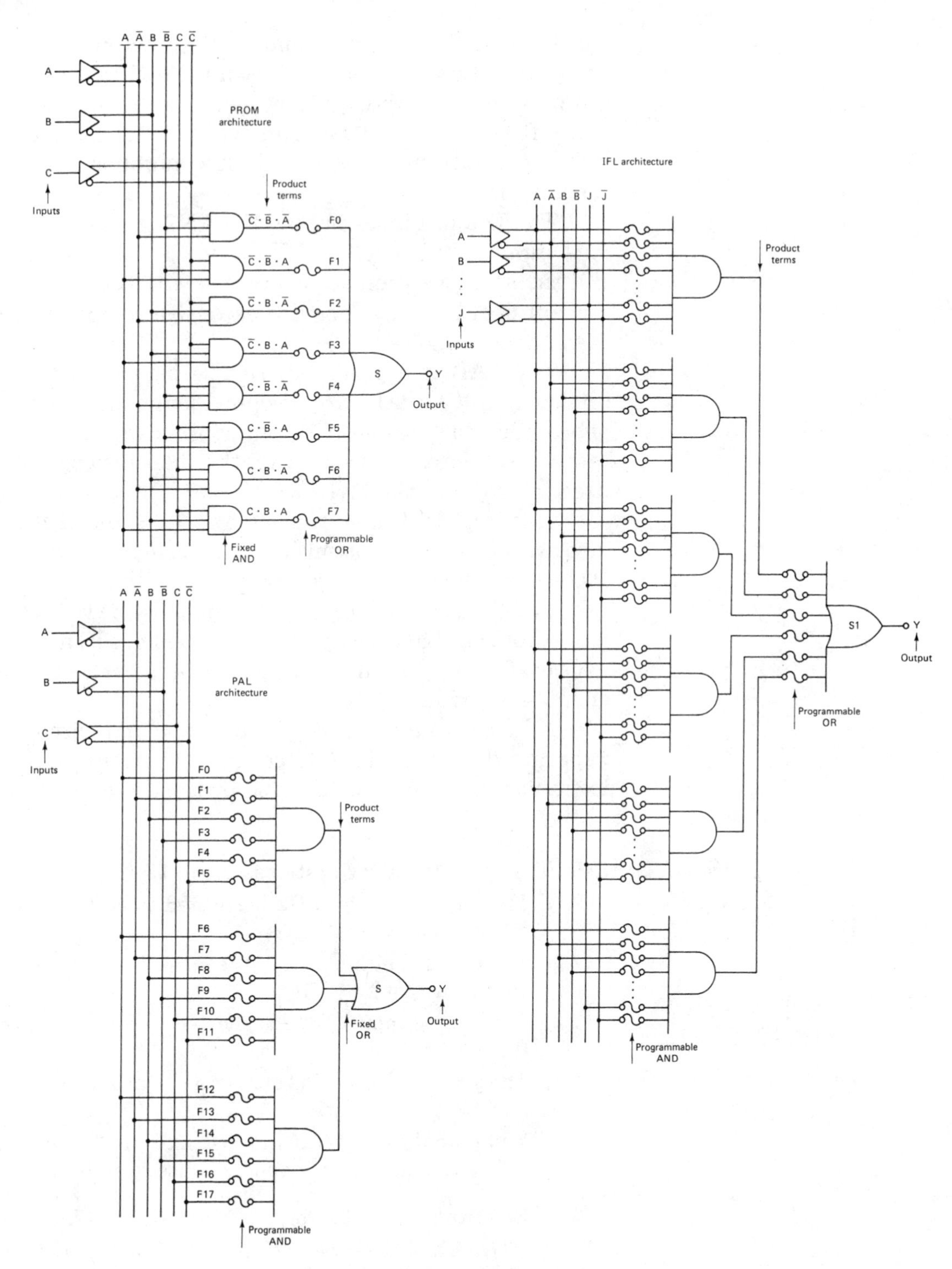

Figure 12.24 PROM, PAL, and IFL architecture.

product of these combinations is applied to the input of the OR gate, via fusible links. A blown fuse will result in a nondynamic level being placed on that OR input. If a fuse is blown during the programming process, the input combination that creates that product will output a logic 0. If a fuse is left intact, the combination that creates that product will output a logic 1.

The PAL architecture is also the standard sum of products.

But instead of fusing the inputs to the OR gates, the inputs to the AND gates are fused. A blown fuse will reduce the number of inputs on the AND gate by one. The PAL has six times as many fuses as the PROM. This will give us the increased flexibility to create an almost unlimited number of 3-bit combinational functions.

The IFL has fuses in both the AND gate inputs and the OR gate input. Obviously, the IFL device is the most sophisticated. But the sophistication does not come without a price. The development of IFL devices require extensive computer-aided design (CAD) programs.

The PAL is a good compromise between SSI/MSI devices and custom ICs. PALs are designed using standard Boolean algebra and truth tables. Computer programs are used to translate the Boolean equations or truth tables into bit maps that are used to program the PALs.

A typical PAL may replace 5 to 10 SSI/MSI ICs. PALs are also available with sequential devices, such as D flip-flops. These PALs are called *logic sequencers.*

The PAL can be programmed on most new PROM programmers. You will treat any programmable logic IC as a semicustom device; If a programmable logic device is suspect, its signature should be verified.

A PAL will appear as a "black box" in a circuit schematic. Separate documentation describing the circuitry that the PAL is emulating should be provided by the design engineer.

QUESTIONS AND PROBLEMS

Refer to Figure 12.1 for Questions 12.1 to 12.4.

12.1. The enable input on G2 is shorted to ground. How does this affect circuit operation?

12.2. The decoded output 2 of U1 is open. What symptoms will the circuit display?

12.3. The enable input of U1 is shorted to the data line. How does this affect circuit operation?

12.4. The ground pin of U6 is bent underneath the IC. How does this affect circuit operation?

12.5. What would happen if the WR line in Figure 12.2 went low before the data were given a chance to stabilize?

Refer to Figure 12.8 for Questions 12.6 to 12.10.

12.6. If SW 1 and SW 3 were closed and SW 2 open, for what range of addresses would RAM 3 be chip-selected?

12.7. If all the switches are open, for what range of addresses would RAM 4 be chip-selected?

12.8. Decoded chip selects for RAMs 1 and 2 are shorted together. How does this affect circuit operation?

12.9. Input C of the decoder is floating. How does this affect circuit operation?

12.10. The A > B input on the comparator is floating. How does this affect circuit operation?

Refer to Figure 12.12 for Questions 12.11 to 12.14.

12.11. Add circuitry to Figure 12.12 so that the 8K bank of memory starts at A000H.

12.12. RAM pair 5 fails to respond. What is the most likely problem?

12.13. Decoded output 7 is shorted to ground with a solder bridge. How does this affect circuit operation?

12.14. The WR line is shorted to MEMRQ. How might this affect circuit operation?

Refer to Figure 12.19 for Questions 12.15 to 12.17.

12.15. How would the circuit be affected if the delay provided by the inverters were only 10 ns?

12.16. The Y outputs of the top 74LS157 appear to be stuck at logic 0 levels. What problem could create this symptom?

12.17. The reset input of the D flip-flop is shorted to ground. How does this affect circuit operation?

Refer to Figure 12.20 for Questions 12.18 to 12.20.

12.18. The RD line is shorted to ground. How does this affect circuit operation?

12.19. MEMRQ is floating. What device would appear to be malfunctioning?

12.20. Why is the 8216 required in the circuit?

12.21. How do the RAS of a DRAM and the CS of a static RAM compare?

12.22. A ROM has 14 address lines. What is its storage capacity in bytes?

12.23. Memory-map the system in Figure 12.23.

13

INTRODUCTION TO DATA CONVERSION

Digital circuits are often used to measure, process, and control analog quantitites. Consider a microprocessor-based system that is used to control a home heating/cooling system. A temperature-sensing scheme is needed. Because temperature is an analog (continuous) quantity, the sensed temperature must be converted to an equivalent digital value. This digital value can then be processed by the microprocessor system. The furnace and air conditioner is controlled with an analog voltage. Some means must be provided to convert the digital output of the microprocessor system into an analog control voltage. This process of converting analog quantities to their digital equivalents is called *analog-to-digital conversion.* The process of converting a digital number into its analog equivalent is called *digital-to-analog conversion.*

An entire book could easily be devoted to the topic of data conversion techniques; this chapter will provide only an overview. Other than in the section on Schmitt triggers, no real-world ICs will be examined. All circuits will be modeled with familiar components. This will enable you to acquire a solid understanding of data conversion techniques without getting bogged down with specific devices.

13.1 THE SCHMITT TRIGGER

The Schmitt trigger is a digital device that converts slowly rising or falling signals, or noisy signals, into clean, precise digital waveforms. Think back to the pulse meter circuit in Chapter 10. We needed a 60-Hz reference for the input to the divide-by-360 counter chain. The power-line frequency in the United States is a fairly accurate 60-Hz. Many digital clocks use the 60-Hz line frequency as a reference. The half-wave rectifier in Figure 13.1 produces a 4-V pulse with a frequency of 60-Hz. (Resistor R_S is a series current-limiting resistor that protects the diode.) Is the output of the rectifier a digital or an analog signal?

It has characteristics of both digital and analog signals. The two predominate levels could pertain to a logic 0 and a logic 1, but it changes between these two levels very slowly. Actual digital signals closely approach infinitely fast rise and fall times.

The sine function is continuous. This means that the voltage of a sine wave is continually changing. A half-wave-rectified sine wave will have a stable logic 0 level, but the logic 1 level will have long rise and fall times. If this signal was used to drive the input of a digital device, the slowly rising and falling edges would cause the digital device to be biased in its linear region of operation. This would cause the output of the digital circuit to oscillate spuriously.

We need a device that will transform this quasi-digital signal into a true digital signal with fast rise and fall times. That is the function of the Schmitt trigger. Schmitt triggers use a phenomenon called *hysteresis*. In the study of physics, hysteresis describes the lagging of magnetization after the magnetic field has been removed. This concept can be implemented in electronic circuitry to form the Schmitt trigger (Figure 13.2). Consider each of the specifications illustrated in Figure 13.2.

Positive-going threshold voltage, V_T+ (Figure 13.2a) As the input voltage into a Schmitt trigger increases, it will reach a point where the output toggles from a logic 0 to a logic 1. This voltage is called the positive-going threshold voltage. Typically, for TTL Schmitt triggers it is equal to 1.7 V.

Negative-going threshold voltage, V_T- (Figure 13.2b) The input voltage is now decreasing from above the positive-going threshold voltage. As the voltage decreases under 1.7 V the output does not change to a logic 0. The input voltage must go below

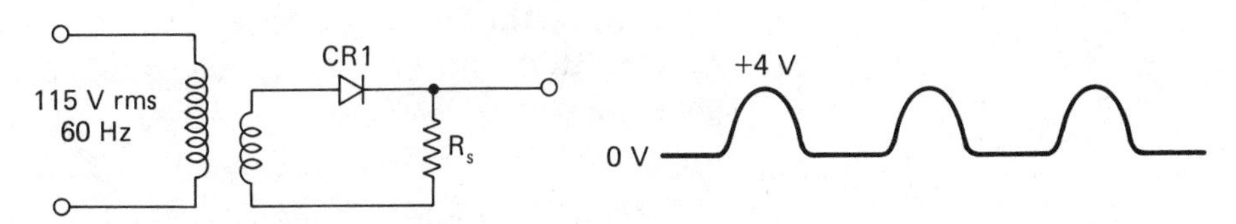

Figure 13.1 Half-wave rectifier with step-down transformer.

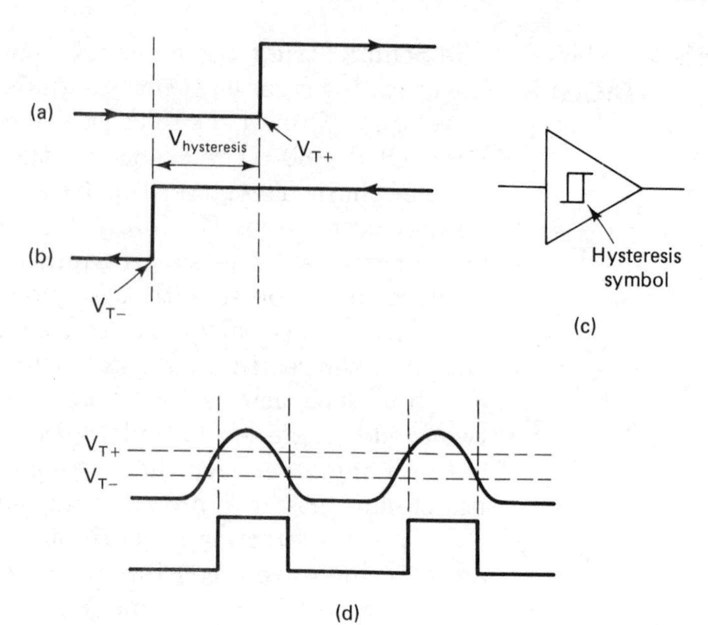

Figure 13.2 Upper, lower, and hysteresis voltages of a Schmitt trigger.

the negative-going voltage threshold before the output toggles to logic 0. For TTL Schmitt triggers this voltage is typically 0.9 V. The difference between the negative- and positive-going voltage thresholds is called the *hysteresis voltage*. Consider the output of a Schmitt trigger circuit just as the input voltage reaches 1.7 V—the output snaps to a logic 1 level. If noise on the input causes the voltage to dip as low as 0.91 V, the output will still stay at logic 1. The hysteresis voltage indicates the noise immunity of the Schmitt trigger.

Schmitt trigger noninverting buffer (Figure 13.2c) Notice the symbol placed in the center of the buffer. This symbol denotes a Schmitt trigger device. The hysteresis symbol is a composite of the upper and lower trigger point graphs. This is how a normal buffer and a Schmitt trigger buffer are distinguished. The 74LS14 is a hex Schmitt trigger that is pin-for-pin compatible with a 74LS04. The 74LS132 is a quad Schmitt trigger NAND gate that is pin-for-pin compatible with the 74LS00. Any logic function can be constructed with Schmitt trigger input circuitry.

Schmidt trigger buffer (Figure 13.2d) The buffer exhibits a half-wave-rectified sine-wave input and digital output. Notice how the output snaps to logic 1 when the input voltage becomes greater than the V_T+ and then snaps to a logic 0 when the input goes below the V_T-.

Schmitt triggers are used to receive digital signals that may have been exposed to noise or that may have degraded rise and fall times because of long transmission lines.

13.2 DATA ACQUISITION AND CONTROL

A Schmitt trigger is used to "clean up" waveforms that are already digital in nature. Physical parameters such as temperature, acceleration, pressure, position, and flow are analog quantities. Figure 13.3 illustrates a typical data acquisition and control system. Devices 1 through 5 are the data acquisition components, and devices 6 through 8 are the components of the control system. Let's investigate each device in Figure 13.3.

1. Transducer: A transducer is formally defined as a device that converts energy from one form to another. An audio speaker is a common example of a transducer; it converts electrical current into sound waves. The transducers used in a data acquisition system convert a physical parameter into an electrical voltage or current. Typical transducers convert temperature, pressure, flow, acceleration, or position into a voltage or current. Good-quality transducers are often the most expensive element in a data acquisition system.

2. Amplifier: Consider a biofeedback instrument that senses alpha and theta brain waves. The brain waves are minute currents that can be sensed on the surface of the forehead. These currents must be amplified before they can be further processed.

A high-gain op amp is commonly used as a voltage or current amplifier. This amplifier must have an extremely high input impedance, which will ensure that the output of the transducer will not be excessively loaded. Excessively loaded transducers will produce inaccurate outputs.

3. Low pass filter: This device is used to filter high-frequency noise from the output of the amplifier. This filter is typically realized with an op-amp and passive components. Active filters

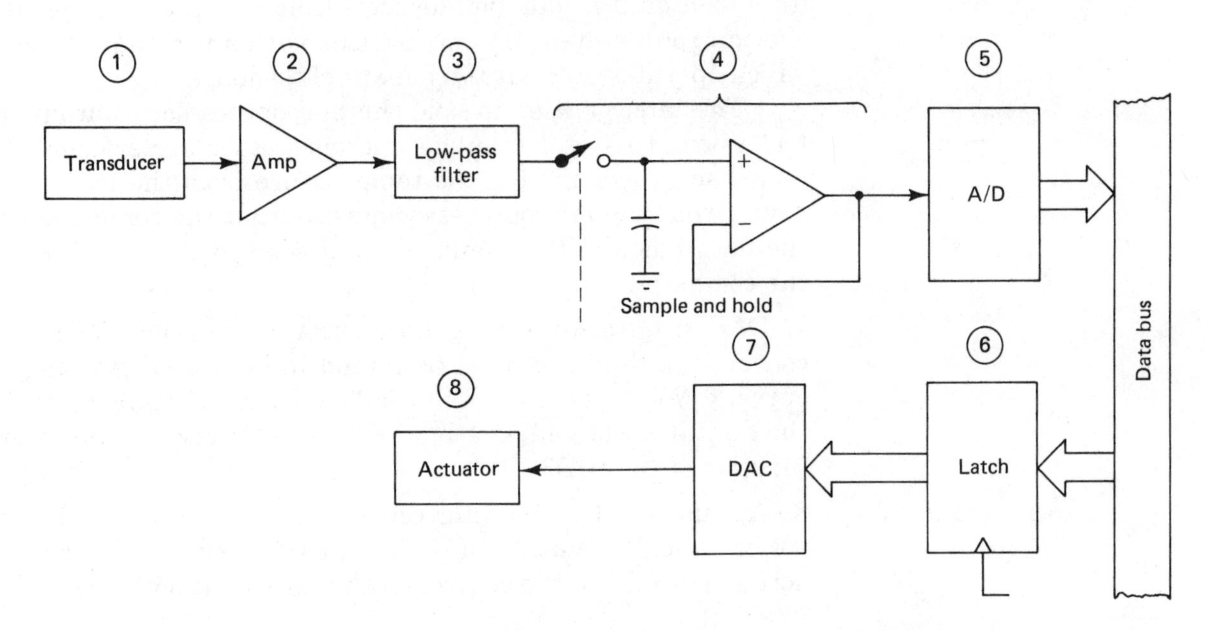

Figure 13.3 Data acquisition and control system.

have sharp response curves and do not attenuate the signal-as do filters constructed exclusively from passive components.

4. Sample-and-hold circuit: This block is constructed from a CMOS analog switch, capacitor, and op amp. When the microprocessor is ready to sample the output voltage from the transducer, it will issue a narrow pulse that closes the CMOS switch for a short period. The capacitor will quickly charge up to the output voltage of the low-pass filter. The op amp is used to buffer the voltage on the capacitor. Like the transducer, the voltage on the capacitor can be easily loaded down. The capacitor functions as an analog memory device. It holds the voltage sample while the microprocessor is performing a conversion. This will ensure that the voltage does not change during the analog-to-digital conversion.

5. A/D (analog-to-digital converter): This is the actual device that converts the sampled voltage into a digital quantity. The digital quantity is then output onto the microprocessor's data bus. A/Ds have a command input that is used by the microprocessor to start a conversion, and a status output that signals to the microprocessor when the conversion is complete.

6. Latch: When the computer has processed the input data, it may be ready to issue a control command to the analog system. Consider a home heating and cooling system that is controlled by a microprocessor. It will continually sample the temperature until an out-of-range error occurs. The microprocessor then must turn on either the air conditioner or the furnace to bring the temperature back into the acceptable range. A system such as this is called a *closed-loop system.* "Closed loop" indicates that the microprocessor continually is given feedback information (from the data acquisition devices) that indicates the result of the control commands. Compare this with an *open-loop system,* which operates without the benefit of feedback.

The latch is used to hold the microprocessor's output control word. This will free the microprocessor to perform other tasks, such as sampling the temperature from the transducer. Notice that the microprocessor must output the command onto the data bus and then send a rising edge to the latch to store the command.

7. DAC (digital-to-analog converter): The DAC is used to convert the digital control command into an analog voltage or current. We have now closed the loop from the analog world to the digital world and back again! The DAC provides the inverse function of the A/D.

8. Actuator: The actuator converts a current or voltage into a mechanical output. In the heating/cooling system analogy, the actuator would be the control on the air conditioner or furnace.

Many of the devices illustrated in Figure 13.3 are optional.

A weather monitoring system that measures temperature and rainfall will use only the data acquisition circuitry; output control circuitry would not be required. A robot arm that is moved by the output of the control circuitry may not require the data acquisition circuitry. If the microprocessor knows the starting position of the robot's arm, it can calculate the resultant position after each command. That would be an example of an open-loop system.

If the output of the transducer is a clean, slowly varying voltage, the low-pass filter and sample-and-hold circuits may not be required. One A/D converter may be shared by many transducers. An analog multiplexer (such as the 4051B) is used to steer the selected transducer output onto the analog input of the A/D converter.

DACs and A/Ds require an external precision reference voltage. Special-purpose ICs are designed to supply temperature and load stable reference voltages. The A/Ds reference voltage will define the largest analog input voltage that can be converted to an accurate digital value. The reference voltage on a DAC defines the maximum output of the device when the digital inputs are all logic 1's.

Resolution refers to the number of digital inputs on a DAC or the number of digital outputs of an A/D. It is also expressed in parts of the full-scale output: an 8-bit DAC is said to have a resolution of 1 part in 256.

DACs and A/Ds are described by many complicated specifications. Those specifications will not be discussed in this text. The aim of this chapter is to familiarize the reader with the fundamental concepts of data conversion.

13.3 THE DAC

Digital-to-analog conversion is not a complex process. A binary word will be converted into an equivalent analog voltage. The concept of "equivalence" between the digital and analog worlds is interesting. The equivalent analog output is equal to the reference voltage multiplied by the ratio of the input binary word to the full-scale binary word. For this reason, DACs that use external reference voltages are called *multiplying DACs.*

13.3.1 Basic Operational Concepts of DACs

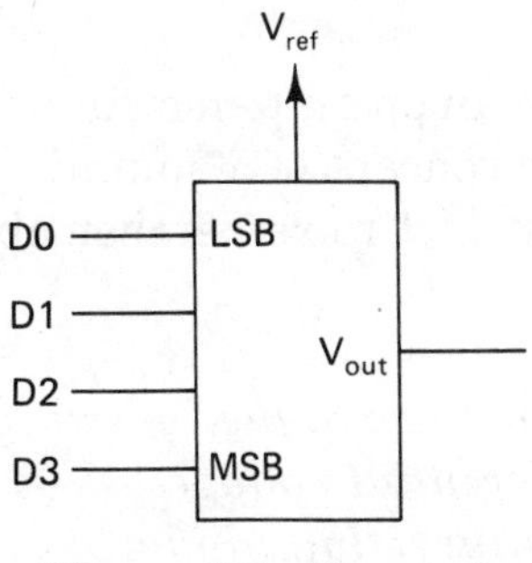

Figure 13.4 Pinout of a 4-bit DAC.

The inputs to a DAC will be a weighted digital word. Depending on the DAC, this word will be 4 to 16 bits wide. Let's examine the pinout of a 4-bit DAC (Figure 13.4). D0 through D3 are the four data bits that will be converted to an analog voltage. The output will be the analog voltage that is equivalent to the data inputs. This range of the analog voltage will be set by the pin labeled V_{ref}. Assume that the reference voltage is 7.5 V. When the digital input is equal to 0000, the output voltage will be equal to 0 V; when the digital input is equal to 1111, the output voltage will be equal to the reference voltage of 7.5 V. A digital input between the minimum (0000) and the maximum (1111) will be equal to a proportional part of the reference voltage.

Digital input	*Analog output (V)*
0000	0.0
0001	0.5
0010	1.0
0011	1.5
0100	2.0
0101	2.5
0110	3.0
0111	3.5
1000	4.0
1001	4.5
1010	5.0
1011	5.5
1100	6.0
1101	6.5
1110	7.0
1111	7.5

The resolution of a DAC is defined as the smallest incremental change of the analog output. A 4-bit DAC has 16 possible input states. The first input state of 0000 will output 0 V. That leaves 15 input states to be divided between the reference voltage of 7.5 V. The resolution of this 4-bit DAC with a reference voltage of 7.5 V is 0.5 V. Notice that the analog output voltage is proportional to the weighted binary input.

Refer to Figure 13.5. The output of a 4-bit binary counter is driving the digital inputs of the DAC. An oscilloscope connected to the analog output will display a stair-step waveform. When the count begins at 0000 the analog output voltage will be 0 V. Each time the counter is incremented, the analog voltage will increase by 0.5 V. This process will continue until the count of 1111, when the reference voltage of 7.5 V is reached. The next rising edge of the clock will reset the counter to 0000 and the analog output voltage will follow by returning to 0.0 V.

13.3.2 A Summing Op-Amp DAC

A simple DAC can be constructed using an op amp in a summing configuration. Let's quickly review the concept of a summing op amp (Figure 13.6). The most important fact to remember about the operation of op amps is:

The output will feedback sufficient voltage to the inverting input to assure that the differential voltage between the inverting and noninverting inputs is equal to 0 V. An op amp operating as a linear device must employ negative feedback.

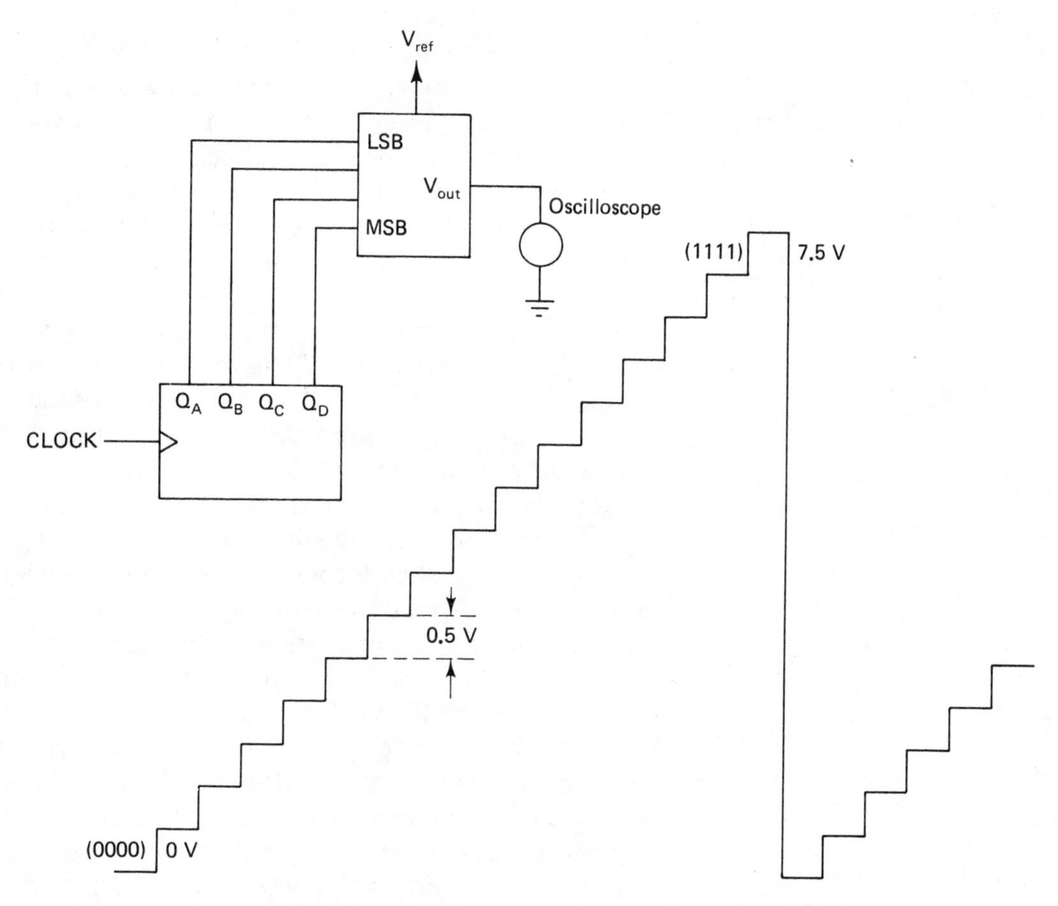

Figure 13.5 4-bit binary counter driving the inputs of the DAC.

Because the noninverting input is tied to ground, the output of the op amp will source just enough current through the feedback resistor to hold the inverting input at ground. The inverting input is said to be at *virtual ground.* That means the common node of the input resistors appears to be tied to ground.

The input currents will be summed at the virtual ground at the inverting input of the op amp. Because one end of each resistor appears to be tied to ground, the individual currents will equal the input voltage divided by the input resistance. The

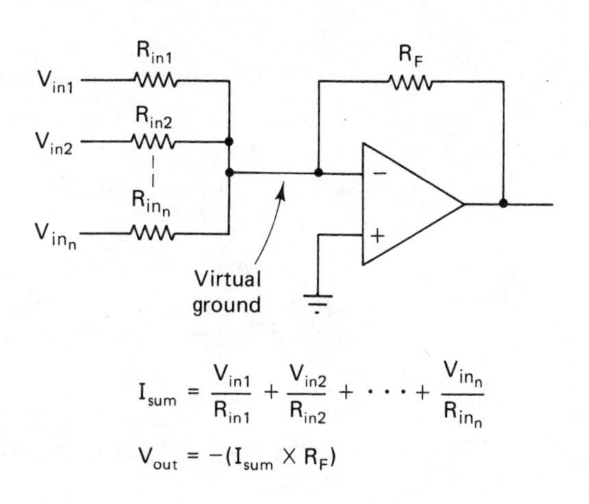

$$I_{sum} = \frac{V_{in1}}{R_{in1}} + \frac{V_{in2}}{R_{in2}} + \cdots + \frac{V_{in_n}}{R_{in_n}}$$

$$V_{out} = -(I_{sum} \times R_F)$$

Figure 13.6 Review of summing-op-amp configuration.

summed current will then be passed through the feedback resistor, which provides the negative feedback path and functions as a current-to-voltage converter. The output voltage is equal to the voltage drop across the feedback resistor: $I_{sum} \times R_f$. Because this op amp is configured as an inverting amplifier, the polarity of the output voltage will be negative. The summing op amp is useful in applications where many signals must be mixed.

We can use the summing op amp with binary weighted resistor values to construct a simple DAC. Consider a 4-bit binary number. The place value of the least significant bit of a binary number is equal to 1. The place value doubles for each successive bit. In a summing op-amp DAC, the least significant bit should provide the least amount of current, and each successive bit should provide twice the current of the previous bit. Remember that the current is being summed, not the voltage. Because the input current of each bit will be inversely proportional to resistance, the input resistance of the least significant bit should be the greatest; the input resistance of each successive bit should be half of the preceding bit. This will double the current for each successive bit. Refer to Figure 13.7. The reference voltage of 7.5 V and an input resistor will create a constant-current source. Switches 1 through 4 are CMOS transmission gates (digitally controlled analog switches; refer to Chapter 7). Each input bit will control one constant-current source. A logic 1 will close the CMOS switch and the constant-current source will feed the summing node; a logic 0 will open the CMOS switch. Data bit 0 controls the least amount of current; each subsequent data bit controls twice as much current as did the preceding data bit.

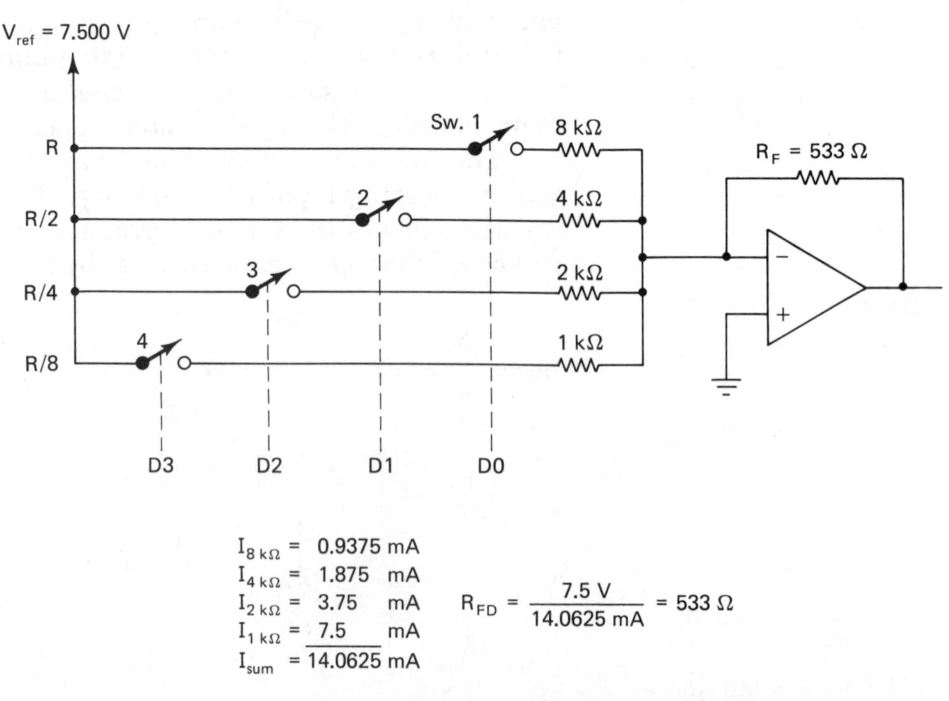

Figure 13.7 4-bit summing-op-amp DAC.

When the input value of 0000 is placed on the controls of the CMOS switches, no current will flow and the output of the op amp will be 0 V. When the input value of 1111 is placed on the controls of the CMOS switches, the maximum current will flow (14.0625 mA) and the op amp will output −7.5 V. Another op amp can be used to invert the negative output to a positive output.

13.3.3 An R-2R Ladder-Type DAC

The accuracy of the DAC in Figure 13.7 is dependent on the precision of the input resistors and the reference voltage. A 16-bit DAC will require 16 different value input resistors. This creates a problem: It is not practical to build DACs from discrete parts, and precision resistors cannot be manufactured within ICs. We must discover a method of creating constant-current sources that does not require a large number of different-value resistors.

We can construct a DAC of any input width using only two different resistor values (Figure 13.8). The absolute value of each resistor is not important; the critical factor is the 2:1 ratio between the resistors. The mathematical analysis of the R-2R ladder is long and tedious. What you must understand is that the R-2R ladder functions exactly as the binary-weighted resistors did in the previous DAC. The major advantage of the R-2R ladder is that it requires only two different resistor values. All monolithic DACs (DACs contained in an IC) use an R-2R ladder.

The dashed lines in Figure 13.8 indicate the part of the DAC that is actually contained in an IC—the CMOS transmission

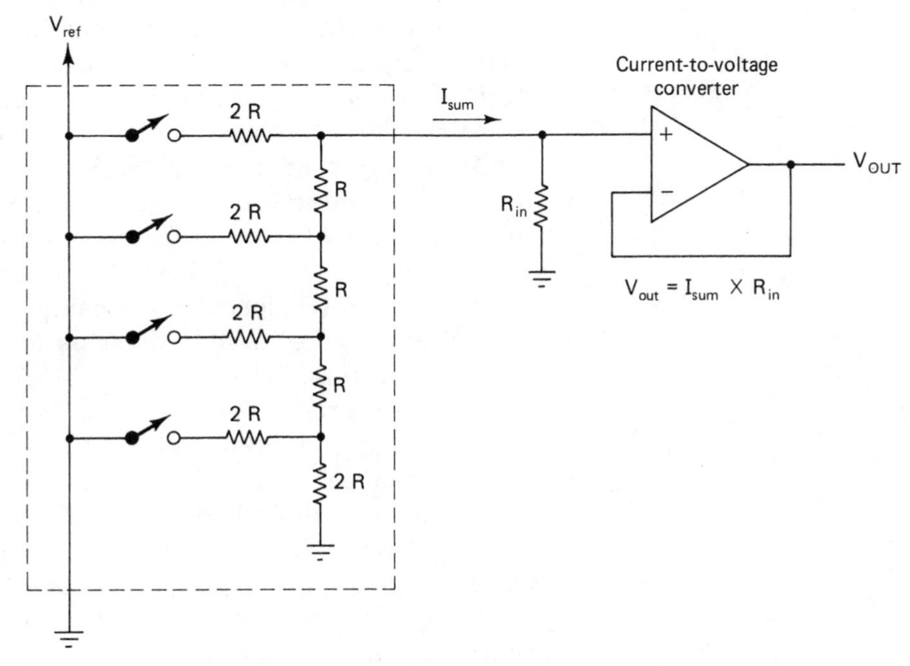

Figure 13.8 R-2R ladder-type DAC.

gates and the R-2R ladder. The output of most DACs is not voltage, but current. An external op amp with an appropriate feedback resistor must be added to the circuit. Most DACs do not contain an internal voltage reference, so this must also be added to the circuit. DACs that do not contain voltage references are called *multiplying DACs.* This is because the output is always equal to the reference voltage multiplied by a fraction denoting the input digital word.

13.4 THE ANALOG-TO-DIGITAL CONVERTER

Analog-to-digital conversion is much more complex than digital-to-analog conversion. A digital word is a known quantized value. Here the word "quantized" means a quantity that is composed of discrete values, unlike an analog voltage, which is continuous. Converting a digital word to an analog equivalent is a simple process. Analog-to-digital conversion begins with a continuous voltage. The A/D converter must quantize this continuous voltage into an exact digital value.

There are four popular methods of converting analog voltages to digital values. Their trade-offs are speed, accuracy, and cost. Each method of A/D has an appropriate area of application.

13.4.1 A Digitized Waveform

A sine wave is a continuous function. One cycle of a sine wave is constructed from an infinite number of points. When we digitize a waveform (convert it from analog to digital), we are assigning a digital value to selected points of the waveform. If we sample enough points, the digitized waveform will be a recognizable copy of the original. If we sample too few points, the digitized waveform will appear as nothing more than a series of unrelated points (Figure 13.9). The intelligibility of a graph is proportional to the number of points sampled. Because the sine wave is a smooth and continuous function, even the graph with only seven samples is fairly recognizable. A graph depicting the acceleration of a revolving body would be highly complex and impossible to reconstruct with only a few samples. In general, the sampling frequency should be at least 100 times greater than the frequency of the waveform that is being digitized.

Every analog sample must be assigned a numeric value. That is the function of an A/D. An A/D will have one analog input and a group of digital outputs (the number of which depends on the particular A/D). The precision of an A/D is directly proportional to the number of digital outputs.

After a sample has been digitized, it can be stored into memory or manipulated in many other ways by the microprocessor. Conventionally, music is recorded in an analog format and degrades with use. Music stored as a sequence of digital words cannot get noisy or degrade; the digital words are a quantized value, not an analog representation of the original music.

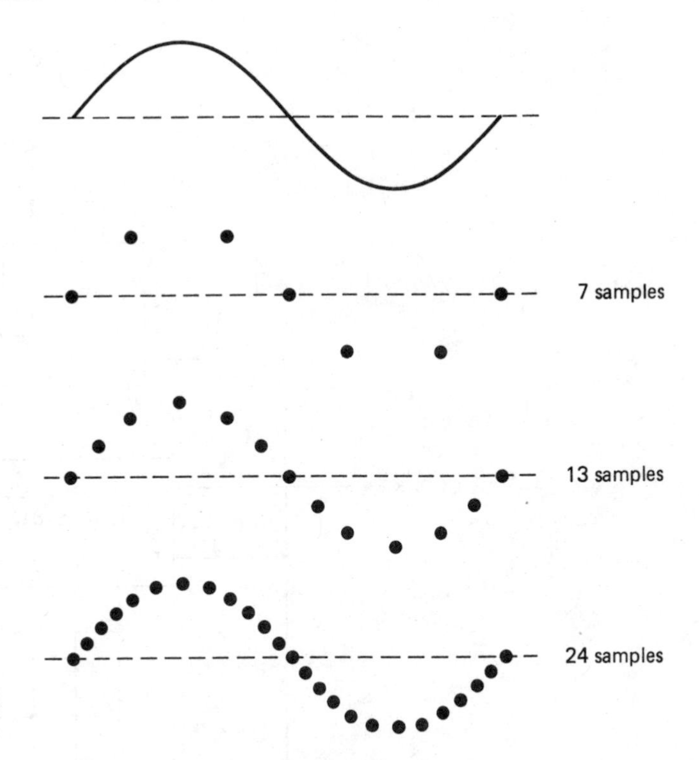

Figure 13.9 Digitized sine waves.

13.4.2 A Counter/Ramp-Type A/D

A DAC is often one of the components in an A/D system. The counter/ramp-type A/D uses a counter, voltage comparator, and a DAC to convert an analog sample into a digital word (Figure 13.10). You have accumulated enough electronics knowledge to completely analyze the operation of the circuit shown in Figure 13.10. Before continuing, take a moment to try your hand at creating a theory of operation that describes the counter/ramp-type A/D converter.

Now let's analyze the function of the individual devices before attempting to describe the system interaction.

Voltage comparator When the analog input voltage is greater than the voltage output of the DAC, the output of the comparator will be a logic 1. This logic 1 will enable the 74LS161 counter. When the voltage out of the DAC becomes greater than the analog input voltage, the output of the comparator will go to a logic 0 and the counter will be disabled. The falling edge of the voltage comparator will also fire the one-shot. The output of the one-shot is a negative pulse, which indicates that the A/D conversion is complete.

Counter The counter will be enabled when the output of the voltage comparator is a logic 1. The 4-bit output of the counter is applied to the inputs of the DAC and the three-state buffer.

DAC The 4-bit DAC will reconvert the digital output of the counter into an analog voltage. The output of the DAC and the analog input voltage will be compared. This is an example of a

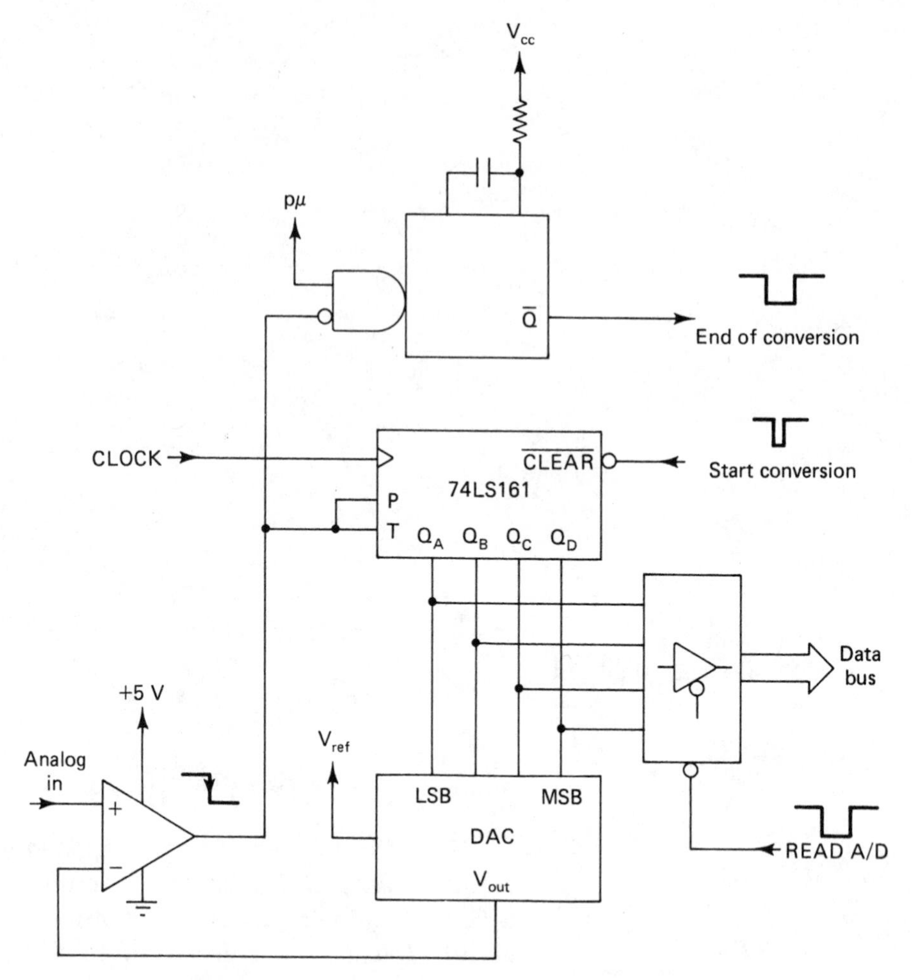

Figure 13.10 Counter/ramp type of A/D converter.

closed-loop system. The DAC feeds the output of the A/D converter back to the input. The counter will continue to count until the output of the DAC becomes slightly greater in amplitude than the analog input voltage.

Three-state buffer The output of the three-state buffer is connected to the microprocessor's data bus. Like any bus talker, the output of this three-state buffer will stay at high-z until it is enabled by the microprocessor.

One-shot As indicated in the description of the voltage comparator, the one-shot will output a negative pulse when the comparator outputs a falling edge. The negative pulse will indicate to the microprocessor that the A/D conversion is complete.

Let's consider the system interaction as an analog-to-digital conversion is processed.

Event 1 Assuming that the analog voltage to be converted is on the noninverting input of the voltage comparator, the microprocessor will issue a negative-going "start conversion" pulse. This pulse will clear the Q outputs of the counter.

Event 2 When the outputs of the counter are cleared, the analog output of the DAC will go to 0 V. This will cause the output of the comparator to go high.

Event 3 The counter is now enabled. At each rising edge of the clock, the counter will be incremented. This updated count will be converted to an analog voltage by the DAC, and then compared with the analog input. The counter will continue to increment until the output of the DAC becomes greater than the analog input voltage.

Event 4 When the output of the DAC becomes greater than the analog input, the output of the comparator will go low. This action will freeze the counter and fire the one-shot. The output of the counter is the digital equivalent of the analog input voltage.

Event 5 The one-shot will fire a negative pulse to alert the microprocessor that the A/D conversion is complete. Remember that the output of the counter is frozen. As long as the analog input voltage does not increase, the counter will stay frozen with the digitized value.

Event 6 The microprocessor will place a negative pulse on the enable input of the three-state buffer. This will momentarily pass the outputs of the counter onto the data bus. The microprocessor will store this value in a memory location.

The A/D conversion cycle is now complete, and the circuit is ready to perform another conversion.

Now that we understand how the system functions, we must consider some of its operational concepts. The conversion time is directly proportional to the amplitude of the analog input. If the amplitude of the analog input is greater than the full-scale output voltage of the DAC, the counter will never be disabled and the A/D conversion one-shot will never fire.

The counter/ramp A/D derives its name from the output of the DAC. The DAC will output a linear voltage ramp whose peak amplitude is slightly greater than the analog input. If more resolution is required, another 4-bit counter and an 8-bit DAC can be used to increase the output of this A/D to a full byte.

13.4.3 Dual-Slope A/D Conversion

One of the most accurate (but also slowest) A/D conversion methods is the dual-slope A/D. It is widely used in products that do not require fast conversion speeds, such as digital voltmeters.

The key to dual-slope A/D is the integrator. An integrator circuit continuously sums a current and stores it as a charge across a capacitor. The op amp is the major component in an integrator. Figure 13.11 is a review of the op-amp integrator. The integrator illustrated in Figure 13.11 is an inverting integrator. A positive V_{ref} will drive the output voltage negative and a neg-

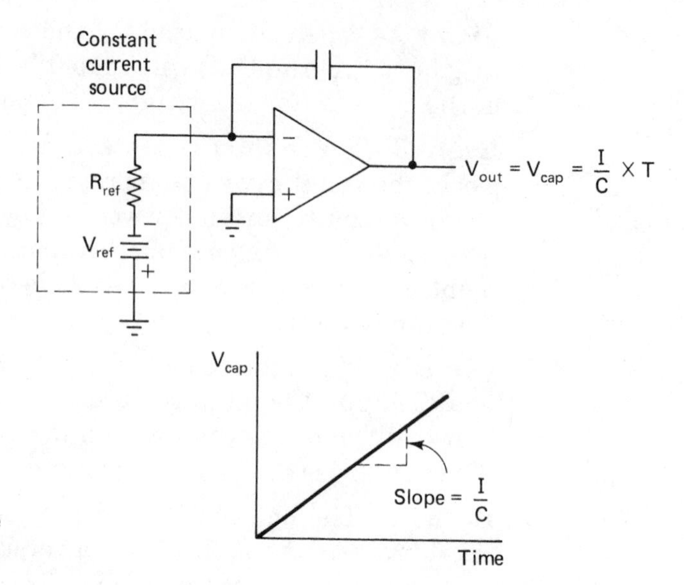

Figure 13.11 Review of the op-amp inverting integrator.

ative V_{ref} will drive the output voltage positive. A reference voltage and resistor are used to create a constant-current source. The output of this constant-current source is used to charge the summing capacitor in the feedback loop.

The voltage on the capacitor is proportional to the magnitude of the constant current and the length of time that it has been charged, and is inversely proportional to the value of the capacitor. This means that the greater the magnitude of current from the constant-current source and the longer the capacitor is charged, the greater the voltage on the capacitor. Similarly, the larger the capacitor, the more time it requires to charge up to a particular voltage.

Refer to the graph in Figure 13.11. The slope of the graph is equal to the constant current divided by the capacitance. The slope can easily be modified by varying the current: if the current increases, the slope gets steeper; if the current decreases, the slope becomes more gradual. This is the basis of the dual-slope A/D (Figure 13.12). The first half of the graph in Figure 13.12 illustrates three different constant currents integrated for

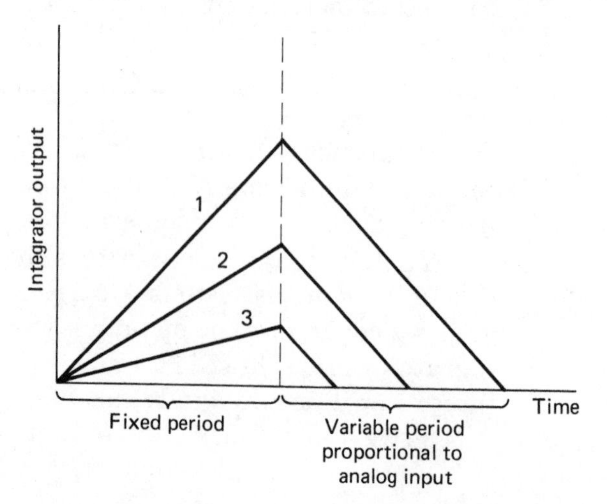

Figure 13.12 Integrator with fixed charge and variable discharge intervals.

a fixed period of time. The second half of the graph illustrates the output voltage as it is discharged by a known reference current until V_{cap} is equal to 0 V.

Remember that the slope of the charging graph is proportional to the magnitude of the constant-current source. The largest constant current is labeled number 1 and the smallest current is labeled number 3. At the end of the fixed-charge period, the voltage across the capacitor is proportional to the magnitude of the constant charging current.

What would happen if we discharged the capacitor with a known reference current of the opposite polarity? The length of time required to discharge the capacitor to 0 V would be proportional to the magnitude of the original charging current. The graph indicates that the discharge slope of all three lines is equal. The largest-magnitude constant current required the longest discharge time, and the smallest-magnitude constant current required the shortest discharge time.

Finally, consider what would happen if we enabled a counter from the beginning of the discharge time until the capacitor was discharged to 0 V. The digital output of the counter would be proportional to the original constant current. We have just completed an analog-to-digital conversion. Figure 13.13 illustrates a highly simplified model of a dual-slope A/D. The schematic symbol for a current source is an arrow enclosed in a circle. Figure

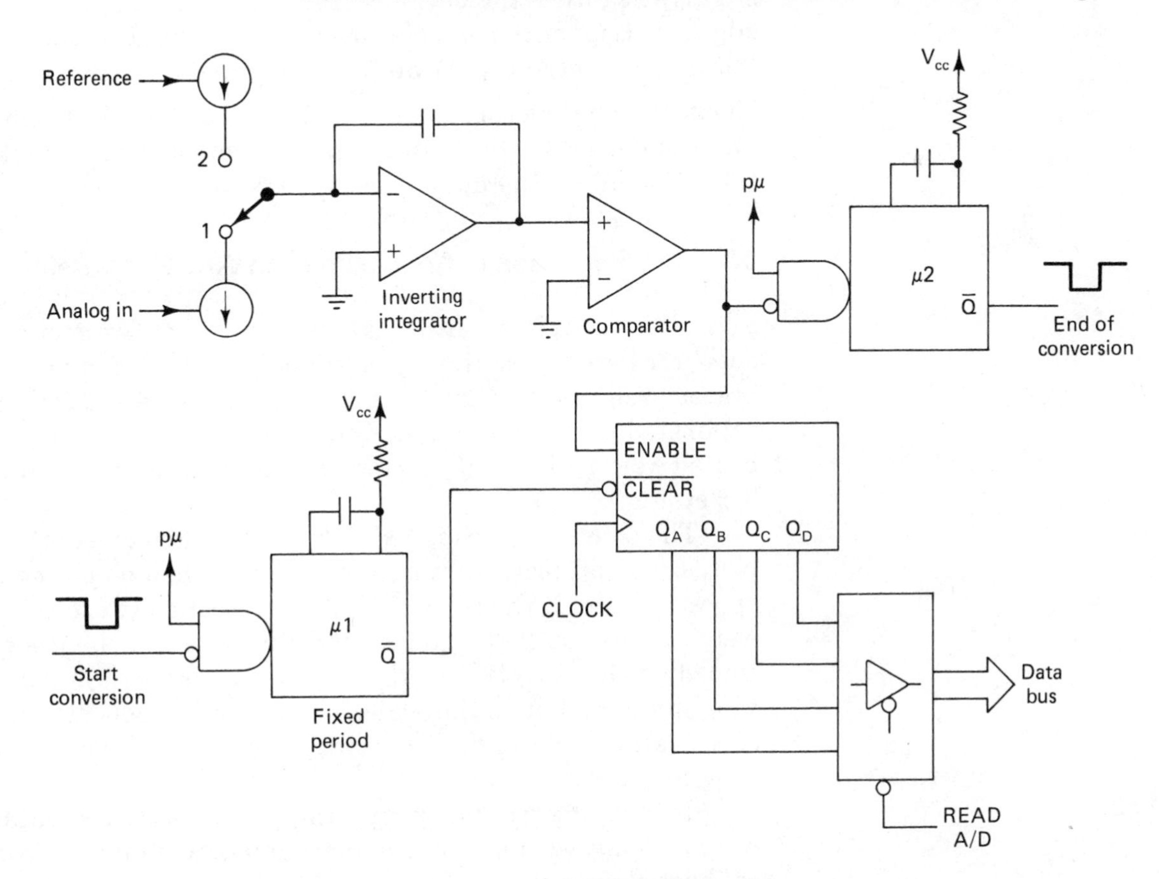

Figure 13.13 Model of dual-slope A/D.

13.13 depicts two current sources: the output of the top current source is proportional to a reference voltage; the output of the bottom current source is proportional to the analog input voltage.

Let's consider the operation of this dual-slope A/D.

Event 1 The microprocessor will initiate the conversion by outputting a start-conversion pulse. The falling edge of the pulse will trigger the one-shot U1. The $\overline{Q}$ of the one-shot will go low for a specified interval. The length of the negative pulse will be the fixed interval during which the integrator will sum the current from the variable-current source. When the $\overline{Q}$ output of the one-shot is low, the counter will be cleared and inhibited from counting.

Event 2 $\overline{Q}$ of the one-shot U1 will go high at the end of the fixed period. The CMOS switch will toggle to position 2, and, because the integrator was charged to a positive voltage, the output of the comparator will be a logic 1. The counter is now enabled.

The reference current source will discharge the integrator to 0 V. The counter will convert the period required to discharge the integrator to 0 V into a digital value.

Event 3 When the noninverting input of the comparator goes slightly negative, the output will toggle to a logic 0. This falling edge will trigger the end-of-conversion one-shot U2, and the logic 0 level will freeze the output of the counter.

Event 4 The microprocessor will issue a read A/D command, and the output of the counter will be passed onto the data bus, thus concluding the dual-slope A/D operation.

13.4.4 Successive-Approximation A/D Conversion

Consider a simple balance scale. It consists of two metal plates supported in the center by a vertical post, and a set of calibrated weights. For this example the balance scale will have calibrated weights of 1, 2, 4, and 8 ounces. Notice that the calibrated weights increase in powers of 2 (just like the binary number system).

The object to be weighed will be placed on one of the metal plates. The largest calibrated weight will then be placed on the opposite plate. If the scale tips in the direction of the calibrated weight, it is too large and must be removed; if the scale stays tipped in the direction of the unknown weight, the calibrated weight will be left on the scale. This process is repeated for each of the calibrated weights, starting with the heaviest and continuing to the lightest.

To find the weight of the unknown object, the calibrated weights that were left on the plate must be summed. With the calibrated weight set of 1, 2, 4, and 8 ounces, an object up to 15 ounces can be weighed with a resolution of 1 ounce.

The act of weighing an object can be thought of as an analog-to-digital conversion. This means that we are assigning a numerical (digital) value that describes the weight (analog parameter) of the object. The block diagram in Figure 13.14 illustrates an A/D converter based on the concept of the balance scale. Figure 13.14 depicts a successive-approximation A/D converter. It is the electronic version of the balance scale. Like the counter/ramp-type A/D, the successive-approximation method uses a DAC and a voltage comparator. Operation of the successive-approximation A/D centers around the SAR (successive-approximation register). The SAR contains a shift register and control circuitry.

After the microprocessor initiates a conversion by pulsing the start conversion input, the SAR will shift a logic 1 into the most significant bit of the output. This digital value (1000) will be converted by the DAC and compared to the analog input. If the output of the voltage comparator stays at a logic 1 level, the SAR will leave the most significant bit set; if the output of the voltage comparator toggles to a logic 0, this indicates that the digital output is too large and the SAR will reset the most significant bit.

This process will be repeated for the remaining 3 bits. The SAR will then pulse the EOC (end-of-conversion) output low, which indicates to the microprocessor that the conversion is complete. The microprocessor will respond by taking the read A/D line low. The three-state buffer will pass the outputs of the SAR onto the data bus. The conversion cycle is then complete.

Unlike counter/ramp-type or dual-slope-type A/Ds, the successive-approximation A/D will always complete a conversion in the same period of time, regardless of the amplitude of the analog input. Successive-approximation A/D is fast and finds applications where many samples must be taken in short periods of time.

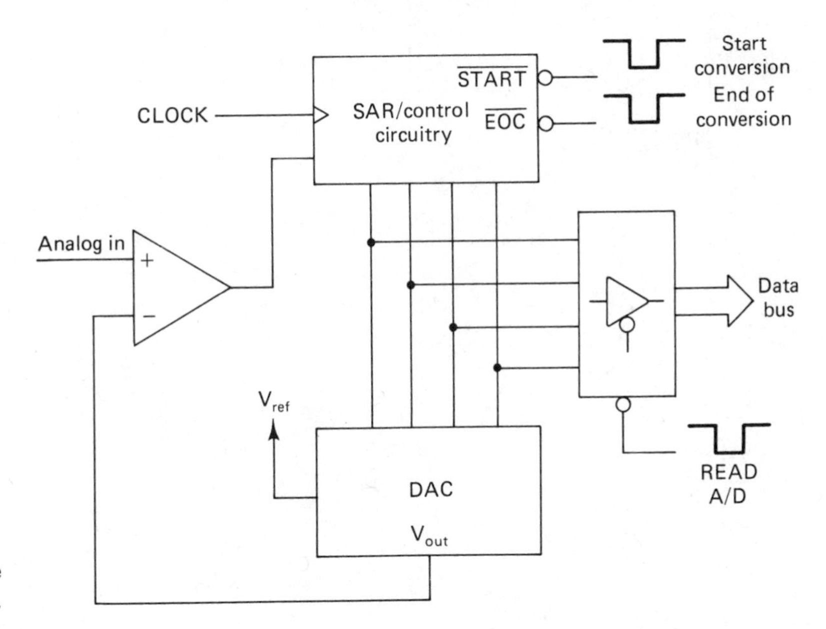

Figure 13.14 Successive approximation A/D converter.

13.4.5 The Flash (Simultaneous) A/D Converter

The fastest A/D converter is the flash type. The other methods of A/D require several passes or clock cycles to complete the conversion. The flash A/D operates in parallel, so all bits are converted simultaneously. Figure 13.15 illustrates a 3-bit flash A/D. For an output of N bits, 2^N resistors and 2^{N-1} voltage comparators are required. The 3-bit flash converter in Figure 13.15

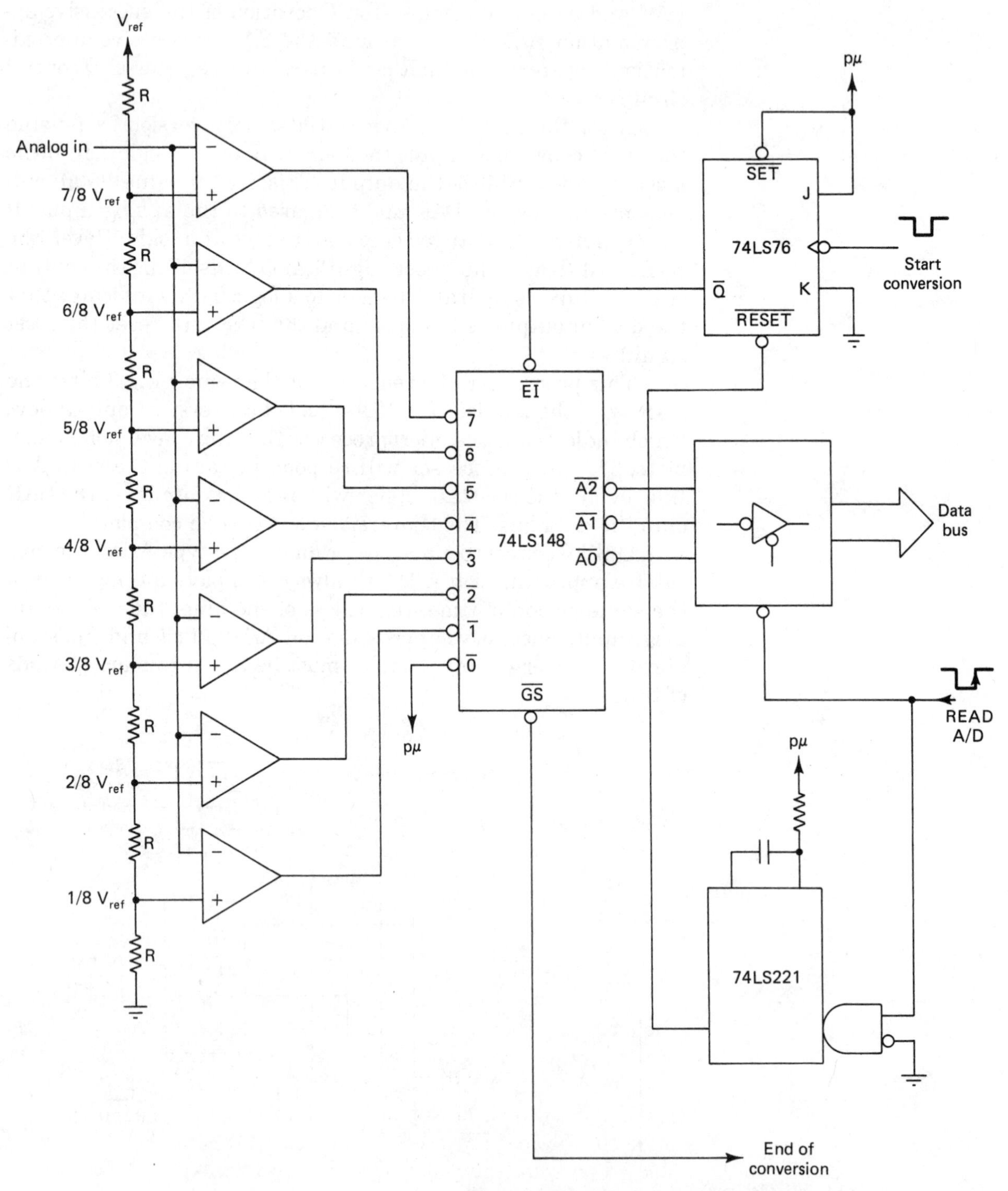

Figure 13.15 Flash-type A/D converter.

has eight resistors and seven voltage comparators. The output of the seven voltage comparators drive the inputs of an 8-line-to-3-line priority encoder. The active-low outputs of the priority encoder are inverted by the three-state buffer.

Each of the eight resistors is equal. The absolute value of each resistor is not important, but it is extremely important that all the resistors be closely matched within a few tenths of a percent. The eight resistors constitute a precision voltage divider. The noninverting input of each voltage comparator will be a multiple of one-eighth of the reference voltage.

The analog input voltage is connected to the inverting input of each voltage comparator. If the analog input voltage is greater than the reference voltage of a particular voltage comparator, that comparator (and all other comparators with smaller reference voltages) will go to an active-low output level. The highest-priority active low will then be encoded into a 3-bit binary output by the 74LS148.

Let's assume that the reference voltage is 8.000 V. The divided-down reference voltages will be 1.000 (at comparator 1) through 7.000 (at comparator 7). We will apply a voltage ramp to the analog input of the flash A/D. The ramp will start at 0 V and rise to 7.999 V.

	Output of comparators:							*Output onto data bus:*		
Ramp voltage	*7*	*6*	*5*	*4*	*3*	*2*	*1*	*B2*	*B1*	*B0*
0.000–0.999	0	0	0	0	0	0	0	0	0	0
1.001–1.999	0	0	0	0	0	0	1	0	0	1
2.001–2.999	0	0	0	0	0	1	1	0	1	0
3.001–3.999	0	0	0	0	1	1	1	0	1	1
4.001–4.999	0	0	0	1	1	1	1	1	0	0
5.001–5.999	0	0	1	1	1	1	1	1	0	1
6.001–6.999	0	1	1	1	1	1	1	1	1	0
7.001–7.999	1	1	1	1	1	1	1	1	1	1

Note: The inputs to a voltage comparator theoretically can never be equal. For this reason the table does not indicate any ramp voltage that is equal to a reference voltage.

Notice that the output of the inverting three-state buffer is the digital equivalent of the analog input. A priority encoder is needed because the output of a less significant voltage comparator does not toggle back to an inactive state when the next significant comparator goes active low.

A conversion will be initiated by the microprocessor pulsing the clock input of the J-K flip-flop. The J-K flip-flop is configured to set on the falling edge of a clock pulse. The $\overline{Q}$ output will go low, driving the enable input of the 74LS148 to an active-low level. The outputs of the comparators will be encoded into a 3-bit active-low code by the 74LS148. After the inputs are encoded, the group select output will go low to indicate that the conversion is complete. The microprocessor will then pulse the read

A/D line. The reinverted binary code will appear on the data bus and the rising edge of the read A/D signal will fire the one-shot that resets the J-K flip-flop.

13.4.6 Summary of A/Ds

Practical A/Ds are not constructed from discrete devices. They are available in two different packages: monolithic (integrated circuits) and hybrids. *Hybrid devices* are sealed packages that contain a combination of integrated circuits and discrete components.

Let's summarize each type of A/D converter:

Counter/ramp Practical counter/ramp A/Ds use an up/down counter. This enables the device to continually track the analog input voltage. Tracking counter/ramp A/Ds are used in medium speed/medium performance circuits, and are usually manufactured in hybrid packages.

Dual slope Although dual-slope is the slowest form of A/D conversion, it is also one of the most popular. Dual-slope A/Ds are simple, accurate, and inexpensive. They also have exceptional noise immunity. Most digital voltmeters use a dual-slope A/D.

Successive approximation Successive-approximation and dual-slope A/Ds constitute over 90% of A/Ds in use today. The conversion speed of successive-approximation A/D conversion is limited by the settling time required for the feedback DAC. High-performance, two-step successive-approximation A/Ds change two bits at a time. This effectively doubles the conversion speed. Successive-approximation A/Ds are used in moderate-to-high-speed applications.

Flash (known also as simultaneous or parallel A/D) An 8-bit flash converter requires 255 voltage comparators and 256 resistors in the precision divider, and the equivalent of a 255-line-to-8-line capacity encoder. This complexity results in such a high cost that flash converters are used only in super-high-speed video applications in excess of 20 MHz.

QUESTIONS AND PROBLEMS

13.1. Complete the timing diagram for the noninverting Schmitt trigger.

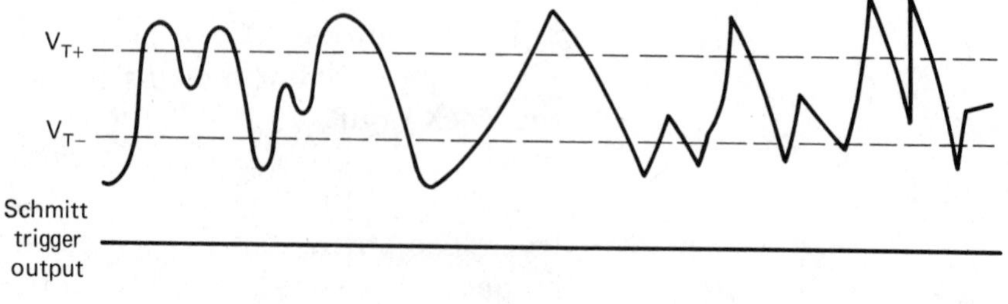

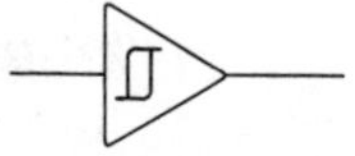

13.2. Describe the output of the circuit shown. *Hint:* Remember that the capacitor will appear as a short to ground when V_{cc} is first applied.

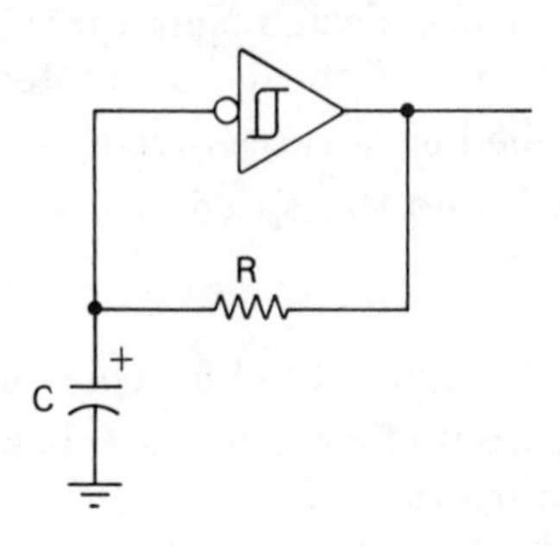

13.3. Describe the output of the circuit shown when SW 1 is closed.

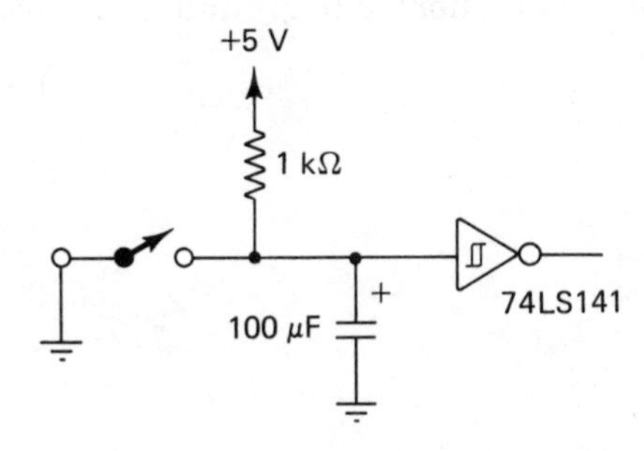

13.4. The diode in Figure 13.1 is installed backwards. What is the output of the circuit? The diode is shorted. What is the output now? What would the output be if the diode were open?

13.5. Define the specifications V_{T+} and V_{T-} as they pertain to the Schmitt trigger.

13.6. In a data acquisition circuit, how would a thermistor be classified.

13.7. Name two devices in which you have seen a capacitor used as a storage element.

13.8. Refer to Figure 13.5. Q_d is shorted to ground. Draw the output of the DAC.

13.9. Explain the concept of a virtual ground as it pertains to op-amp circuits.

13.10. How is the op amp in Figure 13.8 performing a current-to-voltage conversion?

13.11. A square wave with an amplitude of V_{ref} is applied to the input of an A/D converter. What is the output waveform? A triangular waveform with a peak amplitude of V_{ref} is applied to the input of an A/D converter. What is the output waveform?

Refer to Figure 13.10 for Questions 13.12 to 13.14.

13.12. Draw a timing diagram that describes the events that constitute an A/D conversion cycle.

13.13. The microprocessor issues a start-conversion command, and the end-of-conversion pulse never occurs. What is the most likely problem?

13.14. The enable input of the three-state buffer is shorted to ground. How does this affect system operation?

13.15. Where does the dual-slope A/D derive its name?

13.16. Refer to Figure 13.13. The one-shot that provides the fixed period has too much drift for this high-precision circuit. Given a 1-MHz clock, design a circuit that provides a 256-μs fixed time period. *Hint:* Consider a counter.

13.17. Describe how a successive-approximation A/D functions.

13.18. What limits the speed of a successive-approximation conversion cycle?

Refer to Figure 13.15 for Questions 13.19 to 13.22.

13.19. The output of comparator 6 is stuck low. How does this affect circuit operation?

13.20. The EI of the 74LS148 is shorted to the A2 output. How does this affect circuit operation?

13.21. EI is shorted to ground. How does this affect circuit operation?

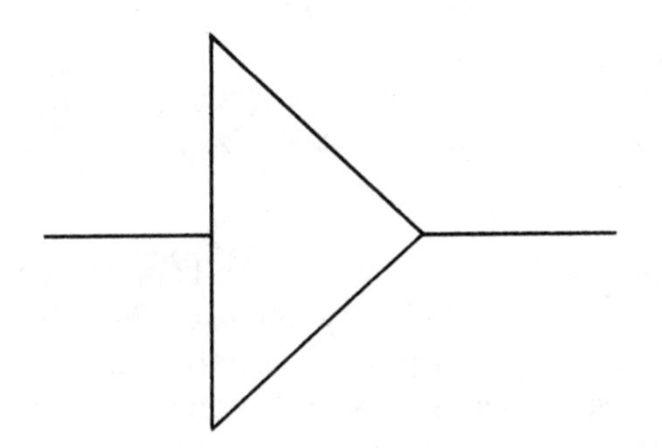

ANSWERS TO SELECTED PROBLEMS

CHAPTER 1

1.1. **(a)** 1101 1010 **(b)** 1001 0001 **(c)** 1010 0000

1.3. **(a)** 62 **(b)** 241 **(c)** 90
(d) 30 **(e)** 51 **(f)** 255

1.5. **(a)** 0001 0011 **(b)** 0110 1001 0011 **(c)** 0011 1001

1.7. **(a)** 8 **(b)** 16 **(c)** 256 **(d)** 65,536

CHAPTER 2

2.3. **(a)** logic 1 **(b)** logic 0
(c) Both levels are unique; they each only occur once.

2.5. **(a)** The OR logic symbol indicates a dynamic input gate.
(b) The AND logic symbol indicates a unique output gate.

2.7. If an input is stuck at a dynamic input level of logic 0, the output of the AND will also be stuck at logic 0.

2.9. A current-limiting resistor must be placed in series with every LED. Assume that the knee voltage of an LED = 1.5 V. The current-limiting resistor will have 3.5 V with 12 mA of series current. The resistor should be 290 Ω ($R = E/I$). 220 Ω and 330 Ω are standard-size current-limiting resistors.

CHAPTER 3

3.3. **(a)** Logic 0—the same as an OR gate
(b) Logic 1—the same as an AND gate
(c) The XOR does not have a unique output level. Logic 0 and logic 1 both occur twice in the truth table.

3.5. Because all other functions can be derived from a combination of NAND or NOR gates.

3.7. One of the inputs is stuck at a dynamic low level.

3.9. It has no dynamic input or unique output levels.

CHAPTER 4

4.3. Notice that the symbols used to represent G1 and G4 no longer indicate the function that they provide. We had used these gates as active-low-input, active-high-output AND gates. Now they appear to be positive-logic NOR gates. The schematic is no longer self-documenting. It is much more difficult to derive a theory of operation from this new schematic than it was from the original version.

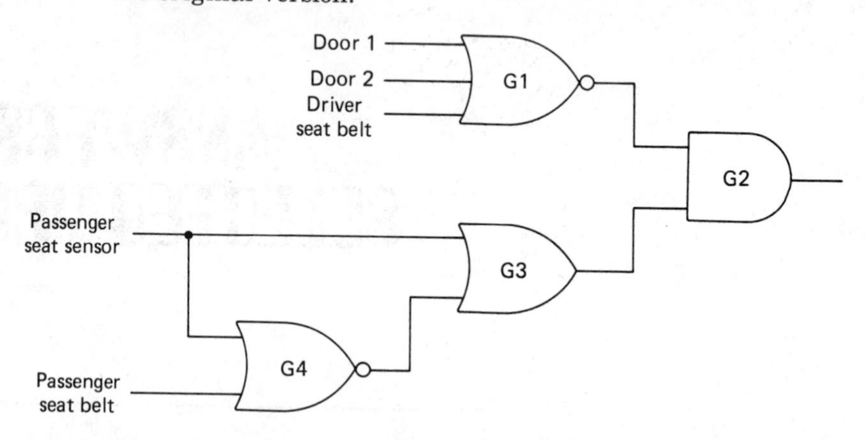

4.5. The XOR function provides a means of controlling the condition of the light from either switch. The XOR gate does not have a dynamic input level. Consider the switch that is not being accessed as the control input. If it were left low, the logic level on the other switch will pass through the XOR gate. If the control were left at a logic 1 level, the logic level on the other switch will be inverted by the XOR gate.

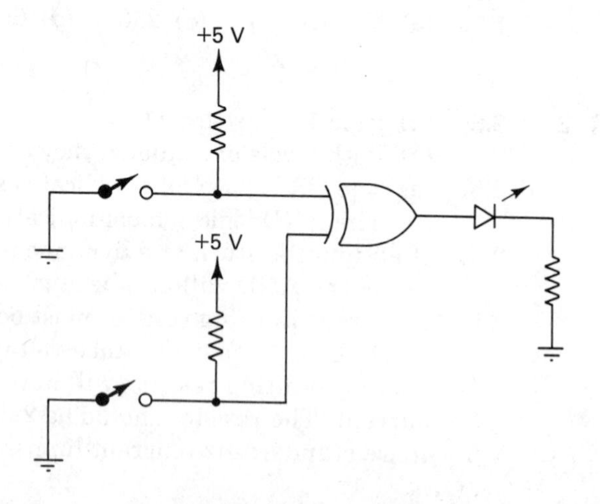

CHAPTER 5

5.1. Notice that two inverters are used to drive the majority and nonmajority LEDs on active lows. The nonmajority LED could be driven directly by the output of the OR gate, but the circuit is easier to understand with the extra inverter.

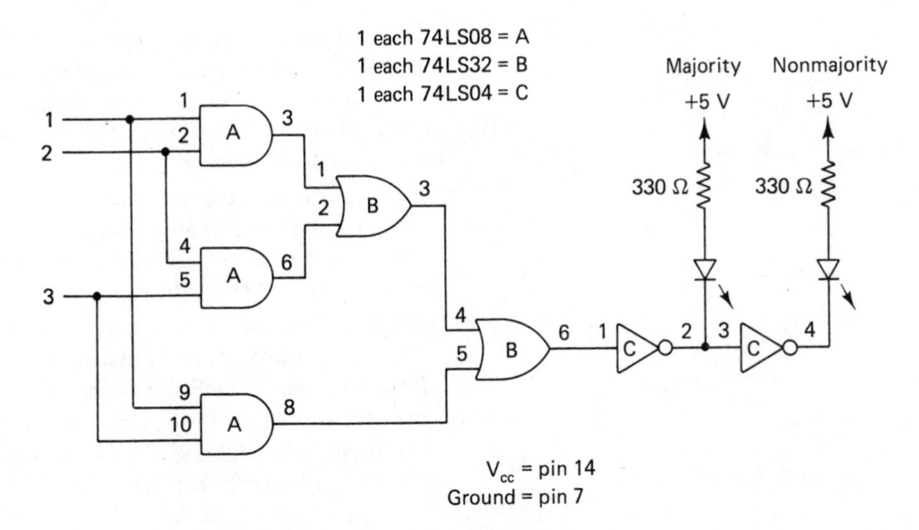

5.3. This circuit is created using only the TTL gates that were introduced in Chapter 5. Once again, notice the use of the appropriate circuit symbols. This concept cannot be overemphasized.

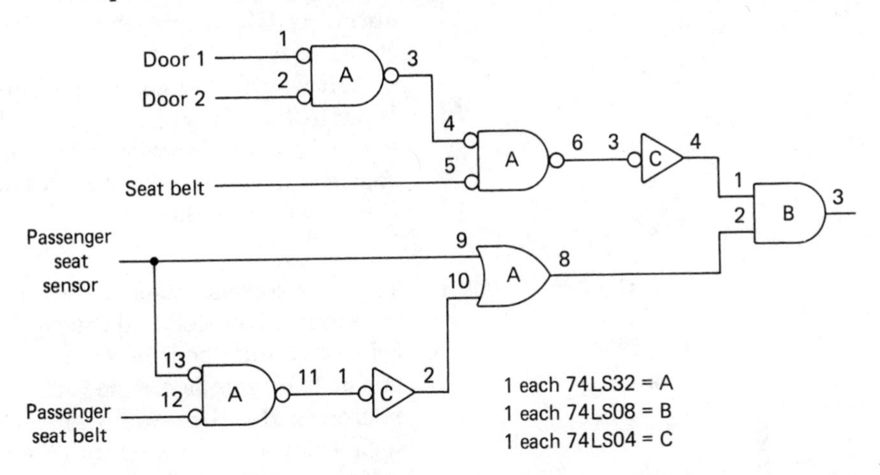

5.5. 3.4 V for TTL, dependent on loading conditions. 11.95 V for CMOS, always 0.05 V from an ideal logic 1 level.

5.7. LED ON = AB + CD. An active-low-input, active-low-output OR gate (AND gate) must be added to the totem-pole circuit. Modern use of open-collector outputs is usually in the area of bus systems, where many devices sharing the same bus pin can bring the node low. Open collectors save the use of an extra OR gate, but they also are slow, noisy, and require an external pull-up.

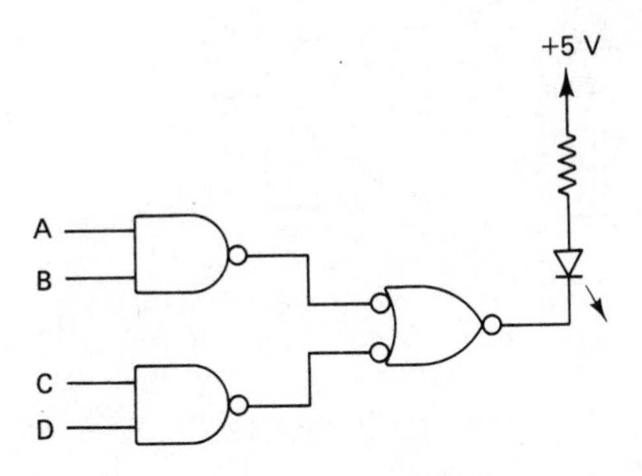

CHAPTER 6

6.3. Bent pins, wrong components, and components inserted backwards can occur during the stuffing process. The easiest way to find these problems is with a thorough visual inspection.

6.5. When the PCB is etched, unetched traces, open feedthroughs, and broken traces can occur. Also, silver-like trace shorts occur during the etching process.

6.7. Bare boards are used to check traces that run underneath ICs, which are difficult to follow on a stuffed unit. Sometimes it is difficult to tell the difference between a valid trace and a solder bridge. The bare board can be used as a reference for this and any other questions regarding traces.

6.21. It is floating. Check for an open.

6.25. Floating CMOS inputs do not appear to be of any particular logic level. CMOS inputs are high impedance and voltage driven, not current driven like TTL.

CHAPTER 7

7.1. With eight inputs this lock has 256 different combinations.

7.3. This is the A = B cascade output of U1. If the A = B input of U2 is floating, U2 will never be enabled. Therefore, the A = B output of U2 will never go active high. (If 74LS85s were used, the A = B input of U2 would float to an active-high level.) The circuit would only notice the inputs of switches 4 through 7.

7.5. Input A3 of U1 will always be at a high level. If B3 is pulled, the circuit will function with the setting of switch 3 as a don't-care condition. If B3 is pulled low, the lock will never open.

7.7. Input D is stuck low.

CHAPTER 8

8.1. The key to designing a switch debouncer is that when the switch has bounced and the pole is between the two throws, the S-R latch must be in a memory state. The dynamic input level of the NOR gate is a logic 1. Therefore, the inputs that are connected to the throws of the switch must be pulled down to a logic 0 level, not pulled up to a logic 1 level. Take care to use relatively small values for the pull-down resistors. If the pull-down resistor is too large, the TTL input will appear as an unbiased, floating input.

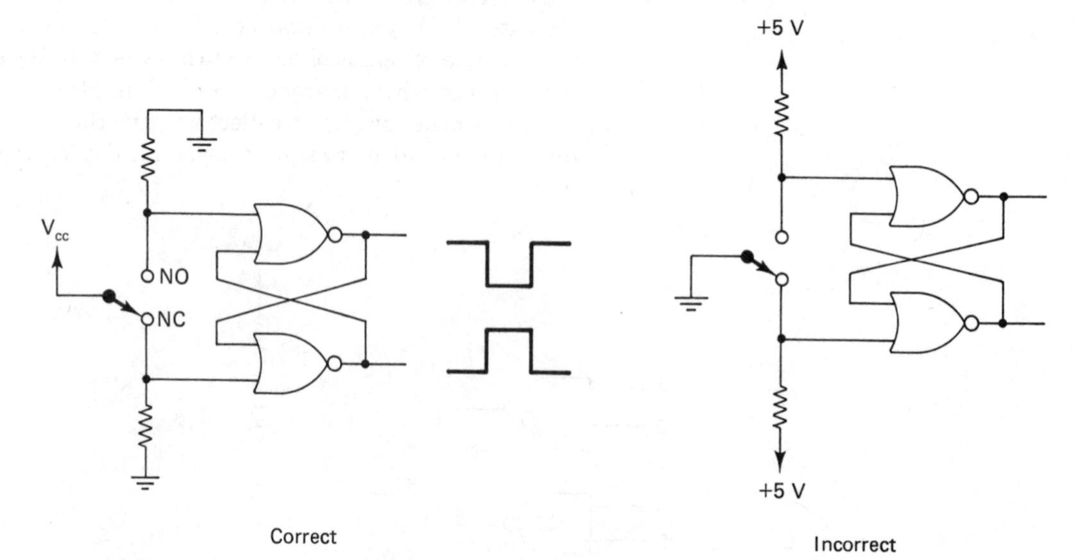

P	G	D	Q	$\overline{Q}$
0	0	0	0	1
0	0	1	1	0
0	1	0	Qn	$\overline{Qn}$
0	1	1	Qn	$\overline{Qn}$
1	0	0	Qn	$\overline{Qn}$
1	0	1	Qn	$\overline{Qn}$
1	1	0	0	1
1	1	1	1	0

8.3. If both of the inputs of an S-R flip flop are at nondynamic levels, the output of the flip-flop is fed back and will maintain the preceding output level. This is the concept of memory.

8.5. By adding the inverter and XOR gate, the 74LS75 can be made to function exactly like the 4042B. When the logic levels on the gate and polarity inputs are equal, the latch will become transparent and pass the D input onto the Q output.

8.7. Complete the following timing diagram:

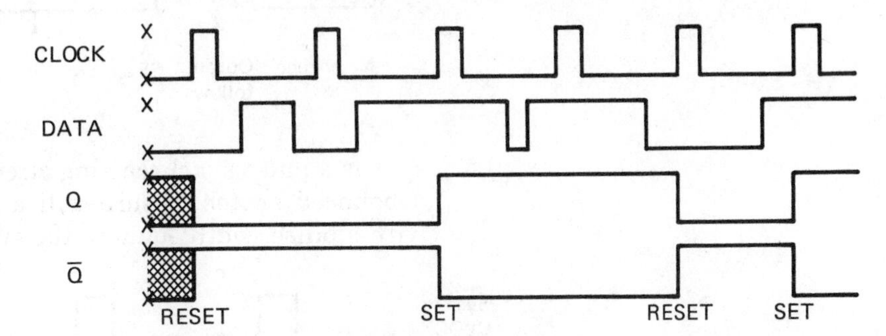

8.9. An indeterminate voltage between 0.8 and 1.2 V would appear at the Q and $\overline{Q}$ outputs and the D input. If the data input was driven by the Q output instead of by the $\overline{Q}$ output, the output would never change. The logic level on Q output would be clocked back into the Q output on every positive edge of the clock. To divide the input frequency by a factor of 4, cascade two flip-flops as shown in figure 8.35. To divide the input frequency by a factor of 8, cascade three flip-flops as shown in Figure 8.35. Each succeeding flip-flop should be clocked by the $\overline{Q}$ output of the preceding stage.

CHAPTER 9

9.1. Complete the following timing diagram:

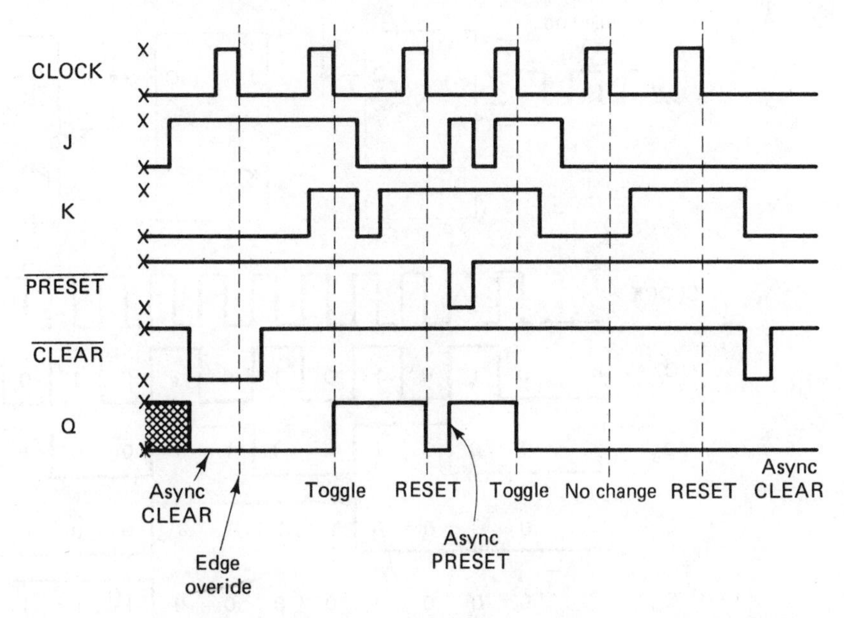

9.3. Complete the following timing diagram. When $\overline{Q}$ is high, the output follows the J input.

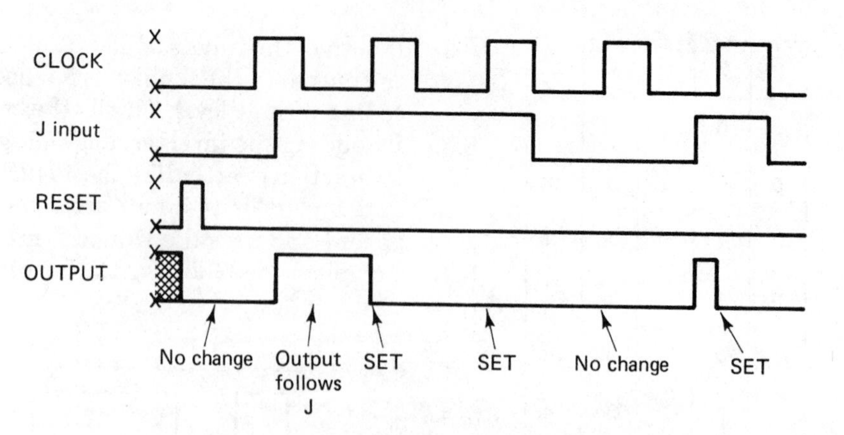

9.5. This is a pulse synchronizing circuit. The pulse created by the debounced switch closure will a synchronized positive pulse with a pulse width equal to the system clock.

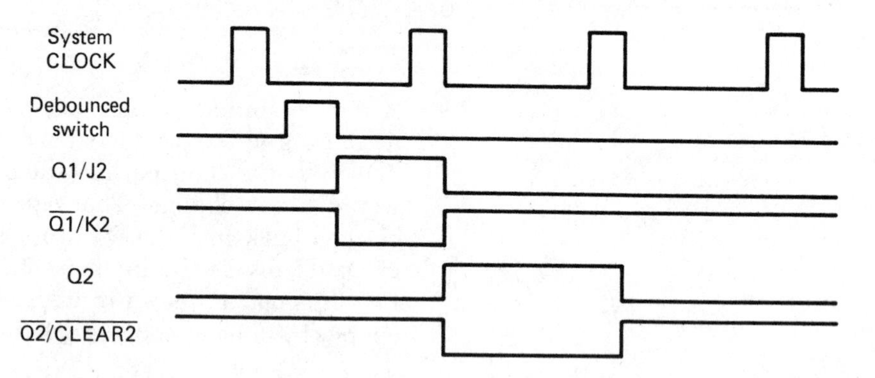

9.7. Figure A9.7.

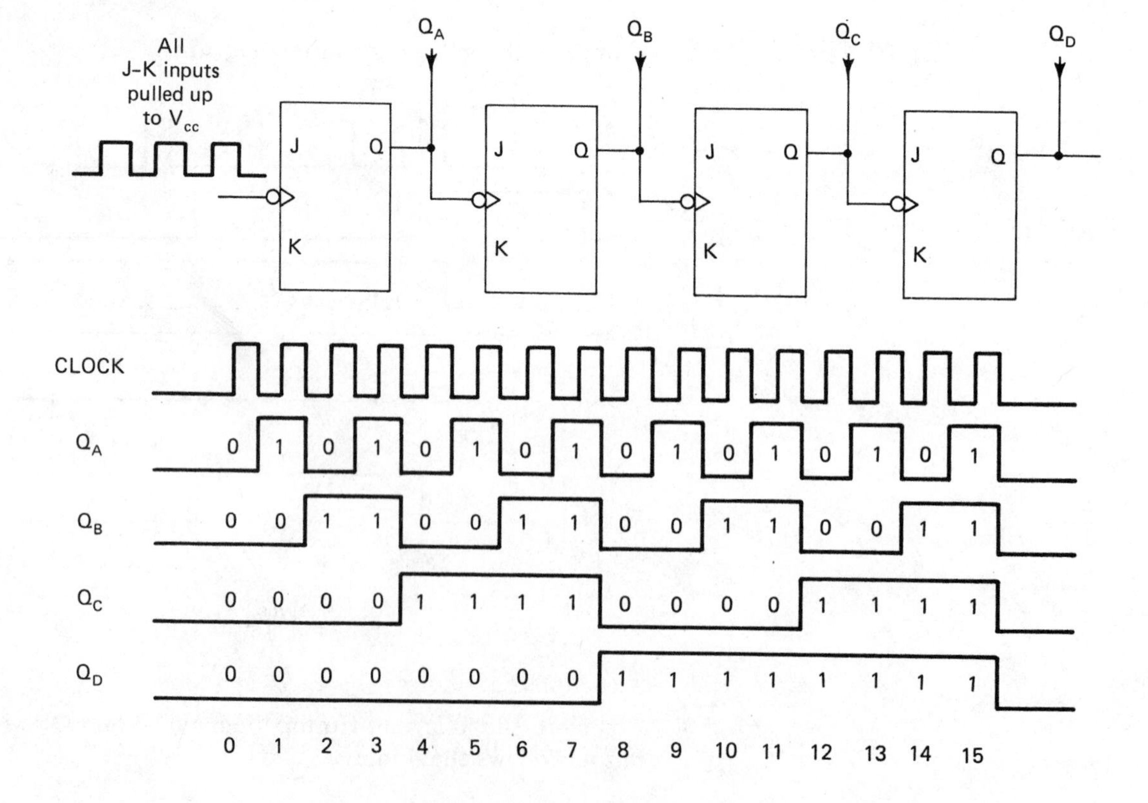

CHAPTER 10

10.1. A device that outputs *one* pulse for every N input pulses.

10.3. The count of 12 occurs only for an extremely short period. That is why the positive glitch occurs on Q_c. When Q_c and Q_d are both high, the clear input goes active. This will reset Q_a through Q_d to logic 0 levels. This, in turn, will cause the clear input to return to an inactive level. This process is indicated by the narrow negative glitch on the clear input.

10.5. 20 ns; one flip-flop and one gate constitute the propagation delay time.

10.7. When Q_a and Q_d are both high, the counter resets to 0. This is a divide-by-9 counter.

CHAPTER 11

11.1. The waveform indicates valid logic 0 and logic 1 levels. The indeterminate level occurs when all the three-state buffers are at high-*z* output levels. Input A of U6 appears to be floating.

11.3. Input B of the decoder appears to be stuck at a logic 0 level.

11.5. It is not being current limited. The cathode of segment a may be shorted to ground with a solder bridge.

11.7. The truth table appears as the truth table of a three-input NOR gate. If A is high or B is high or C is high, the output will go low. The common node of an open collector provides the positive logic Boolean AND function or the negative-logic Boolean OR function. The term "wired OR" best describes the circuit because an active level on input A or input B · · · or input N will take the output to a logic 0 level.

11.9. If the disable input is stuck at a logic 0 level, the outputs will remain at a logic 0 level.

CHAPTER 12

12.1. If the enable input on G2 is shorted to ground, G2 will always be enabled. When the output of G2 and the input data try to pull in different directions, an indeterminate level will result.

12.3. U1 is the row address write decoder. If the data to be written are a logic 0, the enable will stay at a valid logic 0. If the data to be written are a logic 1, an indeterminate level will appear at the enable input of U1.

12.5. The wrong data may be written into the memory location.

12.7. F000H through FFFFH.

CHAPTER 13

13.1. Complete the following timing diagram for a noninverting Schmitt trigger.

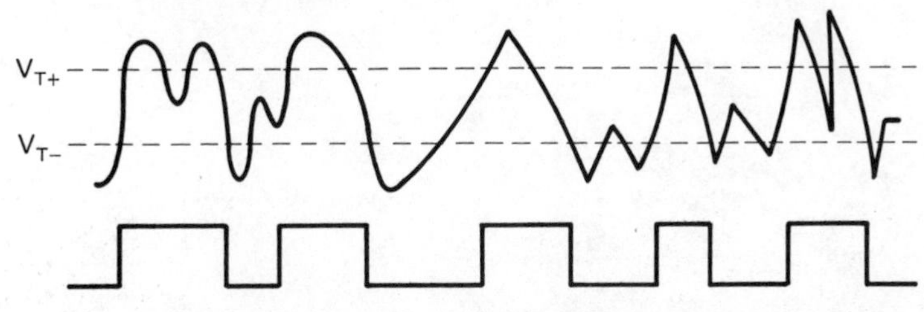

13.3. The capacitor and resistor provide a long *RC* charge time. The capacitor and switch provide an extremely short *RC* discharge time. Normally, the switch is open and the capacitor is charged up to V_{cc}. The output of the Schmitt trigger inverter is a logic 0. The moment the switch is closed, the capacitor instantly discharges to 0 V and the output of the Schmitt trigger goes to a logic 1 level. When the switch bounces open, the capacitor starts to charge through the resistor. Long before the capacitor has charged up to V_{T+}, the switch has bounced closed. The long charge time of the *RC* circuit and the 1.7-V upper voltage threshold of the Schmitt trigger give this circuit the ability to debounce SPST switches.

13.5. The upper threshold voltage is that voltage which causes the output of a Schmitt trigger buffer to switch from a logic 0 to a logic 1 level. The lower threshold voltage is that voltage which causes the output of a Schmitt trigger buffer to switch from a logic 1 to logic 0 level. The difference between the two threshold voltages is known as the hysteresis voltage.

13.7. Dynamic RAMs, sample-and-hold circuits, and integrators all use a capacitor as a storage unit.

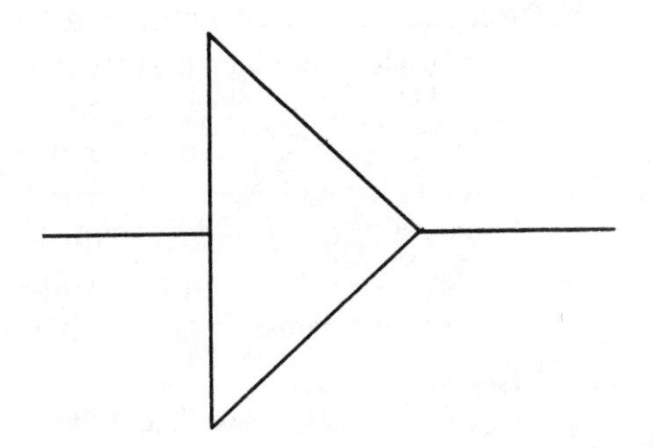

INDEX